Praise for William Leach's BUTTERFLY PEOPLE

"Leach is in pursuit of big ideas about art, science, evolution, collecting, economics and technology." —*The Boston Globe*

"[Leach] uses the rapid expansion of American interest in butterflies after the Civil War as a way into man's arduous process of understanding his place in the order of things. . . . Leach describes the science of this period as a 'fascinating mix of scientific confidence and human longing,' and this book treats both aspects with equal care and well-researched precision." —*The Daily Beast*

"Today's butterfly people will be enthralled. . . . The wonder Leach evokes will captivate all who appreciate the natural world."
—*New Scientist*

"A brilliant work of history." —*Bookforum*

"Masterful and beguiling. . . . A literary cabinet of wonders packed with scientific discoveries, historic artifacts, and artistic revelations to delight scholarly and casual readers alike. No mere flight of fancy, the book is an original consideration of American science, economics and aesthetics set in a time of profound cultural change." —*Washington Independent Review of Books*

"This is truly a remarkable piece of work, and I was totally drawn in. Leach has captured a very important aspect of history as I am sure no one else could have. Along the way it reveals the origins of our obsessions and their trajectory. The breadth and depth of this will stay with me a long time, and I hope the book will be read widely." —Bernd Heinrich, author of *Winter World*

William Leach

BUTTERFLY PEOPLE

William Leach is a professor of history at Columbia University. His previous books include *Land of Desire: Merchants, Power, and the Rise of a New American Culture*, which was a National Book Award finalist, and *Country of Exiles: The Destruction of Place in American Life*.

ALSO BY WILLIAM LEACH

Country of Exiles:
The Destruction of Place in American Life

Land of Desire:
Merchants, Power, and the Rise of a New American Culture

True Love and Perfect Union:
The Feminist Reform of Sex and Society

BUTTERFLY PEOPLE

An American Encounter
with the Beauty of the World

William Leach

VINTAGE BOOKS
A Division of Random House LLC
New York

FIRST VINTAGE BOOKS EDITION, JANUARY 2014

Copyright © 2013 by William R. Leach

All rights reserved. Published in the United States by Vintage Books,
a division of Random House LLC, New York, and in Canada by Random House
of Canada Limited, Toronto, Penguin Random House companies. Originally
published in hardcover in the United States by Pantheon Books, a division of
Random House LLC, New York, in 2013.

Vintage and colophon are registered trademarks of Random House LLC.

The Library of Congress has cataloged the Pantheon edition as follows:
Leach, William.
Butterfly people : an American encounter with the beauty of the world /
William Leach.
p. cm.
Includes bibliographical references and index.
1. Butterflies—United States—History—19th century. 2. Entomologists—
United States—History—19th century. 3. Industrial revolution—United States—
History—19th century. I. Title.
QL549.L43 2012 595.78'9097309034—dc23 2012000389

Vintage Trade Paperback ISBN: 978-1-4000-7692-5
eBook ISBN: 978-0-307-90787-5

Book design by M. Kristen Bearse
Author photograph © Elizabeth Blackmar

www.vintagebooks.com

Printed in the United States of America
10 9 8 7 6 5 4 3 2 1

In memory of Jeannette Hopkins

Beauty has no obvious use; nor is there any clear cultural necessity for it. Yet civilization could not do without it.

—SIGMUND FREUD

I did not dream of there being such splendid things in the world.

—ADRIAN LATIMER *of Lumpkin, Georgia,*
to Herman Strecker, January 27, 1880

Contents

Preface   xi

Introduction   xv

Part One

ENCOUNTERS WITH THE BUTTERFLIES OF AMERICA

1 Yankee Butterfly People   3
2 The German-American Romantics   50
3 Beating Hearts   82
4 Word Power   116
5 The Life and Death of Butterflies   152

Part Two

ENCOUNTERS WITH THE BUTTERFLIES OF THE WORLD

6 In the Wake of Empire   193
7 Butterflies at the Fair   226
8 Death of the Butterfly People   255

Acknowledgments   285
Notes   291
Index   371

Preface

NEARLY ALL MY LIFE I have been drawn to butterflies, among the most beautiful things in nature. As a nine-year-old boy I began collecting in and around a graveyard, in a field that led down to a railroad line, and at the margins of a streambed that ran along the railroad, all within easy reach of my home in a neatly packed lower-middle-class neighborhood. Nothing else so fulfilling or exhilarating marked my childhood. In my mind's eye I see myself running, rushing, falling on the ground, heart pounding, intense beyond belief, lost to an all-consuming purpose. But I stopped collecting abruptly, for want of continuing reinforcement and the absence of informed guidance. I returned to it in my late twenties and, again, in my mid-thirties, in the glorious Cumberland Plateau in Tennessee, where I first saw the Diana fritillary and the zebra swallowtail in all its seasonal phases. Then I stopped for good, although, in spirit, I hardly changed at all—despite the rising climate against collecting and my own misgivings about killing—because collecting took me out of myself and far into nature as nothing else could ever have done. It let me forget time and turmoil and defy many odds, and to see a multitude of living forms. It opened me to the sensuousness of the natural world, which reflected my own inner energies; the color in me met the color outside me. Yet I never took a course on insects or acquired a much greater knowledge until I decided to do this book, the research and writing of which have now cost me many years. I have written it in homage to my own past as well as to great Americans, collectors of all sorts (including some German and English butterfly people) as they came to know butterflies, first in America after 1850, then in the tropics after 1885.

But the book became more than that. It became a story of the American awareness and experience of the beauty of the world, in both its natural and its artificial forms, spanning many years. Soon after the

Civil War, Americans of all classes came in contact with nature—and with butterflies—in pervasive ways. This was a remarkable thing in its own right, for never before had so many people come to know so many butterflies, native and foreign; never before had so many entered "the kingdom of form," as the American philosopher George Santayana put it in his 1896 *Sense of Beauty,* one of the most insightful books of the age. At the same time, the country underwent a massive industrial transformation, which produced a vast number of human-made artifacts, many seen for the first time at the world's fairs, competing with butterflies—and other natural forms—for beauty. This, too, was a remarkable occurrence, opening up aesthetic frontiers of all kinds to more people in an unprecedented way. As Walt Whitman observed, it seemed as if Americans were about to know "nature and artifice" together at once, not as adversaries but as collaborators in the enrichment of life, and such, he believed, would continue to be the case, so long as a balance endured between them. But, as he feared, the balance did not last, and one form of "beauty"—the industrial, technological kind—began to take precedence, and at great cost to nature and to butterflies. So here, then, was my story: how Americans came to know the butterflies and, subsequently, how the butterflies (as exemplary of all natural beauty) began to lose out in the contest with another form of beauty, the industrial-artifactual one, and what that victory meant not only for the well-being of the natural world but for the well-being of American culture as well.

The whole investigation was deeply gratifying for me, to make up for years of ignorance. I'm afraid I never overturned my deficit; nor had I ever expected to, since I had no desire to become a professional expert on insects. Nevertheless, I learned a great deal more about butterflies and about the history of America's attachment to butterflies and of its complex relationship with the beauty of the world as well as of my own. I learned, too, how critical the study of these insects had been to the understanding of the natural world, not merely at the margins of inquiry but at the very core of it; for a brief moment in the 1870s and '80s, some of the most advanced naturalists of Europe and England—Charles Darwin, August Weismann, Henry Bates, and Alfred Russel Wallace, among them—dwelled on butterflies as a way to illuminate the origin and evolution of species, and Americans, too, were at the forefront—leading all the others, in fact, in the depth and range of their insights, embracing Darwin's theory of natural selection as the driving engine of evolution, even as most naturalists at the time rejected or denounced it. I also learned that butterflies captured the imagination of thousands of other men and women (besides the natural scientists), just as butterflies had captured mine when I was a boy, never to lose their hold on me.

Introduction

"THE BEAUTY OF THE WORLD," wrote America's greatest theologian, Jonathan Edwards, in 1725, "consists wholly of sweet mutual consents, either within itself or with the Supreme Being. . . . This beauty is peculiar to natural things, it surpassing the art of man." In the nineteenth century, many Americans, including Edwards's great-grandson William Henry Edwards, encountered the butterflies, among the most evolved in terms of beauty, by some accounts, of all creatures. By beauty here is meant not merely the wings, however beautiful they may be, but the metamorphosis (from the Greek for "changing form") and life history of the insect, from the egg and caterpillar to the pupa and adult, as well as the butterfly in relation to a world full of other life. The encounter took place first with native American species, then with foreign or exotic ones, moths as well as butterflies, the "night" and "day butterflies," as they were called—or, collectively, the Lepidoptera, the order's name referring, in Greek, to the scales covering the wings. By the 1880s, it seemed as if everyone in America was chasing "flying flowers," to quote Augustus Radcliffe Grote, an American expert on moths.[1] People from all walks of life—sheep farmers, shopkeepers, barbers, lawyers, actors, drugstore clerks, housekeepers, wallpaper hangers, priests, Wall Street brokers, glassblowers, miners, and mine managers—had taken up the butterfly net. At the end of the century a new kind of beauty would assume prominence, human-made and artificial; seen especially at the spectacular world's fairs of the age, from Philadelphia to San Francisco, it would expand the world's aesthetic palette, while at the same time challenging the "beauty peculiar to natural things."[2] For a fleeting moment, however, the reign of the butterflies, and of all similar natural life, held sway in the imaginations of many American men and women.

Two phenomena were responsible, in particular, for this history, each ushering people into the natural world with unsurpassed effect. The first was economic and technological and was connected to the rise of capitalism; the second was cultural and institutional and was exemplified by the tradition of Enlightenment natural history, having at its heart a passion for the diversity and beauty of natural forms. One was extractive, the other adoring, and both derived from the thousands of years of experience the Europeans and English imported with them to America. Both existed in the same country and the same people, from Thomas Jefferson to Theodore Roosevelt and beyond. America was founded as a kind of European achievement, the beneficiary, as it were, of centuries of struggle, at once armed with the brilliance to build a new economic empire and endowed with the heritage to understand and protect the natural world.

The American economy had many aspects dependent on nature as landed wealth or property, and that brought Americans close to nature. No economic system had ever done this more thoroughly or swiftly. In what seems a twinkling of historical time, it carried people across the continent in search of land, minerals, virgin forests, and fossil fuels, a situation speeded up by a remarkable series of land surveys and revolutionized by the railroads, which by the 1880s had bound the country into a single market, condensing into only a few decades the expropriation of nature it had taken the Europeans and English many hundreds of years to carry out. As a fortuitous blessing, Americans experienced the country's flora and fauna, its native species of butterflies and moths. Later in the century, the United States joined other Western countries, England and Germany especially, in worldwide commercial trade and butterfly collecting, driven by imperialism and the spread of railroad lines from Canada and Argentina to East Africa and the Asian subcontinent, thereby putting Americans in touch with the foreign parts of the globe.[3] Many individuals made a living finding and selling butterflies on the world market. Most tragic and complex of them all was Will Doherty of Mount Auburn, Ohio, who spent most of his brief life in pursuit of the rarest of species. Extraordinary collections of butterflies in America issued from this descent below the equator. The biggest private one was cobbled together by Herman Strecker of Reading, Pennsylvania, a poor stonecutter with a wild yearning for the "things of endless joy," as he called his insects.

Family farming, America's most widespread economic activity, also exposed Americans to butterflies, even while the majority of farmers had little interest in insects other than to eradicate them. As all the butterfly people came to realize, America had various natural landscapes, from the remote alpine meadows of the Rocky Mountains and the vast blooming prairies of the Midwest to the deserts of the Far West, the freshwater marshes of Connecticut, and the semitropical forests of Florida, alive with "flying jewels," to cite Augustus Grote again. But family farms, much nearer at hand and not fundamentally dependent on the railroad, did perhaps more than any other landscape to convert Americans into butterfly lovers. Farms were distributed throughout the country, and while they sacrificed virgin forests and ecosystems in the short term, they contributed over the long term to nature's vitality.[4] Their distinguishing features were not just plowed fields or barns or silos but also ponds, woodlots, hedgerows, stone walls, open fields along roadsides, and meadows by streams or riverbeds for grazing cattle, all created for human purposes but also serving as likely habitats and hideouts for animals.[5] Renewed by repeated mowing, the meadows, especially, teamed with many kinds of birds; sweet-smelling flowering plants, intoxicating in the summertime, such as milkweed, joe-pye weed, thistle, and clover; and butterflies. Black swallowtails (*Papilio polyxenes*) and monarchs (*Danaus plexippus*), among the most common and handsome of American butterflies, actually *rose* in number in direct relation to the spread of small farms. So did the pretty inch-wide meadow fritillary (*Boloria bellona*), first named by the pioneering butterfly man Samuel Scudder, in the 1870s, because it so often flew in the meadows of New England, and nowhere else. Authorities on butterflies today call these insects "pasture species."[6]

A hallmark of this hybrid rural landscape (hybrid because it intermingled human nature with wild nature), was its "walking through" character, existing before property lines rigidly divided farms from one another, and making nature readily traversable by anyone interested in knowing it. Alpine or mountain meadows of the country's wilderness areas, carpeted by flowers and sometimes staggering in their array of butterflies, were often too far away or frightening for most Americans to visit, but this agrarian tapestry, resplendent with "winged wanderers on clover sweet," was everywhere and usually inviting.[7] Just as the railroad drew people over the horizon to unusual insects, family farming performed a captivating magic of its own. The farm landscape was easily navigated. Hector St. John de Crèvecoeur, a French immigrant and farmer, took delight in it in the 1790s, as, decades later, did the English explorer and collector Edward Doubleday, who witnessed "in Ohio literally tens of thousands" of painted ladies (*Vanessa cardui*), a lovely, wide-ranging species with an intricate lacework of color and design on the underwing, fluttering "on the thistles by the road sides." The writer and reformer Thomas Wentworth Higginson remembered "vividly" as a boy in Massachusetts in the 1830s, "walking along [a] breezy, upland road, lined with a continuous row of milk weed blossoms and white flowering alder, all ablaze with butterflies. I might have picked off hundreds, so absorbed were they in their pretty pursuits."[8] Forty years later, Walt Whitman walked his farm lanes in Camden, New Jersey, and in Brooklyn, New York. "As every man has his hobby-liking," he noted, "mine is for a real farm-lane fenced by old chestnut-rails," along which he saw "butterflies and butterflies, all sorts, white, yellow, brown, purple—now and then some gorgeous fellow flashing lazily by on wings like artists' palettes dabb'd with every color." "In the lane as I came along just now I noticed one spot, ten feet square or so, where more than a hundred had collected, holding a revel, a gyration-dance, a butterfly good-time, winding and circling, down and across, but always keeping within the limits."[9]

Most of America's principal butterfly people would find many new species of butterflies in ecologically wild, relatively pristine places, but they also encountered butterflies living near or in nature that had been modified in some way by farming. Herman Strecker wrote wittily about this environment in his 1878 catalog, *Butterflies and Moths of North America:* "The best time to give [butterflies] chase and try to run them down is under a July sun, with the occasional slight obstacles of fences, creeks, rocks, logs, farmers' dogs and farmers' boys (just as bad)" blocking entry to "a grain or clover field" or to "gardens, marshes and meadows along edges of woods, and above all where plenty of thistles and sumac are growing."[10]

Just as family farms helped Americans see and come to know butterflies, so the natural history tradition created the cultural context for butterfly collecting and study. It gave Americans the means—the language, the interpretative methods and skills—to recognize and understand the living things around them, and, by validating collecting as the cardinal activity, it led curious individuals into a realm of unforgettable sensuous experience. In countries without such a tradition, little existed—save folk taxonomies—to explain what lay within nature's kingdom or to promote collecting or to make of it anything more than a practical activity. In the West, however, natural history had a long lineage, dating from at least Aristotle, who inquired into all nature, from rocks and animals to plants and fossils.[11] Between the late Renaissance and the mid-1700s, it entered a new phase, fresh with purpose and mission, and by 1800 had claimed many thousands of followers, reflecting the Western intrusion into the rest of the world.[12] Natural history institutionalized collecting as a transcendent goal and invented the apparatus of collecting, from killing jars and nets to poisons, baits, and cabinets. It embraced, as well, a systematic approach to the study of nature, with two related aspects, each reliant on collecting. The first was devoted to taxonomy and nomenclature, or to the grouping and naming of organisms; the second, to the study of life histories and the interrelations of all organisms to one another and to the wider environment; this aspect came to be called ecology by the 1880s.

Carl Linnaeus of Sweden, Georges-Louis Leclerc Comte de Buffon of France, Alexander Humboldt of Germany, and, later, Charles Darwin of England, major architects of modern natural history and all scientific autodidacts, commanded, in their books, healthy swaths of nature.[13] Together they helped forge nature as a magnetic fulcrum of transatlantic Western culture, thrilling enough to keep people in its thrall for another 150 years.

Linnaeus was the great exponent of systematic order, establishing a binomial nomenclature, still standard today, that bestowed on all organisms two Latin words, one for the genus, or group, to which the individual organism belonged, the other for the species itself; the two words together represented the complete individual (thus, *Homo sapiens* for "wise man"). Early in his career, Linnaeus seemed to believe that every species was unchanging and God-created, although later he adopted "limited transmutation," and he never failed to root his understanding of species in concrete natural evidence.[14] He looked foremost at plants and flowers, identifying them partly by sexual characteristics, but he also named and described animals, including numerous lepidoptera caught by a brave cohort of young field collectors he sent around the world.

Born around the same time as Linnaeus (1707), the Comte de Buffon, famous for his spellbinding, multivolume *Histoire naturelle,* was determined to cover all animal and plant life, to a degree far beyond Linnaeus. He was critical of Linnaean systematics and of the early Enlightenment, which he thought viewed nature as too fixed in its forms and development. To Buffon, natural organisms or species flowed in "a progression" that changed constantly through almost imperceptible "gradations," self-creating and self-propelled.[15] He considered "Nature" an immense living empire, enfolding everything, animating everything, and urged naturalists to deal always with *living organisms* and to write their "life histories," describing fully and exactly every feature of their existence: their distribution, their outside and inside anatomies, their peculiar habits and everyday activities, their relationships with the life histories of other beings and to the surrounding environment. Species could not be recognized, Buffon argued, only on the basis of single visible sexual features, as was Linnaeus's practice.[16]

Although historians have often seen these two sides of natural history as separate and even at odds, the two so often crossed as to form a whole perspective that grew ever more complex over time.[17] Buffon and Linnaeus shared a belief that humans had a primary right to use nature, first and foremost, for their material benefit; each insisted upon the necessity of describing as well as naming and classifying; both saw the natural world as an interdependent whole; and each was awed by its profusion. "The starting point," Linnaeus wrote, "must be to marvel at all things, even the most commonplace."[18] Still, rightly or wrongly, Linnaeus's fame rested on his binomial nomenclature and Buffon's on his descriptive talents, on the abundance of his engrossing life histories, and on his dramatic ecological vision, which addressed "the great operations of nature" and demanded a "quality of mind that permits us to grasp distant relationships, fit them together, and form a body of rational ideas."[19]

Buffon's version of natural history reached a zenith of influence during the Romantic Enlightenment of the early nineteenth century, most memorably in the mind of Alexander Humboldt, a Prussian aristocrat and the premier explorer of the time, whose multivolume masterpiece, *The Cosmos,* written in his old age and probing nearly every aspect of the universe, had a more profound impact on American views of nature than any other book in the nineteenth century. Humboldt journeyed down the mostly unexamined (for Europeans) Orinoco River in present-day Venezuela between 1799 and 1804, launching the "cartographic penetration of the continental interiors of the Americas." The trip culminated in Ecuador with Humboldt's sensational ascent with his companions—the botanist Aimé Bonpland, his Ecuadorian friend Carlos Montúfar, and an Indian guide—nearly to the top of Mount Chimborazo, an Andean peak then thought the highest in the world. Besides harvesting a huge number of plants, Bonpland and Humboldt collected butterflies, including one delicate but hardy creature above the snow zone—a yellow or sulphur species one inch in diameter (*Colias alticola*)—and two new species later named and described in 1805 by the French butterfly man Pierre Latreille as *Heliconius humboldt* and *Cethosia bonpland*.[20] Despite these discoveries, Humboldt, like Buffon, had no desire, as he put it, to unearth "new, isolated facts" but "preferred linking always known ones together." Also, as with Buffon, he was convinced that "nature, despite her seeming diversity, is always a unity, a whole," and that a "holy, creative, primary force" suffused all.[21] Nature was no "inert mass" but "an inextricable network of organisms," so gloriously fertile in places around the globe that it promised to endure forever.

A half century later, Charles Darwin's theories of natural selection and evolution (along with those of his countryman, Alfred Russel Wallace) would add a radically dynamic element to natural history thinking, conceiving of nature as an always evolving and unstable realm, working from within to create a new abundance of variable shapes, patterns, colors, and sizes: geographic, polymorphic (meaning many forms), and dimorphic (meaning double forms), and encompassing varieties, species, and subspecies—altogether a morphological spectrum never before revealed to naturalists in the same way or to the same degree.

The natural history of these men erected a route to scientific understanding and knowledge. At the same time, it occasionally mixed culture, fantasy, and myth with science; it sometimes treated nature anthropomorphically, told stories, and had literary as well as scientific content; and, most of all, it placed people in touch with the multifarious beauty of the world.

Throughout most of the nineteenth century, naturalists typically learned to draw or paint and, therefore, possibly to see and appreciate better the living beauty around them. We can trace this fruitful relationship to an abiding alliance between art and science, with roots in the late medieval period (if not earlier), when it was commonly held that the ability to pursue art or science depended on the same observational skills and perceptions, the same pair of eyes. This view can be seen in the work of Leonardo da Vinci, Robert Hooker, and William Harvey, among others, and found exponents in the late-eighteenth-century philosophy of Johann Wolfgang von Goethe, Friedrich von Schelling, and many of their contemporaries.[22] It was an outlook, moreover, characteristic of a venerable tradition of artisan craftsmanship practiced by a large community of people outside the elite bastions of science who conjured up images with their own hands, a practice that helped increase the observer's sensitivity and responsiveness to the aesthetic character of nature.[23]

In its heyday, natural history unleashed a stream of pictorial books of insects and other species, with handcrafted plates notable for their costliness as well as for their lively colors, although these sometimes bore little relation to the actual colors of the butterflies and moths depicted. In the early eighteenth century, the German-born apothecary Albertus Seba, then living in Amsterdam, published the four-volume *Cabinet of Curiosities,* a great prototypic catalog, with hundreds of exotic butterflies never before seen. Almost all of the copies were printed in black-and-white; those few that used color were set aside for the tiny royalty of northern Europe who were Seba's benefactors.[24] Other contemporary works with plates, by such people as the Germans Maria Sibylla Merian and August Rösel von Rosenhof and the Dutchman Pieter Cramer, also served a small elite.[25] By the early nineteenth century, both craftsmanship and natural science had overcome the obstacles of poorly executed shades and tints and of an exclusive market. Starting with the rise of lithography in the 1820s, artists and artisans began to depict the real thing in realistic color, and the audience was no longer elite.

Buffon, Darwin, and Humboldt themselves seemed indifferent to drawing lines between art and science. Their prose was often excellent and readable, so much so as to later seem unscientific. They often wrote on behalf of the artistic attractions of nature. "It is certain that natural history is the mother of all arts," Buffon affirmed. "All ideas of the arts have their models in the productions of nature. God created and man imitates."[26] Although no naturalist had more faith in empirical science than Humboldt, he embraced "the ancient bond which unites natural science and artistic feeling." He also rejoiced in the beauty of nature as perceived through the medium of human subjectivity: "For it is the inward mirror of the sensitive mind which reflects the true and living image of the natural world. All that determines the character of a landscape are in antecedent mysterious communion with the inner feelings and life of man."[27] Humboldt believed that anyone without the ability to see beauty would probably also be unable to see the chain of connections underlying nature.[28]

Such a perspective on the natural world, lasting well into the nineteenth century, would be weakened by many changes, ranging from the professionalization of science and the attendant insistence on a rigid separation of art from science, to the invention of photography, which struck a blow at the human hand as the maker and shaper of images.[29] Still, it held up (however hobbled) and finds advocates from all walks of life in the present day.

Natural history, at a very high point in its popularity, made a vigorous claim on the culture of the United States, not least because of the near diffusion of wealth and "comfort" throughout the countryside and because of American educational levels. By the early 1800s, the white population in the United States was already the most literate, the most competent, and the best educated in the world. In the 1830s, for the first time anywhere, women could read and write as well as men could. Both achievements derived from the spread of common schools, the presence of lending libraries everywhere, a vital free press, and an egalitarian ideology that opened up the study of nature to all who cared to pursue it.[30] As a consequence, from the 1810s on, Linnaean societies, natural history clubs, and small college societies appeared in city after city, followed by the rise of natural history museums, where people might see illustrated books of lepidoptera, as well as of other colorful natural life. Improved printing presses aided in the production of books and periodicals on butterflies and moths, while the railroads safely and rapidly delivered countless specimens in the mails and multiplied the locations where butterfly people debated, fought over ideas and membership, and set standards in naming and classification. Photography, in a special way, tied these naturalists into a community.[31] From the 1850s on, but-terfly people mailed thousands of letters to one another, and requests for photographs became a ritual part of their letter writing, a kind of Masonic testimony to a seemingly invisible presence. We "meet in the shadow of the camera," Benjamin Walsh, an Illinois naturalist, put it to his friend Hermann Hagen of the Museum of Comparative Zoology in Cambridge, Massachusetts.[32]

America produced many gifted butterfly people, all of whom deserve today to be much better known, beneficiaries of Linnaeus, Buffon, Hum-boldt, and Darwin and, in time, serious rivals of the English and Euro-peans in creativity. These butterfly people, in turn, depended on an even greater mass of Americans from all stations and classes, many willing to serve the leading figures in virtually any way so that they might have the insects they required to describe, figure, and identify. America was full of such people, their existence a measure of the vitality of American culture at midcentury. Augustus Grote called them "hod carriers," and himself and others like him "boss masons," where today "amateur" and "professional" describe such roles. "Amateur" means, literally, anyone who does something for the love of it; by the 1880s that commanded a lot of ground, since it was true of both professionals and amateurs. It did not imply, however, working for someone else in a self-sacrificing kind of way. "Hod carrier" and "boss mason" came closer to this latter meaning, and closer to the experiences of people involved with butter-flies. At the time, many men actually carried hods (trays of bricks) for boss masons in the building trades ubiquitous throughout the country, and many were on their way to becoming boss masons of some sort themselves, artisans with the skill to give, with their own hands, shape, form, and color to the world around them and, very likely because of those skills, to see beauty in nature.[33] Herman Strecker himself was one of these people, a talented sculptor and designer of decorative objects who started out as a hod carrier, or apprentice, literally bearing bricks for his boss mason father, Ferdinand Strecker. He ended up as an actual boss mason (although he owned a marble yard for only a short period of time) and with a highly sophisticated eye for the beauty of lepidoptera.

In America before 1880, one might begin as a hod carrier who imag-ined becoming a boss mason, despite a lack of formal education, by spending time under the tutelage of someone of more learning and experience—a boss mason, in other words. The terms were so common as to have passed into metaphors by the period in which Grote used them for the butterfly men. "There is no 'hod carrier' who don't think he would be a good 'boss mason,' and as this is a free country, if he wants to try it, who can hinder?" a young George Hulst told Herman Strecker in 1876. (By 1900, Hulst would be a respected authority on American moths.)[34] On the other hand, there were other Americans who were neither boss masons or hod carriers but merely people attracted by lepidoptera, eager to collect them and to know everything about them; these people belonged, very likely, to the biggest and most impressive community of all.

There is no way of accurately measuring the size of this last community, although without it there would certainly have been no hod carriers and no boss masons. The leading butterfly people in America, in fact, emerged out of this community. Each worked hard to reach their goals, and each came from backgrounds shaped in a classic American mold, utilitarian and Romantic about nature at the same time. William Henry Edwards was a coal mine manager who brought the coal business to West Virginia, Herman Strecker a stonecutter who specialized in gravestones for children, and Augustus Grote the musical son of a failed real estate investor and railroad entrepreneur from Staten Island. William Doherty's father founded the trolley car business in Ohio, and Samuel Scudder was the child of a Boston hardware merchant and at first had little interest in "winged wanderers on clover sweet." All of them would come to study the "life and death" of butterflies, observing especially life cycles and ecologies rather than the "dry bones" of taxonomy and nomenclature, although as new claimants to the Linnaean mantle, they would fight bitterly over the merits of naming and classifying. What *is* a name? And who has the right to name or classify? They led other naturalists in the country, too, in absorbing evolutionary thought into their work. And they considered beauty as part of what mattered in nature—ever present, intrinsic, and inescapable, one might suppose, given the splendor they found here and abroad. Edwards, Scudder, Grote, Strecker, and Doherty—these are the key figures of this history, as is another individual, a man far less admirable than the rest. The son of self-sacrificing Moravian Christian missionaries, William Holland was, ironically, a self-serving autocrat, an unsavory opportunist who, after becoming the first director of the new Carnegie Museum of Natural History, in Pittsburgh, would affect the lives of all the other butterfly people. Though he would do some good, he would also help to undermine the promise of a great tradition.

The reigning perspective on nature in America, from the colonial period on, was economic. At the same time, natural history, with its strong aesthetic dimension, existed side by side with the economic approach and, for a short time, made room for many Americans to think and dream about butterflies independent of any material purpose or inclination, and persuading some to make riveting observations concerning the beauty of the world. Today, most Americans come to know nature through technology and the commercial market—through television, computer screens, motion pictures, advertising, and tourism (including ecotourism), or through public institutions that have become wholly or partly commercialized, such as zoos, game preserves, aquariums, botanical gardens, national parks, and even natural history museums. One hundred fifty years ago, family farming, hunting, amateur exploring, and, above all, natural history in its heyday, with its loyalty to the pleasures of collecting, introduced untold numbers of Americans to nature, freely, and in nearly unscripted ways. In our time, butterfly nets have virtually vanished from the American scene, after years of attacks on collecting as insensitive and destructive to nature. This assault may have been a good thing, although nothing seems to be replacing collecting as a way to instill in children a passion for the natural world. In the earlier age, Americans chased butterflies and butterflies coaxed them on, improving their minds, altering the way they lived and what they lived for, weaving into the American cultural fabric yet another thread of democratic life.

Part One

ENCOUNTERS WITH THE
BUTTERFLIES OF AMERICA

ONE

Yankee Butterfly People

C arl Linnaeus of Sweden left a tremendous legacy of naming to American butterfly naturalists. In the mid-1700s, at the peak of the renascence of European natural history, he selected the names for the different stages of insect metamorphosis still in use today and applicable to butterflies. He chose "larva," meaning "mask" in Latin, or a form hiding the true identity of something—in this case, the adult butterfly, and more commonly known as a caterpillar. He adopted "pupa," meaning "doll" in Latin, or an "infant in swaddling clothes," otherwise called a chrysalis, or the stage when the insect "sleeps" until it reemerges as the "perfect butterfly." And he penned "imago," meaning "image" in Latin for the "perfect insect" at the final stage of the life cycle.[1] Linnaeus coined the word "lepidoptera." He also assumed the task of naming American species and, along with his favorite student, the Dane Johann Fabricius, identified a great many butterflies, despite having never visited the places where they existed (their students, whom they sent abroad, bore the brunt of that burden).[2] The original descriptions and Latin names for the monarch butterfly and the tiger and spicebush swallowtails are among their progeny. Pieter Cramer, wealthy Dutch wool merchant with worldwide contacts, especially in port cities, was the first to describe (and to draw in color) the Diana fritillary and the zebra swallowtail. Pierre Latreille, a French priest who studied butterflies in prison during the French Revolution and later left the church to win acclaim as a bug expert, published portraits of America's common clouded sulphur and barred sulphur.

This European domination in the butterfly world held into the next century. At the forefront was the French physician Jean Baptiste Boisduval, who wrote two volumes on American butterflies, the first over an eight-year period ending in 1837, with the help of John Le Conte of Philadelphia, reflecting knowledge gained since Linnaeus's day, and the second, in the 1860s, on Californian insects, with help from Pierre Lorquin, one of his patients. A rugged young lawyer, Lorquin was, like Boisduval, self-taught in natural science, and capable of walking many miles in an afternoon.[3] In the 1840s, what began for Lorquin as a search for Californian gold became a quest for butterflies in the gold fields; many prizes were quickly identified by Boisduval. Among them are the western tiger and pale swallowtails, the Lorquin's admiral, or *Limenitis lorquini,* and the California dogface, a dimorphic sulphur—the male black, yellow, and violet, the female all yellow; today it is the state's insect.[4]

Much of the work of the British and the Europeans was magnificent, the color pictures alone often stunning for their beauty and their power to arouse fantasy. But the systematics (or the complex placing of organisms into relationships or "systems" of arrangement) rested on the study of a paltry few specimens from the field or from private collections, often leading to distortions and myth, with little detailed information or none at all about where insects had been caught.[5]

Among the British naturalists who conceded the dangers and the injustice of preempting the privilege of naming and describing, notable was Edward Doubleday, an aesthetically sensitive young Quaker much enamored of butterflies. In the mid-1830s, when Doubleday came to the United States to collect, he was "struck" by similarities between American and English insects. But after he captured many "very distinct from any known to us," including "a magnificent Luna moth," a soft, light green species with delicate long tails, its caterpillar or larva having a preference for walnut trees, "by far the most foreign Lepidopeterous insect," he changed his mind. Warned in advance that Americans might mock him for chasing after bugs, he soon realized, to the contrary, that he was "actually looked upon with greater respect on the very account of these pursuits." He received "assistance from the neighbors around, with whom I may chance to meet. How I love America!"[6] One of those helping him was W. Wigglesworth of Wilmington, Delaware, who "walked hundreds of miles" to locate for him a "great number of cocoons" of large spectacular American moths, which he then reared to maturity and dispatched to England.[7]

Edward Doubleday wrote an original and influential catalog on butterfly genera (or those butterfly species grouped together according to certain shared properties) and, in a pioneering departure from the English obsession with English insects, dealt at length with the world's butterflies (with most examples taken from the British Museum) and tracked new methods of identification, especially drawing on the veins of the butterfly wings to establish differences among species and genera. Wherever evidence permitted, Doubleday described full life histories, the larval food plants and habits, and the geographical range or distribution, the last, he indicated, "an important branch of enquiry in Natural History, as yet too much in its infancy, for us to venture to draw general conclusions from the facts we possess." He was among the first to discuss seasonal variation in the appearance of butterflies. In the spirit of both Linnaeus and Buffon, he urged butterfly people to "accurately record facts, but guard against the error of making a theory."[8] All of this, along with his formal arrangement of butterfly families—the swallowtails (large colorful insects with tails) first in order and the skippers (usually drab small insects with large bodies and small wings, resembling moths) last—would reappear in the work of William Henry Edwards.

In Louisville, Kentucky, Doubleday fell even deeper in love with America, dumbstruck by "the richest and most verdant pastures" he had ever seen. "The whole country, where it has not been ploughed," he observed, "is like an English park, excepting that the trees are much finer. You see on all sides fine hills, clothed with gigantic trees, valleys with rich meadows, green as emeralds. Many butterflies are out. [There are] masses of flowers; the corn is just browning, and drooping its ripening ears. I never saw such beautiful scenery. Such a sky! Such a sun! Such sunsets! You in England do not know what these things are."[9] He thought seriously of immigrating to Kentucky.

Edward Doubleday, along with his brother, Henry, formally identified many American butterflies, but Edward came to realize that Americans could best describe and understand their own natural world. In an 1839 letter to a naturalist friend in Massachusetts, he apologized for "our collectors" boasting that American species were mere "allied British species," few or none unique to America. He would now leave matters entirely in American hands, he resolved: "I do not wish to take from an American the task of making known the productions of his own country. Little as I am what is called a patriot I know how little I should like a Frenchman or a German to be the first to make known a large portion of *our* insects, and I can from this judge what your feelings must be in a similar case."[10]

Most Americans in the early nineteenth century probably viewed nature as something to dig up or harvest from the earth or to be torn down and transformed into real estate or money. By the late 1840s, they were well on their way to forging a new capitalist civilization. On the other hand, there were many Americans who were, like the mine owner William Henry Edwards, both extractors of natural wealth *and* serious naturalists. Still others resembled Edward Doubleday, turning to nature wholly for insight, and some of these helped create a new American science of butterflies. One such individual was an English émigré, John Abbot, who foraged for insects in the woods, swamps, and fields around Savannah, Georgia; in 1797, he coauthored, with the Englishman John Smith, the first significant volume on American butterflies, *The Natural History of the Rarer Lepidopterous Insects of Georgia,* lauded as "perhaps the best lepidopterological work of the eighteenth century." The book deviated from earlier efforts by illustrating clearly and meticulously the caterpillar, pupa, and adult butterfly, as well as larval food plants and sometimes the egg for each insect observed, most discovered by Abbot himself.[11] Boisduval later adopted several Abbot images in his own first catalog, with full attribution. In his old age, Abbot continued his investigations, working from a small cabin in Georgia, cared for by a slave named Betsey.[12]

Another early enthusiast was Titian Peale, a Doubleday acquaintance and the son of Charles Willson Peale, founder of the country's first natural history museum, in Philadelphia, then a magnet of American natural science, and Titian's birthplace in 1799. Titian's artistic family was fiercely ardent about nature; for him, as for his father, "Art and Nature" were inseparable, and butterflies integrated the two perfectly.[13] At the age of twenty, Titian was hired as a "painter of natural history" on a government-backed expedition to inspect the mostly unmapped Louisiana Territory as far west as Yellowstone.[14] Years later, he sailed recklessly alone to Brazil with a merchant as his sponsor. From the deck of a steamboat on the Magdalena River, he sighted on the sandbars near Buena Vista many hundreds of iridescent, rainbow-colored *Urania fulgens* (from the Latin and the Greek, meaning literally "shining heavenly female"), an unusual moth with swallowtails that flew in the daytime, "their expanded wings, in clusters on the moist earth, rising in such confusion when disturbed that numbers were caught at a single sweep of a collector's net."[15] The sight burned in him a desire for tropical species many decades before others in the United States had acquired a taste for them.[16] Peale had ambitions to write an exhaustive volume titled *Lepidoptera Americana,* a catalog on all known North American butterflies "in the various stages of their existence, and the plants on which they feed," intending to do his own fieldwork and his own plates, and also to prepare another volume on tropical insects; but no market existed, and neither book saw the light of day.[17]

Abbot and Peale, reflecting the impact of Buffon, unlocked the gates for greater American work on life histories.[18] In other ways, their era had limits: no major collections, few journals or societies, no original books, no core group to broadcast the gospel of butterfly collecting and study. But by the late 1860s, these limits were nearly all gone, erased by the rise of a new literature, new clubs and societies for butterfly study, and the brilliant Yankee trailblazers, William Henry Edwards and Samuel Scudder.[19]

Edwards and Scudder were the master figures in the history of the American encounter with the butterflies of North America, building on the achievements of the others. These two men, sons of Calvinists or neo-Calvinists, would tell the stories of more unsung American butterflies than anyone in history. They would go further, too—they would enrich the biographies of all *known* American butterflies, previously described by the Europeans and the English on the basis of only a few specimens, almost always with an eye to the adult insect. In a way well beyond earlier naturalists and, to some extent, later ones, Edwards and Scudder would scrutinize *all* butterfly existence: every stage, what the insect ate, its courting and mating behavior, its "parenting" practices, how it dealt with the surrounding world, how it kept going in face of a grisly army of enemies, and what finally killed it. In books of remarkable depth, figured in unprecedented clarity and completeness, and written in lucid prose, they would deal with butterflies almost completely as living things and with a determination to remain independent from the reigning economic priorities in America.

———

*William Henry Edwards.
Courtesy of the
West Virginia State
Archives, Charleston,
West Virginia.*

Samuel Scudder fought with William Henry Edwards throughout their relationship but still considered him the "foremost student of the life history of American butterflies."[20] That view has persisted up to our own time. Edwards would combine in his life the enterprise of a pioneering coal mine manager with the enthusiasm of a butterfly man; he was an extractor of two different kinds of wealth, both of which he saw as part of the plenitude of the natural world. A direct descendant of Jonathan Edwards, America's foremost theologian and one of its most powerful minds, William was born in 1822, only sixty-four years after Jonathan Edwards's death. In religious terms, the years between them could have been centuries, a void. William hated Jonathan's Calvinism, with its assertion of the sovereignty of God over all creation, of eternal damnation as the price of sin, and of the impossibility of people saving themselves or knowing for sure whether or not they were saved at all. William himself was a secular agnostic and a Darwinian; he believed that most people made their own way in life and had no need for a sovereign God.

Yet William Henry Edwards's feelings about the natural world can be traced back to his renowned relative, who viewed nature not as something to be appropriated as wealth but as "spiritual beauties." "Bodies being but the shadow of beings, they must be so much the more charming as they shadow forth spiritual beauties. This beauty is peculiar to natural things, it surpassing the art of man," Jonathan observed in an essay of his youth, "Beauty of the World." "The fields and woods seem to rejoice, and how joyful do the birds seem to be in it. How much a resemblance is there of every grace in the fields covered with plants and flowers, when the sun shines serenely on them."[21] At the same time, Jonathan delved into the latest scientific discoveries, including those of the equally religious Isaac Newton, which he tried brilliantly to reconcile with his religious vision. Jonathan was fascinated with the small things, such as spiders, especially forest spiders, whose silken threads let them float about through the air. He decided that their "floating" gave them "a great deal of their sort of pleasure," while making them, at the same time, susceptible to the wind which, he thought, carried the spiders off and destroyed them. Such was God's way of checking spider fertility, which, left unchecked, would skyrocket. God used similar methods to restrain all insects, Edwards asserted, so that "taking one year with another, there is always just an equal number of them."[22]

His great-grandson shared his curiosity about the little things, though he specialized in butterflies, not spiders. William Henry Edwards had a comparable conception of an energy coursing through organic life, forming and unforming it, and fostering some kind of harmony over the very long haul. For Jonathan, this power was God power or the "Being" of God, the architect of nature "in the beginning" and throughout time, inherent yet transcendent: "The universe is created out of nothing every moment, and if it were not for our imaginations, which hinder us, we might see that wonderful work performed continually."[23] For William Henry, the deciding impulse was, rather, natural selection, acting in bits and pieces intrinsically, each organism constantly in a perpetual "struggle for existence" that made life and took it, too, checking fertility and creating the fittest specimens. Was this chaotic force but another version of God power, acting ruthlessly and without purpose, just as God often seemed to do in the eyes of Calvinist sinners, helpless to save themselves (like all those spiders dead before their time)? If anyone had told William Henry that his butterfly interests might have been found in the crucible of his forebear's Calvinism, he would have jeered, given his contempt for his ancestor's religion and his own radical take on the "beauty of the world."

William Henry Edwards was born in his grandfather's house, "in the village of Hunter in the heart of the Catskill Mountains," as he put it, nesting ground for many butterflies he would later study. His grandfather, Colonel William Edwards, erected what was at the time the largest tanning factory under a covered roof in the country, standing at the heart of Hunter, with sheds for hemlock bark and cordwood and a boardinghouse for single men. Edwards grew up amid strangers and factory smells, along with the poverty caused by economic cycles that would sometimes afflict his family; he accepted the toxic impact of tanning on nature as a fact of life.[24] William Henry's father, William, cut in the religious mold of Jonathan Edwards, believed in a "personal devil" and "a fiery hell," while his mother, Helen Mann, was a religious liberal. In 1838, the elder Edwards sent his son to Williams College, in Massachusetts, a Congregationalist school governed by strict religious principles, established, in Edwards's words, for "the sons of poor men who had striven hard to get to college." The sixteen-year-old Edwards nearly "froze to death" his first year, during an "infernal winter" that seemed to seep into the beds and books and was worsened by his room-mate, a zealot of twenty-eight who tried, one Sunday, as Edwards later recounted to Scudder, "to get me on my knees to prayers." The zealot tattled to the college president that Edwards had violated a Sabbath law against all secular activity by reading Livy on Sunday, and Edwards was dressed down by the president for violating the ban.[25]

Edwards could not stomach Williams, except for one thing: it was among the first colleges in the country to put natural history at the forefront of its curriculum, unwittingly giving him a secular bridge to the glories of the natural world. Wherever natural history appeared in America, it had this effect: it created and justified the logic of collecting, which was not merely "getting" or "assembling" objects to have or hoard them or to put them on display but a *path into* nature; it was the *thing itself,* the clarion call of Western systematics; countering the economic, extractive approach to nature, it beckoned young men and women to go out into and to feel and touch the living world around them. Under the tutelage of Jeremiah Emmons, a professor of natural history at Williams, Edwards became a collector, later observing, "There is no occupation so delightful as that of collecting examples in any branch of natural history." He read deeply "along natural history lines," admiring, especially, Gilbert White's *Natural History of Selbourne* and Darwin's *Voyage of a Naturalist;* the former set him to keeping a journal, the latter to dreaming of exploration. His first interest, as a freshman, was in the study of birds, one he picked up "all of a sudden," "going out every fine day," aroused by the colors and patterns; he "hunted and stuffed" them, as did many boys throughout the transatlantic world until the end of the century. Edwards sold subscriptions at Williams to Audubon's *The Birds of North America,* the fourth volume of which had just come out in Edinburgh and London. He tried, briefly, to collect butterflies and even made a net, but he had no clue how to pin insects or preserve them from decay. He knew of no one else anywhere who collected butterflies.[26]

Although, at twenty-two, Edwards hoped to become an ornithologist, he had to settle for jobs teaching at middling boys' schools in New Jersey. The work was made bearable, however, by frequent excursions into nature for birds, mostly with other young men, such as a "Mr. Kowalski," who in May 1844 "brought a double barreled gun" with him. "Mr. Kowalski shot a New Warbler. Besides this we shot a Black throat Blue. Black throat Green. Black and white creeping and Blue wing, yellow Warblers, a Red-eyed Vireo. Saw a Scarlet Tanager, but he flew from us."[27] Then, unexpectedly, in 1846, Edwards's resourceful uncle Amory, a businessman near his own age, persuaded Edwards to join him on a nine-month voyage on the Orinoco River, along the very same route Alexander Humboldt had taken and whose *Personal Narrative of Travels* Edwards had read. Once there, Edwards hired soldiers to help kill birds for his collection and journeyed to the breeding grounds of the scarlet ibis and the roseate spoonbill, "the gaudiest birds that fly."[28] Edwards's account of the trip—*Voyage up the River Amazon,* written in 1847, on his return home—made him briefly famous. His only regret was failing "to collect butterflies," "everywhere, and often beautiful."[29]

Also in 1847, Edwards's youngest and dearest brother, Jonas, died at age twenty-four, and William inherited from this budding artist two things: a costly antique gold cross bought partly with a loan (at the time of his Jonas's death still unpaid) and thirty thousand acres of land in West Virginia (then still part of Virginia). William, out of brotherly love, sailed to London to find a buyer for the cross, so the loan could be repaid. Equipped with his explorer's reputation, he had easy access to naturalists.[30] In London, he met William Spence, the elderly coauthor (with William Kirby) of the best-selling 1815 book *An Introduction to Entomology,* the first work of its kind, with lively stories about insects, including butterflies and descriptions of "metamorphosis." Spence and his wife had just finished "reading my book," the young Edwards rejoiced, "and their heads were full of it."[31] He also met John Gould, the "Audubon of England," whose sons had read his *Voyage up the River Amazon* "over and over again," and together they discussed Gould's popular books. Gould showed Edwards his hummingbirds, "exquisitely mounted and arranged in glass cases," a form of display of birds and of insects that had just become popular in England.[32] Edwards "turned the hummers" around, he later recalled, "so as to present their several fronts to the light. I never saw anything half so magnificent."[33] In a visit to his hotel, two starry-eyed young men, Alfred Russel Wallace and Henry Bates, not yet famous, told him they had been inspired by his book to go to the Amazon themselves and were leaving for Brazil the very next day. The many butterflies they caught there later became evidence for their own original evolutionary studies.

Perhaps most stirring of all, Edwards, in the halls of the British Museum, bumped into a thin, intense Edward Doubleday, now the curator of insects, and Doubleday told him of plans to leave the museum for Kentucky, to make his home there. (A year later he would be dead from a spinal tumor.) No record exists of what passed between them, but the Englishman must have given Edwards an earful about what should be done on American butterflies and *why* it should be done by Americans. In his meetings with Gould and Spence, Edwards had been exposed to a new middle-class market in natural history books, but his time with Doubleday, as well as his visit with Wallace and Bates, meant even more, for even though he did not sell his brother's gold cross and was disgusted by London's class extremes—"its gold and rags"—he returned home sold on butterflies, not birds, as his life's work.[34]

The second legacy from Jonas, the title to thirty thousand acres of land bought cheap from a Virginia speculator, would reward Edwards for many years, for under it was one of the thickest veins of coal in the world, enough to make the region the bulwark of the country's coal industry by the end of the century and beyond.[35] Oil, squeezed from cannel coal, had just become profitable as a lamp illuminant, and others in the East—above all, the speculators—were eager to exploit it. (In fact, from the 1790s on, easterners owned more than 93 percent of west Virginia, turning it, in effect, into a colony of absentee American capitalists. It has remained that way ever since, with most residents themselves left poor and landless.)[36] An incipient capitalist as well as a naturalist, Edwards decided to make his living operating coal mines. Several months after Jonas's death and his trip to London, he shrewdly studied real estate law, passed his bar exam in New York (the first man to do so in that state), set up an office on Wall Street, and then traveled to the Kanawha Valley to take possession of his land.[37] After raising capital from London and New York investors, he organized, in 1864, at the height of the Civil War, the Wall Street–funded Kanawha and Ohio Coal Company, with himself as manager at a salary of $3,000 a year. He launched the first coal towboat in mining history, set up cabins for workers and a company store, and blasted the first mines near what would become his home.[38] For four more years, Edwards went by coach and horse, back and forth to West Virginia from Newburgh, New York, on the banks of the Hudson River, where he had lived since 1859 with his wife, Catherine Tappan, the daughter of the abolitionist Arthur Tappan, and their three children.[39] All the while, his pursuit of coal unveiled another reality having nothing to do with coal, and as enthralling to him as it had been to his Calvinist forebear.

Many more people were collecting butterflies since Edwards's days at Williams, a change that helped move him away from birds. He had first begun to "go down the butterfly path," as he put it, around 1856, when he was thirty-three. Probably he picked up the essentials of butterfly learning from John Weidemeyer, a German-American in New York City who, in 1864, wrote one of the first American books on butterflies, the unillustrated *Catalogue of North American Butterflies*—half of which dealt, however, with West Indian and Central American insects. Weidemeyer had a strange notion of "North American"; nevertheless, he raised informed questions about several American species that Edwards later attempted to answer, and in the late 1850s, he generously shared insects with Edwards for study and identification.[40] Another German-American was John Akhurst of Brooklyn, an affable seller of entomological supplies and a taxidermist famous in naturalist circles for his goggles and his flowing shoulder-length hair, sprinkled with the white arsenic he used to stuff birds. Akhurst instructed Edwards in the arts of breeding and preserving lepidoptera.[41]

Samuel Scudder, in his early twenties, would pay visits to Edwards in the late 1850s and early 1860s, once consuming "half a morning to trying to describe the differences" between various fritillaries, as Edwards would later remember it.[42] Edwards often went on trips of his own, one a visit to Washington, D.C., where he learned from Titian Peale how to enclose specimen boxes in glass, thus letting the viewer see insects from nearly any angle, "a great contrivance, for discovery of differences or resemblances," and a reminder of Gould's encasements for hummers in London.[43] Also, in Washington, he met Spencer Baird, busy amassing the collections of the Smithsonian (among the earliest in the country); Baird offered to send him, in exchange for a promise to identify them, all the insects his men had caught on expeditions, a tremendous boon for Edwards.[44] His first interest had been birds, but there were numerous books on birds already, and Edwards knew he could not compete with them. American butterflies, on the other hand, presented nearly an open field.

In 1868 Edwards shed his absentee-owner status and moved south, out of necessity, from New York's Hudson Valley to Coalburgh, the name he himself may have chosen for that tiny town along the Kanawha River, the central artery of West Virginia. Although the state was covered by a vast forested wilderness, Edwards's county had been settled to some degree with family farmers along or near the river.[45] When he cleared land to make way for his dwelling, he left standing elm, sassafras, and tulip trees, the food plants for several butterflies, with moist woodlands near the river, the ideal breeding grounds in particular for the zebra swallowtail, an insect Edwards would soon come to study.[46] Pawpaw bushes, the food plant of the zebra, dressed the lower hillsides along the pathway from Edwards's house to the entrance of the mine on the mountainside facing the river. "As I write," Edwards said in a letter to Henry Edwards, a stage actor, a devoted butterfly man, and one of William's closest friends (although no relation), "I see patches of lavender blue phlox at 200 square, and masses of white trillium, and were I in the spot, I should see millions of violets, blue, yellow, and white. This is the loveliest bit of territory in the United States. I wish you were here to see it." Weeks later he wrote, again to Henry, that "the last rain has made all nature beautiful. I have roses without end. Swarms of Cybele [great spangled fritillaries] are in my clover."[47]

Edwards planted flowers in the garden near the house, especially his preferred zinnias, milkweed, and phlox. In late summer, when he stepped off his back porch, he would see swallowtails "swarming over the Phloxes and Zinnias."[48] Along the wooded peripheries, he put flowering pear, peach, apple, and plum trees, all pleasing to butterflies, and still farther off, well beyond his house, there was a wilder nature, from the narrow swampy lowlands and clover fields along the river below to the sinewy topography above, with its deep hollows, silvery creeks, and rolling mountains, cradling a vast population of butterfly life.[49] Until he was sixty years old, Edwards gathered larvae on the mountaintop behind his home.[50] Weather and season permitting, he went out nearly every day, recording what he learned in a journal he kept from 1859 to the end of the century.

The place he favored most was Paint Creek, a field and streambed that sloped down to the nearby Kanawha River; he decided early on, ironically, that it would also be the site of his coal mining business. In August 1864, he spied a butterfly there he had never seen before, on ironweed—a male specimen of the Diana fritillary (*Speyeria diana*), a bright, handsome orange-and-russet-brown insect, its underwings speckled with silver characteristic of nearly all fritillaries. The insect was feeding in its preferred habitat, an opening at the edge of a mountain forest. Three weeks later, he also saw an even more striking butterfly, larger, dark blue and velvety black, with little white markings along the wing margins, "feeding so quietly on ironweed," as he put it, "as to allow me to stand near it and watch its motions." He soon learned that this was the female counterpart of the earlier butterfly; the two, so visually different, were the same species, European or English claims to the contrary notwithstanding.[51] More secretive and forest-bound than the male, the female was there mainly to find a mate or to lay eggs on violets, its only food plant. "The contrast between the sexes has no parallel among North American butterflies," Edwards wrote.[52] On one April morning, "up Paint Creek," he also saw "myriads of blues," little, jewel-like insects, and caught hundreds in a single arc of the net. The next April, he watched countless female zebra swallowtails "flying through the woods" as hundreds of checkerspots "swarmed" around his feet; a year later, in May, troops of butterflies massed on the wet sandy pathways, forming a sprawling patchwork of color. And then, on one early June day in the late 1870s, he inspected a giant rock slab near the woods "moistened by the drippings from a coal seam over it" and "studded with *Papilios* as thick as they could stand. Allowing one square inch to each butterfly, and this is ample, there were upwards of two-thousand butterflies in that mass."[53]

Such a surfeit of winged creatures, side by side with a mostly buried mineral treasure, defined Edwards's life in the early 1870s. But how did he reconcile these two incompatible forms of energy, one alive and symbolic of nature's beauty, a thing of use value merely, serving as a window into nature's realm, the other dead, attractive for the price it would draw in the market, and capable of severing the link to life and obliterating beauty? Were butterflies and coal the same in his mind, both *property* at the disposal of human beings, or did they stand for opposing ways of being and seeing?

Some Americans at the time feared the impact economic activity might have on nature, among them two New Englanders: Henry Thoreau of Concord, Massachusetts, and George Marsh, a Vermonter and an early advocate of nature protection. Heirs of the Romantic tradition developed by the English and Europeans, both men lamented the demise of the forests caused by railroads and heedless agricultural practices.[54] Thaddeus Harris, a librarian at Harvard College and an all-round naturalist, also worried about economic change, especially in Cambridge, where he lived, "mourning," as he noted in an 1851 letter to Thomas Wentworth Higginson, that the "so-called hand of improvement" had "rooted out many of the beautiful plants and insects that were once found in this vicinity." They have "entirely disappeared from their ancient haunts, driven away, or exterminated by the changes effected therein."[55] Europeans, with so much of their forests already cut down, were even more apprehensive. In the mid-1870s August Weismann, one of Germany's most respected naturalists, painted a sorry picture of an economy that had "caused the extinction" of many "vertebrate animals" and "constantly leads to extermination of many other species of different classes." "When in America hundreds of thousands of acres of primeval forest are annually destroyed," he observed, "the conditions of life of a numerous fauna and flora must be thereby suddenly changed, leaving no choice but extermination."[56] Many years earlier, Humboldt had said the same things, despite affirming, as he frequently did, an inexhaustible nature. "When forests are destroyed," he wrote in his 1815 *Personal Narrative of Travels,* "as they are everywhere in America by the European planters, with an improvident precipitation, the springs are entirely dried up, or become less abundant. By felling trees that cover the tops of and sides of mountains, men in every climate prepare at once two calamities for future generations; the want of fuel and a scarcity of water."[57]

Edwards may have shared these ideas, but no evidence exists that he did, and there are many arguments to the contrary—not least his youthful exposure to the tanning business, which Edwards never faulted, though it stripped the bark of the hemlock trees and polluted streams. Coal was comparable to tanning in its potential to inflict harm, but Edwards never publicly addressed that danger. His vested interest probably prevented his criticizing a business that paid for his butterfly work. So, too, may have been the scale of the mining, overseen by Edwards himself so long as he remained in charge, with never more than one hundred workers, still decades away from the absentee corporate ownership at the end of the century.[58] Edwards, moreover, ran his mines not far from a nearly pristine nature. Who could imagine that mining might decimate the butterflies in what seemed their sempiternal numbers? "You live among a nature not yet wasted," wrote Philipp Zeller, a Prussian lepidopterist, to Edwards, while "our environs [in Europe] are so cultivated as to yield less and less every year."[59]

The tension between butterfly collecting and coal mining seemed not to weigh too heavily on Edwards; nor did it on many of his contemporaries, who, like Thomas Jefferson before them, both worshipped nature's splendor and exploited it. Edwards embodied this tension to the highest degree, but so did other butterfly men, especially in the Far West, where they joined the California gold rush, from Pierre Lorquin, who left France to get rich, to Richard Stretch, an English immigrant to San Francisco and a railroad engineer who collected small lepidoptera, to Henry Edwards, a collector supreme and stage actor, who happily performed Shakespeare for gold miners. In a public lecture entitled "Iron and Its Relation to Civilization," delivered in San Francisco in 1876, Henry Edwards celebrated mining as a "necessity of our being," observing, "what an utter blank the world would be, if iron did not exist—no railroads, no ploughs, no printing presses." He urged America to "develop her mineral treasures," "the grandest foundation on which the prosperity of a country can be based."[60] Nature might be treated both as moneymaking property and as a blessed gift of beauty; take something away from it and it would pop up again somewhere else or in some other form. All is one. Linnaeus, Buffon, and Humboldt believed this. Emerson believed it, too, writing that "man's operations taken together are so insignificant that in an impression so grand as that of the world on the human mind, they do not vary the result." So did Whitman, who affirmed in 1878 "the pulsations of all matter, all spirit, throbbing forever,—the eternal beats, eternal systole and diastole of life in things—where from I feel and know that death is not the ending, as was thought, but the real beginning—and that nothing ever is or can be lost, nor ever die, nor soul, nor matter."[61]

The conception of American nature as endlessly fertile, seen alike in the coal and the butterfly fauna of Paint Creek, hid much wrong in mining. This outlook pervaded the first of Edwards's great three-volume *Butterflies of North America,* issued in ten parts at three-month intervals between 1868 and 1872, published in Philadelphia by the American Entomological Society; it was the first book of its kind by an American. Edwards traveled from Coalburgh to Philadelphia and back again, nursing it through production.[62] The time was ripe for such a work, he believed, since so many species had been unearthed or were better known than since the days of Boisduval. He had hoped to present a "complete" account of North American butterflies but, lacking time, chose "a sufficient number of new, or hitherto unfigured or disputed species, to make at least a moderate volume."[63] Other works could follow later.

Edwards spared "no expense" on pictures, because words alone "hindered" insight into "natural history," he told a friend. "Nothing is more discouraging to the beginner than dry, unillustrated descriptions."[64] Based on specimens he obtained with great care on his own, the images were drawn mostly by Mary Peart and colored by Lydia Bowen, a protégée of Audubon's, who would remain Edwards's principal colorist for the next twenty years, with help from her sister, Mrs. Leslie. The plates were among the best ever done for a work of natural history, unencumbered by fantasies or myths about nature, combining accurate science with a feeling for the beauty of the things they depicted; they nearly approximated the actual color (tints, shadings, suffusion) of the butterfly wings. In 1770, the wealthy English jeweler and butterfly man Dru Drury, in his *Illustrations of Natural History,* imagined "preserving" butterflies from "the ravages of time" by representing them "perfectly" in his volumes; at the same time, he saw the "difficulty" of capturing "the innumerable train of colors, the great variety of tints, the harshness of some, the softness of others, together with the manner of their running into one another." The task seemed insurmountable.[65] But by the mid-1850s, throughout the West, technical advances had overcome many of the hurdles, from the elite market to the poor color. Peart and Bowen reaped the benefits of these changes.

Edwards's volume presented both sexes of many American butterflies for the first time, among them many checkered species of fritillaries, crowned by the grand dimorphic Diana fritillary of Paint Creek and by the nearly equally grand Nokomis fritillary, a "superb" sexually dimorphic species captured in 1871 during a government expedition to Arizona. Edwards was the first American naturalist to systematically study the "blues," the jewel-like insects he found on Paint Creek and its environs. Among the earliest on the wing even as snow still lay on the ground, they gave him a sense of well-being little else could equal.[66]

Edwards wrote many fine descriptions of these butterflies, although there was nothing groundbreaking about them, especially in the early parts of *The Butterflies of North America*. But between 1870 and 1872, he embarked on research that would begin to change how both butterflies *and* the natural world came to be perceived. First, he began to breed seriously from egg to butterfly or imago; second, he became an outright Darwinian naturalist. Edwards insisted, as some of his predecessors had already suspected, that one could not safely identify butterfly species merely by looking at the adult insect, or imago. Imagoes could mislead, and even though the adults of some butterflies might resemble each other nearly exactly, there might be so many differences among them in the early stages as to challenge any designation of them as belonging to the same species. Beneath looks might lurk differences in the early stages; if eggs, caterpillars, and pupae differed, then so did the "butterflies" that succeeded them. One must go, therefore, beyond imagoes (however persuasive as identifiers) to the study of *metamorphosis,* or the "changing of form," at all stages of the life history of the butterfly. To do that, one had to breed systematically, touching every phase, as Buffon long ago had insisted.

Edwards was hardly the first to breed. As long as people had been able to find eggs that had not been parasitized, they'd bred lepidoptera, but almost always to get perfect adult specimens for their collections. In the interest of identifying species correctly, Edwards put breeding on an entirely new footing. In 1870, he discovered how to get eggs from any butterfly, by placing gravid, or pregnant, females in safe containers full of the caterpillar's correct food plant—presto, eggs, the females often "laying all they have at once," not a few at a time as "in Nature." His method, the crucial beginning for knowing any insect's life history, set Edwards apart from the Europeans, who "did not understand how to get butterfly eggs," he told a leading German entomologist, who agreed with him.[67] Edwards converted his own home (basement, porches, clothes closets) and land around it into a nursery, with larvae and pupae stuck on or in virtually every spot—in pots, kegs, barrels, and half-pint jelly glasses with tin tops, and even in his bedroom, where butterflies sometimes pupated, freed by the heat of the fireplaces to fly from room to room. Edwards installed a heated greenhouse in which to grow plants and breed insects.[68]

Shortly after the Civil War, Edwards became a biographer of lepidoptera, with many of his letters and articles repeating the refrain "to get the whole story" from egg to adult, birth to death, with special scrutiny of the larval stage as a way to identify species; no one—except Samuel Scudder, who would one day acknowledge Edwards's contributions—would match him; and it took nearly another one hundred years for life histories to become common practice.[69] Around the same time, Edwards became an ardent Darwinian. Neither God nor any other being made species, he held; rather, nature did that, gradually, through struggle and chance over many years.[70] Edwards believed completely that species existed in nature (otherwise why write books about them): they reproduced only with their own kind and with no other, they reappeared again and again with little significant change, and they could be recognized, indisputably, in breeding. On the other hand, Edwards also believed that species were *not* permanent things, that as some emerged and held their own, others failed to adapt, weakened, and died. Variation and instability, not fixity of species, captured Edwards's imagination, as they did other naturalists who fell under the spell of Darwin's 1859 *On the Origin of Species*.[71] "I am out and out a Darwinian," he wrote Scudder in 1875, long after his conversion.[72] "I think with Darwin," he explained to Henry Edwards, "that greater interest attaches to varieties than to well-defined species. . . . To a Darwinian, there is no such thing as a 'bona species,' except for convenience sake, for the systematist."[73]

America had undergone great changes since the death of Edwards's ancestor Jonathan Edwards. In the mid-eighteenth century, Jonathan Edwards had enlisted Newton's science on the side of God's sovereignty. Now William Henry, his great-grandson, turned to science to banish God from heaven. Certainly William's early estrangement from religion—his disgust, really, with Calvinism—helped prepare the way, as did the larger, ever more secular capitalist culture he lived in, one saturated in change, in capitalist markets replacing older customs, in people moving across continents, in nations formed and forming, in America's becoming. And yet Calvinism itself, inadvertently, may have watered the ground for evolutionary ideas. Wasn't there some similarity between the Calvinist Edwardsian soul, in its quest for conversion always susceptible to backsliding and mistaken interpretation, and the anti-Calvinist Edwardsian species, evolving amid great uncertainty, known for sure only—though never totally—after searching investigation?

Whatever the similarities, Edwards was heir to a new scientific culture, enhanced in England partly by Henry Bates and Alfred Russel Wallace and principally by Darwin, and strengthened in America by Edwards's contact with Benjamin Walsh, an agnostic, like Edwards, whose evolutionary gusto may have exceeded that of all other early American naturalists.[74]

Walsh, Wallace, and Bates all studied insects (and Bates and Wallace, in particular, butterflies) as a way to elucidate the operation of evolutionary change. Their influence was far-reaching on Edwards and on the understanding of insects in America generally, helping to establish entomology as one of the most advanced and dynamic scientific fields in the country.

Walsh was educated at Cambridge University in the same class with Darwin. He emigrated in 1838 and bought a three-hundred-acre farm in Rockville, Illinois, working it with his wife and making every necessary tool or item by himself, from his shoes to his horse's harness. At the age of forty-nine, he began to study insects. He developed a new kind of economic entomology as a service to farmers and until he died, in 1869 at sixty-one, edited the first journal on the subject, *Practical Entomologist*. Walsh singled out "seeing" as the signature of a true naturalist, and with eyes like laser beams, he rapidly became aware of all the various life-forms in Illinois and surrounding places. He learned, for instance, that one of America's most handsome butterflies, the female eastern tiger swallowtail (*Papilio glaucus*) came in two forms, one mostly yellow that flew in the North and another, mostly black, that flew in the South. The Europeans and English considered these two independent species, but Walsh confirmed beyond doubt what two other Americans in the 1830s (John Ridings and George Newman, both of Philadelphia) had already suspected—that they were the *same* species, simply in dimorphic form. Walsh had adopted Darwinism in 1861, or two years after *On the Origin of Species* was published, and while many British and American naturalists denounced the book for its speculative and anti-religious character, Walsh defended it. His work on insects formed a key conduit through which ideas about evolution and natural selection would pass into American thinking about nature.[75]

Edwards and Walsh corresponded from the early 1860s, trading insects and ideas, Edwards sending so many letters that Walsh had no time to respond to all of them.[76] It was from Walsh, and perhaps from reading Comte de Buffon, that Edwards very likely heard of a "new" biological definition of species, one later espoused by the twentieth-century naturalist Ernst Mayr. "My definition of a specific distinctness," Walsh wrote, "is when two forms that coexist do not freely intercross or when we may infer from analogy that if they coexisted they would not intercross."[77] Walsh was fascinated with variation and the mutability, as was Alfred Russel Wallace, the cofounder of the theory of evolution.[78] Focusing on butterflies, Wallace introduced the concepts of polymorphism and dimorphism into natural science, dwelling on sexual dimorphism, or the different forms males and females of the same species take, and on seasonal polymorphism, or the many forms the same insect might take as a result of the impact of changing day length and temperature (today this form of variation is called polyphemism).[79] Both Walsh's and Wallace's thinking on the wealth of natural forms, as well as Darwin's and Bates's on the variability of species, would prepare the way for Edwards's breakthroughs, and when Walsh died unexpectedly in 1869, after being struck by a train, Edwards felt the loss deeply; "well nigh irreparable to American entomology," he wrote.[80]

Volume 1 of Edwards's *Butterflies of North America* marked a new turn in the natural history of butterflies. But for all of Edwards's hymns to caterpillars, almost no "preparatory stages" were depicted in it, since for many species, with food plants still a mystery, successful breeding of females was impossible. "It is a matter of regret," Edwards conceded, "that, in so few instances, I shall be able to say anything about the larvae. Even among our old and common species, the larvae are but little known."[81] Moreover, the full potential of Edwards's Darwinian analysis would not be realized until the publication of his brilliant volume 2, in 1884. Nevertheless, there were at least two occasions in volume 1 where he *did* combine his new Darwinism and breeding to illuminate the complex nature of American butterflies. The first came late in his research for volume 1, so spectacular that he began the book with it: three remarkable color plates by Peart and Bowen of the full life history of the zebra swallowtail, a graceful black-striped species with brilliant vermilion spots at the base of the forewings and vermilion slashes on the underwings. According to Edwards, the zebra swallowtail illustrated seasonal polymorphism in a stunning way: a smaller and slightly green-tinged form in the spring, and two larger, black-and-white forms in the summer. The Europeans and the British, looking only at adults, had assumed they were different species. Edwards, scanning his own property, had doubts; pawpaw bushes, the larval food plant, bedecked the river side of his mountain, with zebras cruising up and down in the spring to visit his gardens and the peach and plum trees near the woods. By 1867, when Edwards had bred them from larvae, he suspected these three forms were one, and in 1870, he nailed it down with his new method and hundreds of eggs—indeed, the summer form of the zebra produced the spring form. A year later, he confirmed that the third form yielded *all three,* thus proving that the zebra, a single species, had three seasonal incarnations.[82]

Edwards noticed the same thing in a common species of angle-wing butterfly known as the question mark (because its hind wings carry a little marking that resembles one). It flew in two forms, one with reddish wings, the other with black, and here again, mesmerized by the wings, naturalists abroad had misidentified the two forms either as varieties of one species or two separate species. Breeding from a single batch of eggs, Edwards realized that the adult butterflies had both red and black wings; hence they were not separate species or varieties but seasonally dimorphic forms of the same butterfly.[83] Another angle-wing had more than two colors. Taking the Darwinian view, Edwards conjectured that the multiple forms, if left apart from one another for long enough and, therefore, unable to interbreed, might result in new species.[84]

Edwards's early years were defined in part by his virtual isolation in West Virginia. Nearly everything moved by stagecoach, with rivers sometimes choked by mud, snow, or debris; he could not send living things (especially larvae) over the Appalachian mountains to Philadelphia, where his artists worked. "In this remote corner of Virginia," he wrote a friend, "the mails are often ten days from New York or Philadelphia."[85] His coal business was a burden. "I have a thousand things to think of, this year more than ever," he wrote Samuel Scudder in August 1871, "and how to find any time at all for the butterflies is a mystery to myself."[86] And the expenses of the book took a toll, more so due to its size (twelve inches by nine, almost folio), its artistic demands (color pictures done by hand), its limited print run, and its consequently high price. He bore the cost himself, paying his artists and the artisans in Philadelphia out of his own pocket and getting subscribers to sign up for installments. He burned all the expense records, lest anyone learn what he had really spent.[87] The labor, he wrote, in the book's preface, was "one of love," and he expected no "remuneration . . . in a pecuniary sense."

Yet Edwards finished his first volume, and he did so with all the strengths he brought to it: his personal wealth (if dwindling), the excellence of his artists and of his own natural science, and his zeal. But there was another reason for his success. Edwards tried to do most of the hard fieldwork himself, especially in very local areas (Coalburgh, West Virginia, and Hunter, New York). But as the demands of his coal business grew on him, he began to proselytize, just as Jonathan Edwards had proselytized, albeit in a different line: William Henry Edwards began to inspire all sorts of people—the earliest of his hod carriers—to work on his behalf.

Thus, John Burke, the Irish-American manager of Edwards's mine workers, found caterpillars Edwards had never seen before, and Burke's daughter, Jenny, captured swallowtails and blues along Paint Creek.[88] The members of the family of Dick Fraser, a prosperous farmer and the owner of what Edwards called Fraser's swamp, a living laboratory for

Edwards's work on local species, netted "perfect" Diana fritillaries in the mid-1860s, as well as an unusual female tiger swallowtail, half black and half yellow, "divided equally down the back," which was later picked by Edwards for an illustration in volume 2.[89] One Fraser son found larvae of the baltimore checkerspot (*Euphydryas phaeton*) in thick clusters on *Chelone* (the food plant of the butterfly) and harvested them for Edwards, helping to launch an investigation that was to last for years. (Edwards announced the Fraser discovery in the *Canadian Entomologist* and concluded with a magnificent life history in volume 2 of *The Butterflies of North America*.)[90] When one of his sisters visited Florida, Edwards armed her with a net and a "bottle

Theodore Mead. Department of College Archives and Special Collections, Olin Library, Rollins College, Winter Park, Florida.

of alcohol" to get monarch butterfly larvae, and he asked another sister, from Clifton Springs, New York, to undertake experiments in her icehouse—"temp 40 degrees all year round and dry"—on the impact of cold on butterfly development. Edwards's oldest daughter, Edith, watched over caterpillars and pupae whenever her father left home and, with her mother, Catherine, drew pictures of insects at his request. He taught butterfly care and breeding to Wesley Bowles, a black stableboy, after observing signs that he was an "incipient naturalist," and in 1870 let him care for "the larvae hatched from the eggs" of a seasonal form of the zebra swallowtail; a year later, Wesley had become "so expert" that Edwards decided to "keep him at it."[91]

The finest hod carriers of all to serve Edwards for volume 1 were so talented as to achieve the status of excellent naturalists. Benjamin Walsh, of course, was among them, but two others—Theodore Mead and Henry Edwards—were nearly as gifted. Born in 1854, Ted Mead was William Henry Edwards's chosen successor; he grew up at 674 Madison Avenue in Manhattan, two blocks east of the newly created Central Park; his father, a wholesale grocery merchant, had moved the family to the East Side from Fishkill, New York, in 1857. Henry Edwards, an

English-born actor, was the most good-hearted individual in the entire history of butterfly people, American or otherwise. The naturalist David Bruce called him "the most unselfish man I knew."[92]

Mead shared with William Henry Edwards a seventeenth-century colonial ancestry, a family enthusiasm for abolitionism and women's rights (he went with his mother to hear a young feminist firebrand, Anna Dickinson, lecture in Manhattan), a like-minded religious heritage, and, as he grew older, a comparable aversion to the church world of his parents.[93] As a teenager, Mead was fluent in French and German, with impeccable manners, but like so many naturalists of this age, he hunted, killed, and skinned birds as if it were second nature to him. Once, he so bungled the taxidermy as to be left with only the skull of the bird, which he "boiled" and mailed to his friend Willie Edwards, son of William Henry Edwards. An amateur photographer, he experimented on domestic cats, killing and beheading them and then dipping their skulls in photographic silver to produce negatives. In 1875, his younger brother Sam accidentally killed himself with a gun loaded with "explosive bullets" of his *own* invention capable of "deranging the internal economy" of a large predator. But before these things occurred, Mead happily joined Sam in target practice, shooting at dead trees in northern New Jersey in the winter to watch the wood shatter "into splinters as long as a man's arm."[94]

Mead collected tropical fish and exotic ornamental plants, especially orchids—all new to the New York scene in the 1870s—but his "essential favorites" were the butterflies. He had begun hunting them at twelve and became such an aficionado that he impressed Albert Bickmore, the director of the new American Museum of Natural History, founded in 1869 and at first located in Central Park within easy walking distance of Mead's home. In 1872 Bickmore appointed Mead, then eighteen, "acting curator" of entomology, a nonpaying position that, nevertheless, invested Mead with the authority to purchase all the insects and to classify them "precisely" as to "date of capture" and "locality."[95] "I am the only person at the American Museum," Mead told another butterfly man, Herbert Morrison, "who knows anything about insects."[96] Three years earlier, Mead had read about William Henry Edwards in the magazines, which inspired him to write Edwards directly, inquiring about where best to catch butterflies within five hundred miles of New York

City. Edwards replied that Coalburgh was the perfect place and invited Mead to visit; Mead soon arrived, stayed a whole summer, and charmed everyone, especially Edwards's sixteen-year-old son, Willie—or "darling Willie," as Mead came to call him, affirming his "love" in letter after letter, while often at the same time boasting of how many birds he'd killed or cats skinned. Mead learned about butterfly breeding from Willie's father, became a skilled tracker of their life histories, and spent nearly all his time on the lookout for butterflies around Coalburgh, through what he called Edwards's "terrestrial and entomological paradise."[97] What began as a service to Mead ended as a bonanza for Edwards, as Mead made one surprising find after another, most notably the chrysalis of the Diana fritillary, a mystery to everyone up to that time.[98]

Mead idolized Edwards, convinced he was the most advanced entomologist of the day. Edwards, in turn, was awed by Mead's talents. "Mead is as great a collector as lives, remarkably quick of eye and hand, and overlooks nothing," Edwards told Scudder. To Henry Edwards he wrote: "I think Mead may make an eminent naturalist, and he begins by being a wonderful collector."[99] Even as he aged, Edwards prided himself on doing his own fieldwork, both locally and in Hunter, New York, where he was born. The thought of going West to "collect for myself" and "to gain more knowledge of the larvae" always thrilled him. But as he got to know young Theodore Mead better, he began to consider seriously sharing with him the excitement of field exploration. "When I can get a young friend like Mead here," he explained to Henry Edwards, "I can see with his eyes, and hunt with his net, quite as well as if I had myself."[100] To Mead himself he wrote, "Tell your father this is my opinion. One of these days I hope you might take up the Butterflies of North America where I leave it, and continue it."[101] Their mutual admiration may have taken a toll on Willie, Edwards's only son, who, with little interest in butterflies, felt left out and jealous, and even subtly vindictive, while professing persuasively to be Mead's "best chum." As he grew older he began to draw Mead away from butterflies, and from biology in general. "We were fond of one another, and Willie had a great influence with me," Mead wrote in his autobiography, "He was very urgent that I should do something more practical than butterfly collecting."[102] With Willie's encouragement, Mead followed Willie to Cornell University, joined his fraternity, and majored in engineering.

In the spring of 1871, Edwards arranged for Mead to join a government exploratory expedition to the Colorado Rockies and places westward, its purpose to case the natural riches of the mountains, butterflies included. The two planned to divide the expense and the insects, but Mead knew the real benefits would accrue to the older man.[103] The summer journey had its dangers—angry Indians near the collecting spots, gray wolves at the higher reaches, and many icy-cold nights. But for the most part the experience was a lark, a joy without pain, for Mead and his companion, his only brother, Sam. There were stage routes dotted with comfortable lodging facilities serving good food, settlers all around and visible, and railroad lines from Chicago to Denver, with sleeping berths and dining rooms to carry the boys to the edge of the wilderness of Colorado. The Meads sometimes rode horses and built their own snug wigwams, wrapping themselves in blankets by the fire. While Sam "popped his Winchester" all over the place, Ted pursued tinier game, catching one specimen at so high an altitude that he "had to lie down frequently to regain his breath."[104] A boon for Edwards, the expedition gushed a stream of butterflies: seven hundred by July 4 and three thousand by December 5.[105] Mead unearthed twenty new species of butterflies, all ending up in volume 1 of Edwards's *Butterflies of North America*, most brilliant of all a fiery orange sulphur Edwards named in Mead's honor, *Colias meadii*, which Mead thought "one of the prettiest butterflies on the continent."[106] In 1875, the ex–cat killer discovered, on Edwards's lands in Hunter, New York, and Coalburgh, West Virginia, the food plants of the pearl crescent and comma butterflies (*Phyciodes tharos* and *Polygonia comma*, respectively), permitting Edwards to breed the full life histories of both. On another trip west, Mead did the same for the western tiger swallowtail (*Papilio rutulus*), "getting twenty eggs of Rutulus, the best find of the season," as William Edwards reported glowingly to Henry Edwards.[107]

In San Francisco, Mead met the butterfly man Henry Edwards, then on the stage in the Manhattan Theater, one of the city's leading companies, and Edwards "loaded him down with butterflies" from his own "magnificent collection," as Mead called it, an unexpected act of generosity.[108] The actor treated others in the same way, bestowing a wealth of "diurnals" (day-flying Lepidoptera) on the young California collector and lawyer Charles McGlashan. "I shall not forget your kindness in robbing your own cabinet to swell mine," McGlashan wrote. After finding out that Samuel Scudder studied skippers, Edwards assured him that "I will do all and anything in my power to be of service to you—with great pleasure I will send you all the species I have."[109]

Henry Edwards was born in England in 1827 and, as a very young man, he had tried to satisfy his parents by clerking in a London commercial house. But he had loathed it and, for relief and with unexpected delight (but to his family's disgust),

Henry Edwards.

had joined the theater. He also studied natural history, learned taxidermy, enlisted in the relevant societies, and may have collected insects for Edward Doubleday.[110] In 1853, fed up with clerking, Edwards sailed off to Australia, lived for a while on his brother's farm near Melbourne, and then joined Melbourne's new Theatre Royal, playing Petruchio in Shakespeare's *The Taming of the Shrew* and Iago in *Othello*.[111] Gustavus Vaughan Brooke, the manager of the Theatre Royal and himself a celebrated British actor with whom Edwards had a complex relationship offstage, drank so heavily that he had to give up the management and also his wife, Polly, who divorced him. Edwards was the gainer, however, inheriting the management of the theater and Polly as well, marrying her soon after Brooke returned to England, debt-ridden and humiliated.[112]

When he could find time, Edwards pursued natural science and even sought permanent work as a curator at the new Sydney Museum of Natural History.[113] A seventy-page letter of 1854 displayed his command to a naturalist friend, John Jones of Gloucester, England; in it, Edwards described the "Natural History" of Australia according to standard "classificatory groups," from the "Marsupials" at the top to the "Diptera," or flies, and other low-life at the bottom, and indicated along the way the life histories and Latin names of numerous species and their interrelationships, with special attention to predators. Edwards spent page after page on cuckoos, cockatoos, parrots, and parakeets, savoring their palette of colors, and concluding his account with the Blue Mountain parrot, a rainbow-colored bird "with the breast of bright orange, passing at the upper parts into scarlet, and at the lower part giving place to a deep blue, the wings green on the outer webs, the inner side black, with a band of sulphur yellow, the head is blue, with a dark streak down the centre of each feather." He liked colorful beetles, too, finding many species among "the most strikingly beautiful" things in nature, "beautiful in the extreme." Surprisingly, the lepidoptera (and of these, most were moths) got barely one page, out of the seventy, just beating out earthworms, cicadas, and walking sticks. "The Butterflies are particularly few," he observed, "only nineteen being known to me, and the most common, is closely allied to [the European] *Cynthia cardui.*"[114] But soon new vistas opened before Henry Edwards, revealing a startling abundance of butterflies spread throughout the entire Australian region, which he began to collect and exchange frantically in 1862, first at home and then elsewhere in the world, promising "300 to 400 specimens of the various orders" in exchange for "*one half* the same number of Lepidoptera of any country, my attention being now devoted particularly to that order."[115] Just as in the case of William Henry Edwards, the beauty of butterflies had trumped that of birds.

In 1866, Edwards emigrated, with his bugs and wife, to San Francisco, a city quite on the make, as Melbourne had been, with a lively theater culture. There he performed in, and helped manage, two companies, the Metropolitan and the California Theaters.[116] He played Polonius in a cast with Edwin Booth and Lillie Langtry and worked the mining circuit, even into the Rockies, at a time when Americans from all walks of life enjoyed Shakespeare.[117] He cofounded two centers for intellectual advancement, the Bohemian Club, where men and women met to hear "original contributions" in "music and literature," and the California Academy of Sciences, where he became the curator of entomology, trustee, and vice president, in the company of other amateur butterfly men (above all the Germans Hans Hermann Behr, James Behrens, and Oscar Baron).[118] Edwards disliked "dry classification" (although he did a lot of it) and admired the "seeing" sensibility or "that faculty of observation even of the meanest things." Anything in nature caught his eye, but mostly the small things—the spiders, beetles, grasshoppers, scorpions, wasps, water bugs, bees, starfishes, and millipedes.[119] But, above all, he was "omnivorous" about butterflies and moths, exotic as well as American, the food for his "dreaming and drifting temperament," as one friend observed.[120]

In San Francisco, Edwards continued collecting and trading exotic butterflies, while joining Americans in their quest for native species, hunting nearly every day he could spare, first nearby, and later in the deserts and on mountaintops; he was proudest of catching a swallowtail (*Papilio indra*) and a satyrid (*Chionobas ivallda*) on a "very high peak" in the Sierras, the northernmost spot for these alpine-zone butterflies. Henry's wife, Polly, also a lepidopterist, accompanied him on expeditions to the Sierras, discovering on her own at Summit Station two fully grown caterpillars of the brown elfin (*Incisalia augustinus*), a pretty little lycaenid with orange-brown underwings.[121] Henry befriended John Muir, the Romantic wilderness man who had come to San Francisco at about the same time and took "a revolutionary moral position" toward nature: that human beings had no right to harm any part of creation for the sake of their own "happiness."[122] Edwards and Muir may have met at the California Academy but more likely in Yosemite, where Muir lived and worked. Muir called for an end to all hunting and hurting of nature, but when it came to butterflies and to Edwards, who shared with him similar Romantic feelings, he betrayed his convictions: "You are now in constant remembrance, because every flying flower is branded with your name. . . . I wish you all the deep far-reaching joy you deserve in your dear sunful pursuits [sic]."[123] Edwards named a butterfly after him, John Muir's hairstreak (*Callophyrs muiri*), a dark russet little California species historically confined to "serpentine and gabbro soils where its hosts, MacNab and Sargent Cypresses, grow."[124] A patriot of a new place, he had begun to call the butterflies and moths of America "our fauna."

Within two years of his emigration from Australia, Henry "pledged himself" to William Henry Edwards announcing that he would send to no one else *all* the unnamed species he found.[125] His loyalty knew no bounds. "I will bow to no one else," he wrote a friend. "He is our best authority on diurnals."[126] He sent William countless native American butterflies, along with many details of life histories that often resurfaced on the pages of both volumes of William Edwards's *The Butterflies of North America*, perhaps the most significant record of Henry's achievement.[127] William put the actor's photo in a "place of honor in my library," and when a letter failed to come as expected, he was miserable:

"I am thirsting for a letter from you like a parched soul in a desert." After it arrived, William wrote that "your familiar hand gave me a sense of pleasure like that of sunshine on a bank of flowers."[128]

The success William Henry Edwards had with volume 1 of *The Butterflies of North America* depended on the likes of Henry Edwards and Theodore Mead, just as it depended on his artists, on his natural science, and, not least, on the butterflies of West Virginia. Edwards's first volume was one of the most rooted-in-place books ever published, saturated in "local coloring of West Virginia," as Augustus Radcliffe Grote observed.[129] And how gratifying it must have been for Edwards to behold the reaction of the naturalist press. On seeing the first plates in 1868, a leading London magazine warned the British to take heed, for "if illustrated works of so much beauty and accuracy as this can be produced on the other side of the Atlantic, it behooves Natural History Iconographers in our old Europe to look to their laurels."[130] There is "no one" in Britain "who can make such drawings with such fidelity to nature," wrote Arthur Butler, the curator of butterflies at the British Museum, in a letter to Scudder.[131] The natural science had the same effect, humbling even France's Boisduval, who, indifferent to American pride, had described so many American species. Writing to Edwards, Boisduval compared him to the German naturalist Jacob Hübner, self-taught in butterflies, who designed cotton textiles for a living but was also the first great "world lepidopterist." In the early 1800s, Hübner had invented a system of classification to accommodate, for the first time, all the known butterfly species since Linnaeus's day; he named many genera of tropical butterflies that still stand today. Boisduval was willing to share his *entire* collection with Edwards. "I have nothing, absolutely nothing I wouldn't let you have," he wrote. "All my collection is at your disposal. I consider you the Hübner of North America."[132] Reporting on Edwards's first installment, Charles Valentine Riley, the respected co-editor with Benjamin Walsh of the *American Entomologist,* observed, in 1868, that "if the future numbers shall bear the same marks of care, correctness, and artistic skill, as does this first number, we shall, without hesitation, declare it the finest work of its kind ever published in this or any other country."[133]

If Edwards's phobia for Calvinism helped drive him toward butterflies, Samuel Scudder's affection for a milder version of it had a comparable effect on him. One of the finest all-around naturalists in America, Scudder was a Yankee, like Edwards, but a life-long Protestant. He believed that the world had been called into being by a God power, as Jonathan Edwards had understood it, as an all-generative force of nature. This state of mind was strengthened by the naturalist Louis Agassiz, Scudder's most significant teacher, who reviled Darwin for think-

Samuel Scudder. Courtesy of the Museum of Science, Boston.

ing that species evolved rather than having been formed, all at once, in a series of complete creations after every climatic catastrophe in the earth's history, by a Higher Reality.[134] Fervent Congregationalism, still bearing the Calvinist imprint, mixed with anti-Darwinism—no recipe could have irked Edwards more. Scudder showed symptoms of both.

Scudder was born in Boston on April 13, 1837, descended, on both sides of his family, from Puritan stock dating back to the founding of the Massachusetts Bay Colony in the 1620s. His father, Charles, a commission and hardware merchant of middling success and a Calvinist Congregationalist, ran a strict "Puritan home" and sent three of his five sons to Williams College to protect them from the "commercial life."[135] Scudder enrolled Samuel at Williams explicitly to place him under the influence of Mark Hopkins, president of the college, and the very same man who had scolded Edwards for studying on Sunday. According to Charles Scudder, Hopkins was a gifted Christian preacher (and more of a religious liberal, in his view, than a strict Calvinist) dedicated to teaching young men to marshal all their powers toward a noble end. "Man can have strength of character," Hopkins said in a Boston sermon, "only as he is capable of controlling his faculties; of choosing a rational end; and in its pursuit, of holding fast to his integrity against all the might of external nature."[136]

Samuel enrolled at Williams at sixteen, the same age as Edwards sixteen years earlier. If Edwards had hated the college on religious grounds, Scudder loved it on those same grounds. Both boys became collectors while at Williams. Unlike Edwards, Scudder did not begin with birds but went straight to butterflies, partly because Williams students collected them, whereas in Edwards's day no one had. Early in his first year, Samuel noticed, in another student's room, a "glazed box of butterflies, perhaps a dozen or twenty in number, artistically arranged and hung as a picture," packed with specimens caught in Williamstown. "Got them right here," the boy explained. "There are lots of them." It was a revelation, a shaft of light. "I had not dreamed," Scudder recalled in 1896, "that such beautiful objects existed, least of all at home, or that so many kinds could be found in one spot."[137]

Scudder began to hunt for butterflies around Williamstown, a landscape in the Berkshire hills rendered passable by farmers. He entered a deep ravine, shut in on both sides by steep mountain slopes, warm in summer and thick with butterflies, where he captured his "first love," the "banded purple," a "showy insect," he observed, with "a broad white bow stretched across rich purplish black wings." He also caught a rare early hairstreak there (*Erora laeta*), a small insect with turquoise hind wings speckled in red, its family name derived, Scudder wrote, from the fine "delicate markings threading the under surface of the wings" in the manner of all such butterflies. It landed "at my very feet," and, in an instant, "my net was over it" and "I triumphantly" blurted out a line from a Shakespeare sonnet: "How *have,* I say, mine eyes been blessed made / By looking on thee in the living day" (Scudder wrote "*have*" in place of Shakespeare's "would").[138]

Also responsible for Scudder's enthusiasm for butterflies was Paul Chadbourne, an all-around naturalist with an entomological bent, who taught natural history at Williams and attracted a small student coterie to study insects with him.[139] Under Chadbourne's wing, Scudder became— like Edwards—a naturalist, but with a difference, since he apparently also studied natural religion with Chadbourne. Chadbourne, no reactionary in science, welcomed all advances in "the means of observing," or in what he called "Humboldt's power of seeing."[140] "In one year a man may see more of the earth than Humboldt could see in ten," he noted, yet he insisted that "man stands alone," above all other animals in his "belief in God, in immortality, in accountability, and in having

the instinctive impulses of prayer and praise *towards an unseen Being.*"
An "Unseen Power," he maintained, circulated through the "structures
of the earth," which humans experience when they study nature. He
was not a strict Calvinist like Jonathan Edwards, but his views were
in the Edwardsian spirit, muffled perhaps on the theological level but
full-blown in terms of the natural world. Chadbourne's double vision
helped convert Scudder into something of an entomological zealot.[141] A
beneficiary (if that is the right word), like William Henry Edwards, of
the waning of religious orthodoxy and the rise of secular natural history,
which exposed many eyes to the beauty of the world in a way Jonathan
Edwards could not have known, Scudder was freed to study butterflies
at the cost of studying God.

Scudder might well have followed his older, much-loved brother
David, who, enamored of Mark Hopkins at Williams, became a Tamil-
speaking missionary on the streets of Madras, India, in 1851, only to
die there in 1852. The news crushed their father, who died two weeks
after hearing it. Samuel chose as his own mission the study of insects—
butterflies, above all, with a preference for skippers, many mailed to him
from Madras by David. At twenty a Williams graduate, Scudder wrote
a renowned Harvard naturalist, Hermann Hagen, that he "intended to
pursue the study of Entomology through as long and short a life as Prov-
idence may grant me, and with as much vigor as soundness of mind and
body will allow."[142] He pledged to enlist all "the reverence and devotion
to faith that had characterized his ancestry" in behalf of this endeavor,
to quote his best early biographer.[143]

Scudder now brought his entomological self to Louis Agassiz, one of
the foremost scientists of the age, and studied with him for six years, as
a kind of postgraduate assistant, at the Lawrence Scientific School and
at the Museum of Comparative Zoology, both at Harvard, from 1857
to 1863. These were tumultuous times in American history, years of civil
war but also of a fierce feud over Darwin's *On the Origin of Species,*
which, in 1860, had erupted in Boston, with Swiss-born Agassiz at the
podium. Trained in natural science at German universities, Agassiz, at
thirty-nine, had arrived in the United States in 1845 with a huge reputation, based on his work on fossil fishes and theoretical investigations of
past ice ages. He crossed the Atlantic to study the natural history of the
continent but, feted by Americans, accepted an appointment at Harvard

and remained there for the rest of his life. Handsome and eloquent, with impressive fund-raising abilities, he was the most exciting thing to happen to the country's natural science in the years before the Civil War. In 1860, Harvard created the Museum of Comparative Zoology as Agassiz's very own museum, with a separate faculty and separate students, and, under Agassiz, it became the country's first high-quality research and teaching institution, with the most complete account of natural history materials in the country. In the first class of students was Samuel Scudder, tuition-free; many, like Scudder himself, aimed to change the character of American natural science.[144]

Agassiz had at his fingertips a commanding knowledge of European science, and he lifted the level of the natural history tradition in America to a greater excellence. He taught students—and anyone interested in nature—how to collect, identify, label, and sort scientific materials. He spread the use of the microscope, and he fostered embryology, comparative anatomy, and paleontology, all signifying a formative process toward greater complexity he had learned in Paris under Georges Cuvier. A legatee, too, of the renowned Comte de Buffon and a close friend to Humboldt, who greatly respected him, Agassiz was indebted to Humboldt's ecological mysticism, awed by "a soul-breathing epos" in nature, fusing "all animals and plants" into "an expression of a gigantic conception, carried out in the course of time." Like Buffon, he was opposed to conventional Linnaean systematics for dealing only with "surfaces," not with the "true" identity of things as they "really existed in nature."[145] He took seriously things in context, the building blocks that formed species and linked them. To classify and identify "the character of species," he insisted in his theoretical tour de force, the 1857 *Essay on Classification,* required knowing their "relations to the world around them, to their kindred, and upon the proportions and relations of their parts to one another." Food, habits, periodicity of changes, metamorphosis—all demanded scrutiny; "descriptions of species ought to be comparative" and "assume the character of biographies and attempt to trace the origin and development of species during its whole existence."[146] In 1853, Agassiz researched the fish of the United States, sending out six thousand circulars to Americans, urging them to tell him what they knew about the fish in the bodies of water in their towns and regions. The response was tremendous, testimony to legions of new naturalists,

Drawn by D. Wiest.   Bowen & Cº lith. & col. Philadª

DIANA. 1. 2. ♀. 3. 4. ♂.

1. The Diana fritillary belongs to a family of butterflies, the fritillaries, one the largest in the world, whose species, male and female, are mostly orange and brown, with crisscrossing lines and rows of dots on the upper surface of the wings, and metallic silver spangles beneath that shine in the sun. Diana is unlike most other fritillaries in its muted markings and radical sexual dimorphism, a discovery of William Henry Edwards, who showed that the male and female butterflies took different forms: only the male is orange-brown, while the female is a dark blue and black. For the more typically patterned fritillaries, see plates 2 and 31.

2. William Henry Edwards identified and described correctly more species of American fritillaries than any other American butterfly person. The term "fritillary" itself is derived from the Latin for "dice box," the tossed dice producing a range of dot patterns that may have reminded a gambling naturalist of the wings of the butterflies. Shown here are several western species, assembled in a plate photographed in color and taken from *The Butterfly Book* by William Holland (New York: Doubleday, 1898). Most of the specimens came from Edwards's own collection, and most of the species, each with its own unique arrangement of lines and dots, were originally described and named by Edwards.

AJAX, VAR. TELAMONIDES. 1. 2. ♂; 3. ♀.

*4 Young larva. 5.Mature larva.
6, 7. Chrysalids.8.Egg, magnified.
Food-plant.- Pawpaw.*

3. William Henry Edwards was awed by all the diverse forms in the natural world. Around 1870, he discovered that the zebra swallowtail, a mostly southern species, appears in three different forms from early spring to late summer, each slightly larger and differently colored than the one preceding it. This plate shows the earliest form in its life history from egg to butterfly, the first complete history ever published of the insect. It was drawn by Mary Peart of Philadelphia, who would become Edwards's premier artist, and colored by Thomas Sinclair's lithographers, also of Philadelphia. The vernacular name speaks for itself.

Drawn by Mary Peart.

L. Bowen, lith. Phil.ᵃ

INTERROGATIONIS. VAR. UMBRO SA. 12. ♂. 3. 4. ♀.

a. Egg, nat. Size b. Eggs magnified.
c. Young larva, d, Same after first moult.
e. Same after second moult; f. Mature larva.
g. Chrysalis.

4. This plate, from volume 1 of *The Butterflies of North America,* reveals the life history of the question mark butterfly, the name derived from the silver sign on the underside of the hindwing that looks like a question mark. The plate was among the earliest collaborations of Edwards's key artists—Mary Peart, who drew the butterflies, and Lydia Bowen, who colored them. Bowen had earlier worked for John James Audubon and was likely related to John T. Bowen, who ran the influential lithography business in Philadelphia under his name. Together, Peart and Bowen would help make Edwards's work on American butterflies famous.

T. Sinclair & Son, Lith. Phila.

5. The White Mountain butterfly (number 9 in this plate) was one of Samuel Scudder's favorite butterflies. An arctic species that flew at the top of Mount Washington in New Hampshire, it was first discovered in 1828 by the American naturalist Thomas Say, who also gave it the Latin name *Oeneis semidea*. Scudder chose what became its vernacular name. He was also the first to describe its life history, from egg to butterfly, with a fullness unrivaled by any naturalist then or now. The insect's adult form, both forewing and hindwing, appears at the bottom of the plate in Scudder's volume 3, right beneath the monarch and viceroy butterflies, insects critical to Scudder's discussion of mimicry in his first volume.

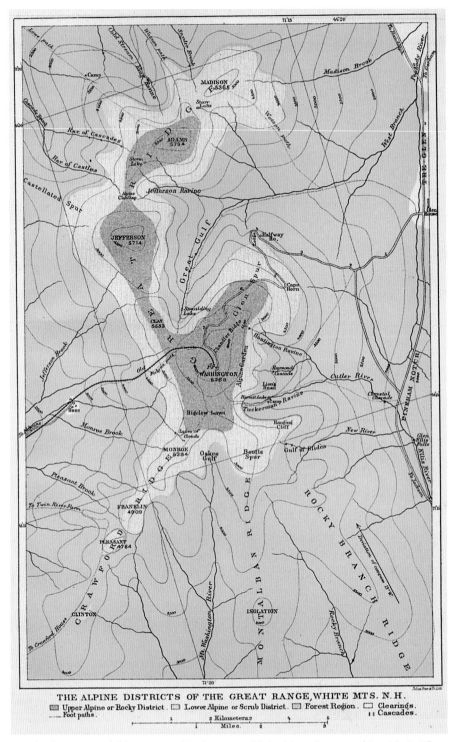

THE ALPINE DISTRICTS OF THE GREAT RANGE, WHITE MTS. N.H.

Upper Alpine or Rocky District. Lower Alpine or Scrub District. Forest Region. Clearings.
Foot paths. Cascades.

Kilometers.
Miles.

6. At the center of this topographical map, published in volume 3 of Scudder's *The Butterflies of New England and Canada*, is Mount Washington in the White Mountains of New Hampshire, the highest mountain in the East, with arctic air at the top, an essential symbol of Scudder's imaginative life as a naturalist. He often climbed the mountain alone or with his friends, but did so, above all, with his only child, Gardiner, from boyhood on, in all sorts of weather. It is one of the four peaks in the White Mountains named for the Founding Fathers.

7. This plate shows many underwing moth species, all drawn and colored by Herman Strecker. The underwings are called such because, when the insects are at rest, their forewings are folded back to cover their hindwings, thereby "converting" them, with their bright colors, into the *under*wings. The sweetheart underwing (number 15), which Strecker fell in love with as a child, appears on the far left at the bottom of the plate. Published in Strecker's *Lepidoptera* (1878).

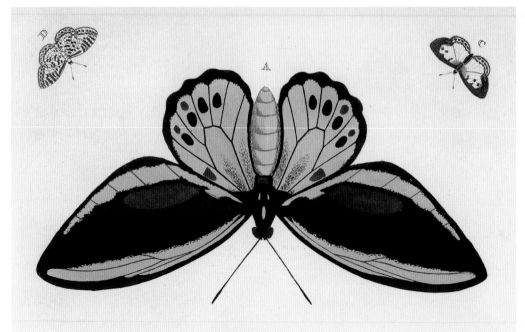

8. Strecker longed to possess a rare tropical butterfly from southeast Asia after seeing this depiction as a young boy in a grand 1779 catalog, *Papillons exotiques* by the Dutch butterfly man Pieter Cramer, in the basement library of the Academy of Natural Sciences in Philadelphia. It was the "first bird wing" species ever discovered (in Latin, *Ornithoptera priamus*), so called because it looked like a bird high up in the trees. Linnaeus named it in 1758, and Cramer drew and colored it, on the basis of a single specimen. *Courtesy of Craig Chesek, © American Museum of Natural History.*

although Agassiz never finished his fish book. (Two others, one on turtles, the other on jellyfish, were completed, however.)[147]

Agassiz's reputation was also shaped by his hostility to evolutionary thinking, inherited in part from his Paris mentor Cuvier, who, in the 1830s, had disparaged his countrymen Jean-Baptiste Lamarck, Comte de Buffon, and Geoffroy Saint-Hilaire for imagining that "species" could "pass insensibly into one another" or might have derived from simpler earlier forms. Agassiz, as a young man, had condemned such thinking himself in the German *Naturphilosophie* of Goethe and Schelling.[148] The 1859 publication of *On the Origin of Species* raised the ante terribly for Agassiz, forcing him, throughout the 1860s, to fight the Darwinian genie.

A sensational moment occurred late in the fall of 1860, when Agassiz joined one of the first public debates on evolution in America, at the Boston Society of Natural History, the reigning such society in the country, founded in 1830 as an outgrowth of the earlier Linnaen Society of New England and funded by wealthy Bostonians.[149] Scudder was sitting in one of the front rows, with his fellow students, as Agassiz took the floor to confront William Barton Rogers, a professor of geology, who defended the pro-Darwin position. What an emotional charge this must have carried for Scudder, his teacher at the very zenith of his fame and in the year the Museum of Comparative Zoology had been dedicated. Vain, with noble bearing, Agassiz seemed invincible, and he might have proved so had the times been different and had some quick-thinking colleagues— Harvard's botanist Asa Gray and geologist Rogers, for example—not been recent converts to Darwinism.[150] There were other apostles to Darwin in America, too, not on anybody's faculty—men like William Henry Edwards and Benjamin Walsh, among the first naturalists in the United States to adopt natural selection as the propulsive engine behind evolution.

In the Boston auditorium that day, Agassiz defended his categorical conviction that species issued ready-made from God, in their totality, and *exactly* where they now lived, wholly bounded organisms invulnerable to variation and hybridization and to the impact of migrations. According to Agassiz, a continent like North America consisted of independent "islands" set off from other "islands."[151] Except in limited, confined ways, change did not exist as a general phenomenon in nature.

As one of Agassiz's students in the audience observed later, "Prof. Agassiz stated that he knew no such thing as variety in the animal kingdom, except such as are stages of growth, within the limits of species." When the geologist Rogers countered with fossil data to demonstrate real change from primitive to more advanced forms, Agassiz did not budge, asserting that, in the end, his views would prevail, no matter what Rogers or others might claim.[152]

Agassiz's opposition to evolutionary thinking flowed from deep within him, from a fear that all who espoused the theory of evolution aimed to banish God from the universe and invoke the Void. "It was the icy gloom of atheism Agassiz feared," a modern historian of science has written.[153] A religious liberal with a passion for inductive science and contempt for those who relied on the Bible as a source for scientific truth, Agassiz nonetheless believed in something on the order of a Divine Being or a Transcendent Creator. The notion that species might mix, or move from place to place, or that varieties might be more interesting than species, or that species as a category might not exist at all, as William Henry Edwards (following Darwin and Wallace) believed, was anathema to him because he thought it denied the existence of the Supreme Intelligence. God never stumbled. Natural selection, in all its sloppy small steps and variations, did not create nature; the Creator did, and to perfection. Animals were separate from the mediums in which they existed, as far as their origins were concerned.[154] Other people might endorse Darwinian ideas and still believe in God; not Agassiz. But articulate as he was, Agassiz did not come off well in this debate and, in fact, offended many who had heard him argue similarly before. To some he seemed simply stubborn. Agassiz himself felt vindicated.[155]

William Henry Edwards would never meet the men who influenced him most, Darwin and Walsh. Scudder, on the other hand, as a young man had nearly daily contact with Agassiz, who watched over him like a brooding hen, expected his allegiance, and bestowed on him a priceless feeling of being at a golden moment in the dawning of natural science in America.[156] Like all great naturalists, Agassiz taught Scudder to depend on his own eyes when in nature's gardens, to "look and look again," until he figured out exactly the characters and conditions that underlay the identity of living things. In their very first meeting, Agassiz removed a fish from alcohol and told Scudder to study it until he could decide what it was that made that fish a fish. He left his student alone in the room for hours, until Scudder came up with something and passed the test; the memory of even the smell of alcohol would never leave him.[157] Still, Scudder was not entirely loyal to Agassiz; like others, he was dismayed by Agassiz's performance in Boston. He also disliked being dictated to, especially from one who demanded independent thinking. In 1863, Agassiz's most ambitious students, including Scudder, refusing to toe the line any longer, began to publish articles on their own and to seek employment without his consent. Furious and hurt, Agassiz claimed that whatever his students did at the Museum of Comparative Zoology was his "intellectual property" and threatened to expel the guilty students. Whereupon all walked out the door, Scudder among them.[158]

Nevertheless, Scudder, during his six years' dose of Agassiz, learned to specialize in the eggs, fossils, and morphological structure of butterflies; to use the microscope adeptly; and to follow the ecological approach to insect life. He became more Humboldtian under Agassiz, who sent him off, in 1860, to Canada, on a thirty-five-hundred-mile round-trip, to observe an eclipse of the sun, just as Humboldt had famously done in Cumaná, Venezuela, in 1789.[159] Although tall and strong, Scudder suffered from asthma all his life—it may have kept him out of the Civil War—yet off to Canada he went, eager to please Agassiz, with 650 pounds of chronometers, sextants, telescopes, and containers of alcohol for specimens, which he helped lug around himself, gasping all the way. (After all that, he saw little of the eclipse because of overcast weather.)[160] Further, he accepted Agassiz's views on Darwinism, in part because Agassiz confirmed in him what he had already taken from his father, as from Mark Hopkins and Paul Chadbourne at Williams: that an Unseen Power authored the natural universe. Species were "created independently" and had "a separate existence," Scudder wrote Benjamin Walsh in 1864. This only ignited Walsh's caustic wit: "I could easily conceive of yellowness or virtue or patriotism having had an actual independent existence *before* matter and men were created. The idea that a species has a separate existence of itself reminds me of Paddy's recipe for making a cannon—'Take a large hole, and poor melted iron around it.' Do you believe that holes existed before matter had any existence?"[161]

Agassiz's dedication to the study of *American* nature shaped Scudder's own desire to study his country's butterflies above all others. Ever since his epiphany at Williams, he had continued to hunt lepidoptera around Williamstown, but he had also found a better place: the White Mountains of New Hampshire, distinguished by Mount Washington, at all levels inhabited by butterflies, even at the iciest elevation in the East, in swirls of arctic air. He climbed it all seasons and, at the top one day, came upon the caterpillar of a rare arctic butterfly, *Oeneis semidea*, a hardy refugee of ancient glaciers retreating northward; proud of his discovery, Scudder would record *O. semidea*'s life history in book after book on American butterflies. He collected along the wagon road and rail route up and down both sides of the mountain, highways for both butterflies and people. On an old wagon road in the woods, he saw banded purples floating up and down or in dense groups on the ground, iridescent blue-and-purple butterflies flecked in red and banded in white. "It is one of the delights of camp life in northern New England to meet this butterfly," he wrote.[162]

In 1861, Scudder married a wealthy descendant of Puritans, Ethelinda Blatchford, and they settled in Cambridge, a short walk away from the museum, where he studied with Agassiz and near the Boston Society of Natural History, which hired him—first as recording secretary and later as curator of entomology and librarian; in the 1880s, he became its president. He gave many papers there, once assuming the persona of Humboldt to explain the "distinct zones of life on high mountains," with insect illustrations from the White Mountains.[163] Scudder read the letters of Thaddeus Harris, Scudder's predecessor at the society as curator of insects, which were stashed away in the society's archives. Harris was also the first teacher of natural history at Harvard, and along with Titian Peale and John Abbot, a pioneer in the study of the life histories of American insects. A friend of Edward Doubleday's, he defended the right of Americans to identify their own species and had the "climax of his life" in August 1840, when he found the caterpillar of one of America's loveliest butterflies, the iridescent blue pipevine swallowtail, named for the food plant it fed upon, crawling about on a shrub in Harvard's botanical garden. Harris was convinced that he was the "only person" in the country who cared about butterflies and other insects for their own sake. He confided to Doubleday that "you can never know what it is to be alone in your pursuits, to want the sympathy and the aid and counsel of kindred spirits." At the same time, Harris's reputation rested largely on his *Treatise on Some of the Insects Injurious to Vegetation*, the first influential book in America on economic entomology, which treated insects not for their own sake but mostly as pests to be destroyed. Harris wrote the book for farmers and never seemed inclined to criticize the economic thrust of his work, although one of his students at Harvard, Henry David Thoreau, whose two years at Walden Pond derived partly from Harris's teaching, minced no words about his opposition to economically driven natural science. On seeing the "Beauty of a blue butterfly" one day in the woods, Thoreau observed in his 1859 journal that "the only account of the insects which the State [of Massachusetts] encourages is 'Insects Injurious to Vegetation.' We are not interested in birds and insects as they are ornamental to the earth and cheering to man. Come out here and behold a thousand painted butterflies and other beautiful insects, then go into the libraries and see what kind of prayer and glorification of God is there recorded. We have attended to the evil and said nothing of the good."[164]

Samuel Scudder admired Thaddeus Harris's innovative life histories, and, on these grounds, he decided on the spot to edit and publish Harris's letters, ending with the last about desolated pockets of nature in Cambridge left behind in the wake of "improvements." It is very unlikely he was ever aware of Thoreau's attack on Harris's *Treatise*. One thing is certain, however: before the late 1860s, Scudder had no plans to concentrate on butterflies, though they were his first love; he thought of himself as a generalist, interested in "the structure and development of [all] insects."[165] Time spent with Harris's papers, along with Agassiz's advocacy of all things American, and—perhaps most propelling of all—William Henry Edwards's plans to publish his *Butterflies of North America*, helped shift Scudder completely toward butterflies. His attraction to their beauty kept him there. He would never write a thing on economic entomology.

In 1869, the year the Harris letters appeared, Scudder announced that he was beginning a major work on American butterflies, many not yet described or figured. He called the work-to-be *The Butterflies of New England* but planned to examine *all* species in the eastern United States and Canada. (Later, he would publish the book under the title *The Butterflies of the Eastern U.S. and Canada*.) Like Edwards, he, too, began what he thought would be a new kind of book, dealing not just with adults but with every aspect of butterfly life and death, peppered with "abundant colored plates."[166] In line with Buffon and Agassiz, he made no exception: every insect would get its life history or not be published, a stance Edwards did not take at first, since many of the butterflies in his 1872 volume did not yet have life histories.[167] To ensure thoroughness, Scudder sent out circulars, as Agassiz had done for his books on fish and turtles, exhorting amateur naturalists everywhere to catch and rear butterflies for him and then send him the results; he promised (and would give) full attribution.[168] "No similar work has ever been attempted," he wrote, ". . . but I believe I can succeed by pressing into service the young naturalists of the country."[169]

In one month alone, Boston men and women inundated Scudder with dead and "living butterflies (!)," many packed in cotton wool. They sent him lists of butterflies found in different localities, with exact data on the time of "first appearance and the duration of each species," and even material on the *parasites* of butterflies "with observations about time and manner of attack." From Clarissa Guild of Walpole, Massachusetts, Scudder learned about a little ichneumon fly so hungry as to nibble away by itself nearly the entire larva of the painted lady (*Vanessa cardui*), and from Caroline G. Soule, of Jamaica Plain, Massachusetts, that the American copper butterfly "slept" by "clinging" to blades of grass "with drooping wings."[170] Gene Stratton-Porter, a well-known nature writer, claimed that Soule and her partner, Ida Mitchell Eliot, had "in all probability raised more different caterpillars for the purpose of securing life history, than any other workers in our country."[171]

In the interest of getting data on "insects from every possible quarter to arrive at a definite knowledge of their habits," Scudder cast a wide net over the continent. "I am anxious to obtain the larvae of *all* the American species," he told Henry Edwards of San Francisco in 1869. "Don't you think you could obtain some of the California species? . . . I do not believe that the larva of *one* of the distinctively California species has been described. Here is an open field for you. Why don't you occupy it? We of the East would be pretty pleased."[172]

Though 1869 might have been a thrilling butterfly year for Scudder, personal misfortune robbed him of this glory. His wife, Ethelinda, became gravely ill after giving birth to a son, and they sailed to Europe and North Africa for her recuperation, traveling as far south as Alexandria and Cairo and settling for most of the time in Montreux, Switzerland, on Lake Geneva at the foot of the Alps, where many naturalists lived. (One hundred years later, Vladimir Nabokov, Scudder's modern champion, would live and die there.) Scudder hated being far away from the butterflies as well as the naturalists at home, principally from Edwards, and Edwards was unhappy to see him leave, partly because he felt Scudder needed him. "You are unfortunate in having to be out of country and dependent on the observations of others," he wrote. "I could aid you a good deal if you were easier within reach, for I am always observing something." But he needed Scudder, too: "I have some puzzling things to consider of occasionally and I am sorry you are so far away that I cannot write you as frequently as I would for consultation."[173]

To compensate for being abroad, Scudder scaled the high Alps to catch butterflies, and he studied fossil species recently found in Europe, in time emerging as a great authority on fossil Lepidoptera.[174] He spent hours in stuffy local libraries cogitating on butterfly genera, with results that would upend conventional notions of butterfly groupings and, eventually, infuriate the likes of William Henry Edwards. Known, in the jargon of the times, as a "lumper," Scudder made few species but many genera, grouping species together by shared "conspicuous details of structure," such as the wing veins or the genitalia. (On the other hand, a naturalist who made many species but fewer genera was known as a "splitter," a term, along with "lumper," still used today.)[175] Following Darwin and Walsh, Edwards viewed genera as arbitrary and artificial constructs, with no actual existence in nature, and serving mostly as convenient devices for classifying organisms. Scudder, on the other hand, believed, with Agassiz, that genera were God-designed and perfect, just like species. "I find them in Nature," he told Edwards in 1871, "and it is my business, undertaking the work I have, to weigh and publish what I see."[176]

He was concerned, however, that too many genera had been made over the years by uninformed individuals who'd used indefensible evidence and fostered a mayhem of names. To establish order, Scudder daringly revised older descriptions, applying in each case "the law of priority," a "law" invented in England governing the classification of *all* butterflies and enforcing stability in naming (for example, the name chosen by the first describer sticks, so long as it is accurate).[177] "There is hardly one of the genera which I can leave in its old limits," he told Edwards. On "systematic" grounds, Scudder rejected an older ordering that had put the big showy swallowtails, or the Papilionidae, at the top of the butterfly hierarchy; he placed, instead, the medium-sized, low-flying, and dull brown satyrs and wood nymphs, or, collectively, the Satyridae, at the top, beginning with the genus *Oeneis,* and put the species *Oeneis semidea,* Scudder's own anointed favorite from Mount Washington, the very first in line! He added numerous new genera, enlisting names penned by Jacob Hübner, the man Boisduval thought Edwards most resembled and whom Scudder believed had "priority" in the naming of many butterflies. In the end, Scudder created nearly one genus for every two species![178]

Scudder sent a copy of his essay to Edwards, followed by another unsettling paper, this one co-written with Edward Burgess in Cambridge, demonstrating how the genitalia of male butterflies (the most complex of the organs visible on male butterflies, with a character often unique to each species) could be relied on to identity butterfly species. Some Europeans had earlier shown how genitalia might be so enlisted, but Scudder may have been the first to look at them under a microscope, at magnifying powers of 125 times and up.[179] Inviting Edwards's reaction to both papers, he sought an exchange of letters that would, in its own right, cut down the distance between them at this critical moment when both were working on their books at the same time, on the same subject, and for the same publisher. "I am delighted to hear that Hurd and Houghton are to publish your book," Scudder wrote Edwards in November 1871. "I should have told you before that mine was to be in their hands. I am sure we shall assist each other in this new way."[180] Their letters to each other give a glimpse into their temperaments: Edwards, older by sixteen years, paternalistic, arrogant, and mincing no words; Scudder, guarded and well mannered but just as unbending on matters dear to him.

Sniffing the scent of Agassiz in Scudder's analysis, Edwards challenged his colleague, "I do not believe in such a thing as purity of genera. A genus is fluid as water or elastic as rubber." Two years before, Edwards had asserted nearly the same about species, based on his research on polymorphism, much of it unpublished at the time and shared with Scudder (who was then in Cairo with his sick wife and must have been excited to receive it). "There are some curious things about this species," Edwards explained about seasonal forms of the zebra swallowtail, "not paralleled by anything I have read of." He added, "I don't know what to say of such a strange complication. It confounds one's notions of species. I don't believe in the immutability of species anyhow, and I study these variations with great interest."[181] Edwards, known in his own day (and in the present) as a splitter, not a lumper, ever the careful Darwinian, sometimes seemed neither of these.

Yet Edwards did not oppose making new genera, so long as "there is really a reason for it," he advised Scudder, and especially if one could show some "natural" justification. "And if you follow the butterfly from egg to imago and form your genera upon differences in all stages, nothing can be said against them, that is within any reason." Still, Scudder, he complained, was himself making too many genera, and relying too much on Hübner for names. Although Boisduval had flattered Edwards by calling him the "Hübner of America," he had unwittingly praised the wrong man; Edwards actually detested Hübner and his system, thought him uneducated and barbarian, and despised his "tribal names" as nonsensical. Edwards said nothing against Hübner in his letter to Scudder, but he did argue that the established arrangement of butterflies was sound. "The system was a good one," he maintained, the Darwinian supposedly more radical than Scudder. Nor was Edwards alone in this belief; few labeled *him* a radical or troublemaker (although they might well have), whereas many were certain that the younger Scudder, a believer in the God-ordained, was a bull in a china shop. Edwards himself accused Scudder of betraying "the fathers." "I don't allow your right to deviate from usage and to attempt introducing an entirely new mode of identifying species, and that no man in a hundred could follow," he wrote. "Microscopic examinations are very well as a mode of distinguishing difficult species and so an adjunct, but you should not cut adrift from the fathers."[182]

Edwards was unhappy that Scudder had "put his name to such a paper" and suggested that he pay less mind to classification. Taxonomy and nomenclature were important, Edwards conceded, and he certainly did his share of it, but they often proved a waste of time. In echoes of Comte de Buffon's harangue against Carl Linnaeus in the 1750s, Edwards urged Scudder to stick to natural history or biology. "I cannot go with you in this barren field you have entered upon. Life is too short."[183] He criticized Scudder, too, on morphological grounds, disputing his use of male genitalia—or what he half-jokingly called "the tails"—to identify species.[184] And please remember, he added, when you address "the tails," that butterflies come in two genders; "your scheme makes no provision" for "a female. She is a superfluous individual."[185] Scudder reacted defensively: "I do not intend to make the genital organs the prime feature of my book—I only meant that it would be a noticeable one because heretofore neglected. I shall simply describe those of each species in connection with the description of the other parts of the body; and I shall figure each because they are difficult to comprehend without illustration."[186] Privately, Edwards also disparaged Scudder's hierarchical arrangements—that one butterfly family was "higher" or "lower" than any other. Nature, Edwards maintained, on Darwinian grounds, did not make such distinctions, and nor should we.

For all their testiness, Edwards's letters to Scudder were warm and friendly, as were Scudder's to him. "I have been daily watching for the first butterfly," Edwards wrote, soon after a last winter snow storm had blanketed Coalburgh, "which will be Violacea (an azure blue), and till that happens, I will not allow that spring has really come."[187] Once, learning that Scudder and his wife were going to Paris, he asked the younger man to get Boisduval's photo from Boisduval himself "for my Album" (a ritual among butterfly people), and Scudder did so with pleasure, as when he'd been in Cairo and Edwards had requested Egyptian stamps for his children.[188] Edwards may have disagreed with Scudder's generic distinctions, but he wanted the two of them to present the same face to the world. "Now give me your genera," he wrote in December 1870; "it is better that we be in accord." And, three months later: "I would rather be uniform with you, if I know your determinations."[189]

Both men complained that Europeans snubbed Americans as ignorant about their butterflies. In June 1871, Edwards wrote Scudder about the German naturalist Otto Staudinger, one of the most sensational men in the history of butterfly science, whose catalog of European species was so influential, another leading naturalist had argued, as to induce "all collectors to use the same names." Staudinger had asserted in a letter, Edwards told Scudder, that most of the American fauna were "the same as the European," and to find out if the German knew what he was talking about, Edwards had mailed him three butterflies that looked alike and that Staudinger had never seen. "I innocently asked him to tell me what they were," and Staudinger had replied that they replicated exactly a single European butterfly. How ludicrous, Edwards told Scudder, since the two Americans knew how different the larvae of each insect was from the other. "Suppose I undertake to pronounce on the butterflies of Europe with one or two of each or none at all in hand. They would feel insulted by the suggestion."[190] (A year later, Edwards stated this categorically in volume 1 of *The Butterflies of North America*: "The truth is, the sooner the theory of the identity between the European and North American fauna . . . is exploded the better. . . . The number common to the temperate regions of both continents can be counted on one's fingers.")[191]

Scudder responded in complete agreement: "The view of Staudinger and others about our species seem to me inadmissible on general grounds. They seem to think that our fauna in its general aspects differ but little from theirs, but I shall be able to show that it is far from being the case."[192] How ironic that both these men—one a Darwinian who believed that butterflies might migrate or hybridize freely, the other an anti-Darwinian certain that the boundaries of species could not be breached—were arguing for the independence of American butterflies. As the targets of Edward Doubleday's advice, they wanted to affirm to the world a common sense of American destiny and identity.

In August 1872, Scudder's wife, after much sufffering, died in Montreaux, leaving him broken and with the care of their son, Gardiner. The following year, in April 1873, he came home to begin, in earnest, his work on butterflies. In the summer, with help from his wife's legacy, he traveled to the Smithsonian in Washington to inspect the butterfly collection there, followed by a visit to Edwards in Coalburgh, West Virginia, and ending in Reading, Pennsylvania, to see the insect cache of the weird and intense Herman Strecker, a German-American stonemason in love with the beauty of butterflies as no one else in America.[193] Scudder's purpose was to examine all the specimens he could, and in the greatest number, so that he might distinguish species correctly. Later in the year, he presided over the Cambridge Entomological Club, the first such group in America and one of the two important clubs he either cofounded or led in the mid-1870s. In the Cambridge club's early years, members convened at a midsummer campsite on the slopes of Mount Washington. In 1874 it published a new journal devoted to insect study, called *Psyche*, meaning in Greek both "butterfly" and "soul." Still available today, it was edited by Scudder for a time, with articles by both men and women.[194] Around the same time, Scudder further upset Edwards by defending, in *Psyche* in 1874, the common names for butterflies. Edwards preferred the Latin names of Linnaeus, again the "radical" Darwinian loyal to "the fathers." Although Scudder, too, employed Latin or Greek nomenclature, he also thought that Americans might prefer more everyday names. "It is my belief," he wrote, echoing a view prevalent in England, "that the study of butterflies would be far more popular, if they . . . had common names."[195] He invented many names himself, still used today, such as "the Blue Swallowtail," "Regal Fritillary," "Great Spangled Fritillary," and, most celebrated, the "Monarch," so called because it "ruled a vast domain" and had a life span longer than any other butterfly on the continent.[196]

A second club, cofounded by Scudder in 1878, the Appalachian Mountain Club, was the country's first enduring mountain-climbing group, modeled after a British prototype, and responding to a wave of Romantic recreational mountaineering begun in the 1840s.[197] Scudder edited the club's journal, naming it *Appalachia*. The club, also chiefly Scudder's handiwork, had a remarkably mixed membership, thirteen women and eleven men, most of them from Williamstown and nearby towns, its purpose to know "the natural history of the localities." Familiar with every inch of the countryside and the group's leading spirit, Scudder conducted one exhausting "tramp" after another, "scouring the whole region," scrutinizing ponds, glens, meadows, basins, and springs. On one occasion, he supervised a spectacular three-day hike of eight men and women to the top of Mount Washington, on a pathless route through a maze of scrub, soft mosses, leaves, and hazardous rocks, ending at twilight. The Cambridge and Appalachian clubs were partly excuses for Scudder to get back to the White Mountains. "It is a good thing," he later said, "to come into direct contact with Nature in the fields and woods."[198]

In 1874 Scudder wrote an eloquent chapter on insects for *The First Volume of the Final Report upon the Geology of New Hampshire*, devoted principally to the distribution of butterflies. He laid out beautifully the three distinct zones of life along isothermal lines (forest district, subalpine, and alpine) in his favorite place—the White Mountains, which he explored up and down in the Humboldtian manner, tracking what he had come to believe was the most exceptional butterfly fauna in the United States.[199] He began, of course, with an account of his discovery, O. *semidea*—the longest and most affectionate description in the volume—followed by a parade of fritillaries, skippers, angle-wings, swallowtails, purples, and the monarch. But it was only a taste of the work that in 1869 he had hoped to finish in a few years' time. He had misjudged: his grandest achievement lay in the future, coming twenty years later than he had planned. Edwards's best work, too, lay in the future, and would emerge, surprisingly, out of the sudden wreckage of his coal business.

TWO

The German-American Romantics

Throughout the nineteenth century, Germans migrated to America in larger numbers than any other group, introducing into the country a mature scientific tradition. They brought with them the oldest butterfly culture in Europe, dating back perhaps to the late seventeenth century.[1] From San Francisco to Newark, often in rooms above saloons, they either founded or cofounded America's first societies for the serious study of insects. In the early 1870s, many of Brooklyn's butterfly people, most of German extraction, met upstairs at a saloon on Broadway in Williamsburg to talk butterflies and other insects, afterward stepping downstairs, where the beer flowed freely. Once the same group gathered in another Williamsburg saloon, John Kramer's on Graham Avenue, each man with a raffle ticket giving him a chance "on a collection of butterflies to be drawn for." "The event was an orderly jolly gathering of entomological enthusiasts," one ticket holder recalled, "who discussed collecting experiences and drank vast quantities of wholesome lager beer."[2] Germans also emigrated to Philadelphia and to the Baltimore region, among them Karl Zimmerman, an aristocratic German who in the 1840s collected butterflies and other insects in the southern United States and in Brazil, financed by German merchants eager to sell exotic wildlife, including butterflies.[3]

Another Baltimore resident and lepidopterist was John Morris, a Lutheran minister and the son of a German mother and a Prussian-German doctor who'd served in the American Revolutionary War. Born in 1803, he briefly taught at the Pennsylvania College in Gettysburg, and, with the help of thirty students, organized one of the country's earliest natural history societies. "I am the only person in Baltimore," he told a friend, "who pays any scientific attention to butterflies."[4] In 1859, he completed the first catalog devoted entirely to American butterflies by an American; unlike the German-American John Weidemeyer of New

York City, whose later 1864 book, *Catalogue of North American Butterflies,* also dealt with Central American species, Morris kept his sights clearly on nature above the southern border of the United States, treating nearly two thousand insects, which he arranged along the German model.[5] It had no pictures and was unoriginal in its entomology, since all its species had already been described by Europeans. Nevertheless, as William Henry Edwards later observed, it was not until Morris's book came out that one knew "whether a given insect had been named and described or not. Its publication gave a start to many collectors, and the work of describing caterpillars and moths went on rapidly thereafter."[6]

Herman Strecker and Augustus Grote came out of this German immigrant world, beer and all, and, along with the Yankees Edwards and Scudder, belonged to the advance guard of American butterfly people. In the 1870s Strecker would write one of the ablest (and most egomaniacal) books on American lepidoptera and would amass the largest private collection of butterflies in American history, now housed in the Field Museum in Chicago. When Grote was fifteen, in the mid-1850s, no one in the United States could identify any native moth species, but twenty-five years later, thanks largely to his labors, hundreds of species had been described and identified, usually based on specimens lent or given to him by others; many still stand, including nearly one-fourth of all Noctuidae, a great and diverse family of "night butterflies," its name derived from Latin meaning "owl."[7] At the end of the century, Grote would make original contributions to the evolutionary study of butterflies and moths.

Both men were German-American Romantics. Both followed Goethe and Humboldt in affirming—in Strecker's words—"that all that is great and sublime in nature and art is more or less intimately connected" and that beauty was an elemental feature of the natural world, not contingent but basic. Strecker believed that Humboldt "far exceeds in rank all of earth's potentates, of whom a monarch of Europe once said, '*Der grösste Mann seit Noah*' ['The greatest man since Noah']."[8] Grote's awareness of the relationships among all natural phenomena derived in large part from Humboldt. He practiced "aesthetic entomology," a discipline rooted in German traditions.[9] "Entomology," he wrote in 1886, "combines Art and Science in a peculiarly seductive manner.

Even in flowers we have no more beautiful patterns and colors. Tints which we do not find in Art often brought together, are here harmoniously blended."[10] He considered "the larger species" of moths "the most beautiful objects one can wish to see."[11] Grote and Strecker also shared a *Sehnsucht* characteristic of the German lyrical tradition, Grote expressing his literally in his own poetry, and Strecker in his lifelong craving of butterflies. "I have such a terrible *Sehnsucht*," he would say to his friends, meaning literally longing (*das Sehnen*) to addiction (*die Sucht*), but more poetically a "heart yearning" for some unreachable object, an exotic butterfly or rare moth.[12]

To a degree matching the British influence, the German Romantic naturalist outlook had an impact on *all* American naturalists, not merely on the Germans who came to American shores. William Henry Edwards owed a lot to such men as Weidemeyer, Akhurst, and Morris, and he also absorbed Humboldt's *Personal Narrative* before retracing Humboldt's trail down the Orinoco. Scudder got his feel for instruments of measurement and exploration and for the kindred affinities of all natural phenomena directly from German-trained Louis Agassiz, who, in turn, got them in part from Humboldt, his own great hero. Before the Civil War and for some time after it, many American naturalists— perhaps the majority—believed that the study of art and science were not only compatible but should be integrated. In places like Philadelphia and Boston, many an evangelist preached the aesthetic distinctiveness of nature—Titian Peale for one, Audubon, for another.[13] At the same time, in the early 1870s, Grote and Strecker felt a stronger loyalty to German Romanticism than did most of their contemporaries. Romanticism helped bring them together, as did their desire to make major contributions to the American science of butterflies and moths. And they, too, had to meet the European charge that America really had no unique species of its own and not much by way of a science of butterflies, either.

Herman Strecker was besotted with the beauty of butterflies. He was an antinomian of butterflies, willing to break laws to capture the fire inside the wings. The family religion was Lutheranism, but he had no interest in it or in any church. Neighbors knew him in his maturity as a "blatant agnostic," and sometimes, in jest, he called himself "one of the Devil's favorite children."[14] He was born on March 24, 1836, in Philadelphia, the son of an expert stone carver, Ferdinand Strecker, who, the year before, had immigrated from Stuttgart, the home of many artists and artisans, among them Herman's uncle Wilhelm and his two daughters, "artist-painters by choice," who remained in Germany, where they practiced a venerable European tradition that connected a variety of visual expertise—from glass and window design in churches and cathedrals to skilled drawing using brush and color—to nature study, including lepidoptera.[15] Herman's mother,

Herman Strecker. Courtesy of the Field Museum of Natural History, Chicago, Illinois.

Anna Kern, was a second-generation German-American, the daughter of the deputy collector of the port of Philadelphia, with three brothers, Benjamin, Richard, and Edward, explorers who joined John Frémont on his missions to the Far West. Edward and Richard Kern excelled as artists, were self-taught in topography and in the use of instruments of exploration, and were able to identify a wide range of plants and animals.[16] Yet despite these enticing uncles, whom he admired but would never meet, Strecker himself had mostly a "dreary" childhood with no close or loving kin. "At the death of my parents," he wrote later in life, "we all scattered—brothers and sisters—one to California, another down South and so on, and for years I have heard no word, none having had any sympathy in common with the other."[17]

When Herman was eleven, his father abruptly moved the family— Herman and two brothers and two sisters—from Philadelphia to Reading, Pennsylvania, a tinier world where he might stake a bigger claim. He built a marble business. He made Herman an apprentice stone carver and then a full carver, or *Bildhauer* (maker of images), work he would do all his life, usually under other master stone carvers, except for a brief time after his father died in 1856, when he owned his own business.[18] "I make my living," he told a friend, "by sculpturing statues, monuments, and other memorials for the dead."[19] He specialized, as he put it, in cutting marble angels for children's graves. He took pride in the work, but it was demanding, exhausting, and dirty. He sometimes dropped heavy marble on his feet or got flint in his eyes.[20] At the same time, he did artistic pieces of high quality, such as diminutive illuminated color pictures, inspired by medieval prototypes and only fully viewable with a hand lens, which he sold in the Philadelphia art market.[21] In 1867 Joseph Drexel, a prosperous commercial banker in Philadelphia and an avid butterfly collector, and his wife, Lucy, commissioned the young Strecker to sculpt a bas-relief of Poe's raven for their new home, urging him to "do something better than tombing."[22] Strecker would also make a special art out of his butterflies, sketching and coloring pictures of them to integrate into what became one of the most provocative catalogs on lepidoptera ever done by an American.

He started collecting at five. His father tried to beat it out of him with a strap, but he stood his ground, befriending a Philadelphian taxidermist, Charles Wood, popularly known as the "butterfly man," who taught him about lepidoptera and later sold him butterflies as well as stuffed birds and birds' eggs, which Strecker collected with the same fervor as had William Henry Edwards.[23] Christian Sproesser, a chronically drunk German apprentice of Strecker's father, in a sober moment gently placed in young Herman's hands a small board on which was pinned a moth specimen of *Catocala amatrix,* vernacularly called the "sweetheart underwing." (The genus name is Greek for "beauty," the species Latin for "mistress"; together, "mistress of beauty.") Strecker rotated it around in his hand, marveling at its long gray forewings and hind wings slashed with scarlet; he treated Sproesser to lemonade (but the apprentice wanted more beer) and always remembered him. "If living, I hope he is well," Strecker later reflected, "or, if dead, has gone to where he belongs. For all have had their day / the grave and the gay, / Then blow to the devil and vanish away."[24] Sometimes, too, along the docks of the port where his grandfather worked, Strecker would get to see the many boxes of butterflies from China, brought by merchant ships from the port city of Canton. They "seemed to come over in loads" and "sometimes with good things in them," he would later say.[25]

Strecker never got beyond grammar school and, unlike the Yankees Edwards and Scudder, never had a natural history course, but Philadelphia was the mecca for natural history in America up to the 1850s and the home of its first natural history society, the Philadelphia Academy of Natural Sciences, founded in 1812. At ten already a seasoned butterfly collector, Herman visited the academy's basement library, admitted by a curator, Joseph Leidy, and gawked at the butterfly pictures in the mighty catalogs from England and Europe, among them Pieter Cramer's *Papillons Exotiques* (1779), Edward Donovan's *The Epitome of the Natural History of Insects of India* (1800), Jacob Hübner's *Sammlung exotischer Schmetterlinge* (1806), and Dru Drury's *Illustrations of Exotic Entomology* (London, 1770, 1837). Each had hand-painted illustrations—in Donovan, the iridescent blue *Papilio ulysses;* in Drury, the "exceedingly rare" *Papilio antimachus,* which proved to be the largest butterfly in Africa; in Hübner, an exquisitely drawn and colored South American *Morpho achilles;* and, most stunning of all, in Cramer, the iridescent gold-and-green *Ornithoptera priamus* (Latin for the "first birdwing," called such because it looked like a bird when European collectors first saw it, high up in the forest canopy). The earliest known of all the birdwings of southeast Asia, *priamus* was "the Ur-birdwing" of the entire genus. Strecker may have actually seen it years before he visited the academy's library, when he was only five years old, and it would become his lodestar, sinking into his consciousness, always there to be savored, as if it were the most beautiful thing in the world.

Today, photographs of butterflies—and of other natural forms— have become so seductive as to sometimes serve as substitutes for the real things, interrupting or even blocking contact with the living natural world, a counter-world against which the real one is measured or ignored. In the nineteenth century, pictures were more often allies of contact, inspiring the collector or explorer to pursue nature. Humboldt was perhaps the first modern naturalist to recognize the way pictures of nature drew people into nature (in his case, landscape pictures). "By showing all the diversity of form of the external world," he wrote in his *Cosmos,* "landscape painting incites men to a free communion with nature." Strecker probably knew and understood this passage well. "Great God what a Heaven opened to me!" he later wrote of his time in the basement library. "How I gazed wonder-struck on [these insects] depicted by the old authors, never dreaming that I should ever become the happy possessor of such treasures."[26] The pictures exposed him to the aesthetic-sensual core of natural history, and with enough primal force to make him "think of nothing else, dream of nothing else," for the rest of his life.[27]

In Reading, Herman hunted butterflies in neighborhood fields and especially on Neversink Mountain, a perfect natural paradise of living things for boys and girls, rising only nine hundred feet but covering many hundreds of acres, with evergreens at the top and dense vegetation, streambeds, and the Klapperthal Glen at the bottom, the whole of it "undeveloped" until the railroads were driven through in the 1890s, bringing the vacation hotels.[28] Memorable butterflies and moths were everywhere, and sometimes Strecker took an ally with him, Russell Robinson, and the two would collect together. Robinson was a grandson of the engineer who built the Philadelphia and Reading Railroad, the full line completed in 1839, among the first such roads in the country. Strecker "fired" him up "with the collecting zeal," according to Robinson's grandson Wirt Robinson, and the two became buddies—Herman, at sixteen, the individual in charge, and Russell, at ten or so, an apprentice assistant. Their paths would diverge, then cross again at the end of their lives.[29]

In the early 1850s, still in his teens, Strecker ran off to Mexico and Central America and, for a month or so, rejoiced in fantastic Aztec temples and unknown butterflies, heightening his already refined sense of the exotic. Back in Reading, he slipped the bounds further and may have fathered children out of wedlock, alienating his parents as well as his two sisters; after they married and returned to Philadelphia to live, neither sister had anything to do with him. In 1855, he began courting a Louisa Roy in nearby Altoona, Pennsylvania. Even Strecker's mother warned her against marrying such a "crazy man." He did look bizarre, apparently, with his high brow, long neck and torso, irregular beard, and brown, gentle, bespectacled eyes, glittering with energy. In a letter to Louisa, fuming with grievances, Strecker called his mother "a devil spirit" for messing in his affairs; defiant, he married Louisa in 1856, the year his father died, and his mother left Reading for Philadelphia to join her daughters.[30] Strecker's marriage was passionate but blighted, beginning with the deaths of two baby boys and ending with Louisa's death, from malaria or miscarriage, on the same day in 1869 that his mother died. (Samuel Scudder lost his own wife around the same time, and so, too, would Strecker's fellow German-American Augustus Grote.) Strecker was now alone, wifeless, childless, and kinless. As he said to a friend, "I am an isolated old fellow"—he was only thirty-four when he wrote this—"with no family or ties whatever."[31] Although matters of death and disease were constants in Strecker's life, he wrote of them seldom, and when he did, only offhandedly, as if they concerned no one.[32] The effect of it all was to propel him decisively into his collection, where he stayed, his erotic energies radiating mainly through the butterflies, despite a new marriage two years later, this time to a practical "hausfrau" from Prussia; two children were born, and both survived.[33]

Around the time of his second marriage, Strecker began to follow a daily regime that would last until he died. "I get up at 6 a.m.," he told a friend, "dress, eat, go to the Post Office"; then "I work at the shop cutting marble angels, etc., until 1 o'clock—to the Post Office, eat dinner, write letters, back to the shop at 2 p.m."[34] The U.S. Post Office in Reading was to him what Paint Creek and the White Mountains were to William Henry Edwards and Samuel Scudder: the source of his butterflies, sent by hundreds of people from around the country and the world; without it, his butterfly life would have died. At six p.m. he went home from the marble yard and his shop to begin the rest of his ritual regime, climaxing between eight or nine and midnight, when he turned wholly to his "things of endless joy," pursuing them with the fervor of a crazed monk.[35] If the day often felt like night, with all the dead children to sculpt for, the real night was alive. Here, groggy and worn out in his butterfly room, Strecker examined the packages from the post office, arranged his insects, wrote his letters (usually unpunctuated and unedited, flushed out of his brain like mere smears on the page), and read those sent to him, many thousands over the years, from all sorts and classes of people, united by a common delight in butterflies.

"Mine has been such a life of ups and downs," Strecker wrote a friend in 1872, "that I take things rather philosophically, for after all there is in our favorite science a balm for all worldly ills, and when night comes and I can get to my butterflies, cities and nations may rise and fall for all I care." Three months later: "What a life of pain, toil, and trouble this is, in my own case. Were it not for the butterflies, it would have been indeed almost 'one drear night long' but the delight to be found in the study of nature compensates for the many miseries." And two years later: "my circumstances are wretched now, but I don't want to complain, as the powers that be allow me more pleasure with my beloved butterflies, than most men could have with their money."[36]

He differed from Edwards and Scudder not only in his lack of privilege and formal education but on account of his unbuttoned *Sehnsucht.* "Unusual in that he was interested in everything," as a friend said of him, he wanted *both* moths and butterflies—"omnium," he said, no distinctions.[37] He loved butterflies principally because of their beauty and specialized, therefore, in the adult insect. "I only care for the perfect butterfly or moth," he wrote a fellow butterfly man, "and do not want ova or caterpillars."[38] Unlike the others, he wanted exotic insects as well as American ones, a taste he nursed as a boy when he kept a loosely bound notebook, recording the names and habits of moths and butterflies, American and foreign.[39] Thus: "Papilio swallowtails are first class diurnal Lepidoptera. The larvae have sixteen feet. Chrysalides are always naked and attached by tail and commonly angular." Or: "There are about 250 species distributed over all the world" (but this was the 1850s, and some 530 species were known, so Strecker was off by 280). The notebook contained his drawings based on butterflies he may have seen in Mexico and Central America (or in pictures), such as one of his favorites, the morphos of Brazil, "a magnificent genus in which the superior surface of the wings is invariably of some shade of brilliant blue and the under surface is generally dark and invariably ornamented with numerous ocelli" (or eyespots).[40]

Around 1870 Strecker began systematic collecting, at the very time when Edwards and Scudder had launched their own investigations (Strecker did not fight in the Civil War; why I do not know). In a few years he had "an immense collection of Lepidoptera," upwards of thirty thousand specimens, patched together mostly through exchanges that stretched throughout the country and abroad. He was no longer isolated in Reading, as a host of naturalists —including Samuel Scudder and Theodore Mead—traveled the Philadelphia and Reading Railroad to see his acquisitions.[41] He exchanged as well as bought and sold for cash with others via his post office, strategies that brought him in contact with butterfly people from San Francisco to Dresden, Germany.

Buying and selling outright was always treacherous for Strecker, forcing him to walk a fine moral line, as well as to deny himself everything else, including food (after his first wife died, he didn't mind this suffering; things changed a bit once he remarried).[42] He dreamed (it was only a dream) of never going into debt and never buying on credit in lieu of cash payment. The better way was to dispense with money altogether by exchanging his insects for the insects of others, so long as he gave as much as he took; if it turned out he didn't, then he had to face the heat and, sometimes, the moral fury of others, but at least money was not directly at stake, something he feared and could not easily finesse. Nevertheless, changes in the character of the economy, coupled with his appetite for the exotic and tropical, made it harder and harder for him to escape money transactions. In 1876, Strecker's friend George Hulst, a "beginner" collector longing for specimens from the "far West and pacific Coast," urged Strecker to "sell on the 'one price' system, so there can be no occasion for haggling."[43] But how *did* one set a standard price for a butterfly? Increasingly, as a result of the rise and development of the railroad and of faster ocean transportation, one could get lepidoptera from great distances, and Strecker found market forces inescapable, though at every step they compromised his quest for beauty, whose value transcended price. "The law of supply and demand," he would later write, "regulates the prices" under conditions where "there is no set value; a species that you may get today for a dollar may be worth, in a week, five, or what may be five today may in a short while fall to one."[44] In any case, all prices offended many naturalists; they refused payment for nature's progeny, viewing monetary transactions as inherently degrading. "Buying seems like desecration," wrote H. Landis, a German-American physician from Columbus, Ohio, to Strecker, "when compared to catching butterflies in one's own net."[45]

Strecker no doubt agreed, given his preference for exchange. Yet he had little choice but to rely on prices or paying in cash and by credit. From the start, as he built his circuit of collectors in Europe and America, he paid in money whenever feasible, especially for large lots, and then resold those lots piecemeal at a profit in the United States, to finance his own collection. Strecker would always distrust the modern market economy, but, as his desire grew in relation to the growth of that very economy such that he could not, at times, master his desire, it turned out he was willing to accept some degree of "desecration." He never seemed concerned that the intrinsic value of a butterfly might be subverted by its price. Nor did he worry about losing face or status because he sold and bought butterflies; Strecker never sought a profit to meet his own material needs: *everything* he did was in behalf of his collection, *con amore.*[46]

Strecker's key European contact was Otto Staudinger of Dresden, the most influential butterfly man of the nineteenth century, an entomological Svengali whose spell over many Americans would grow stronger year by year. Born in 1830 to upper-middle-class Prussians, Staudinger was educated at the University of Berlin in medicine, then in natural history, specializing in insects. From his thirteenth year, like so many German youths, he chased butterflies, hunting throughout Europe, from Iceland to Granada; in Sardinia, he discovered the larva of *Papilio hospiton,* the loveliest and rarest of Europe's three swallowtails. He became one of the most respected butterfly authorities in Europe, with enough firepower to infuriate Scudder and Edwards, who objected to his uninformed judgments about American butterfly fauna. Staudinger, however, did more than study butterflies or write catalogs about them; to help pay for his travels, he also created a butterfly business, formed, at first, from his own collection. In Dresden, he converted it into a successful enterprise, at first buying and selling only European lepidoptera, then slowly dipping into the tropical trade, with the capacity to determine price levels— a skeletal standardized market—throughout the transatlantic world, just as he had tended to dictate European naming. Edwards and Scudder disagreed with Staudinger over his ideas on American butterflies, but Strecker's relationship was entirely of another kind. Around 1870, Strecker was trading American butterflies for European, but, as with his other business relationships, paying cash when necessary, at great personal sacrifice, and always trusting that Staudinger had given him an honest price; sometimes he borrowed money to get what he wanted. In time, he would get snared, almost helplessly, within Staudinger's seductive webs.

A few Americans shared Strecker's desire for exotics, and Strecker did everything within his power to court them. One was Tryon Reakirt, a manufacturer of lead products in Philadelphia, an excellent naturalist briefly rich enough to collect exotics as well as domestic species (many of which he actually named and described); he even wrote up descriptions of African and South American butterflies, which Philadelphia's Academy of Natural Sciences published in its proceedings. He bought many from Pierre Lorquin, the man who collected for Boisduval, and Reakirt thought he'd discovered new species but ended up only redescribing many already identified by European naturalists. In 1870, Strecker tried desperately to obtain the butterflies of Reakirt, after he put them up for sale. Generously, Reakirt let Strecker study the collection and even allowed him to take it home with him for a while, provided Strecker help find him a buyer; having little money of his own, Strecker tried unsuccessfully to persuade the banker Joseph Drexel to purchase the butterflies. Sometime in early 1871, Reakirt fled the country, never to return, leaving Strecker in full possession of his treasure, for which Strecker apparently never paid a cent.[47]

Strecker's most regular American supplier of exotics was the actor Henry Edwards of San Francisco, a man willing to exchange specimens—always Strecker's preference and less open to potentially dishonorable dealings. Edwards's influence on Strecker, at least for a short time, matched Staudinger's. Edwards had his fingers on an enormous reserve of American specimens, as well as on exotics from the Australian region, where he had lived before immigrating to America in 1867. In 1869, after hearing from a friend about Strecker's obsession, he promptly opened trade with him, ready to "drop all my correspondents in the Atlantic States, and send all and everything to you."[48] "I am perfectly omnivorous as regards Lepidoptera," Henry Edwards said. Strecker countered, "I am as omnivorous regarding Lepidoptera as you possibly can be" and "I want omnium."[49] Sprinkled with such yearning phrases as "I have longed for," "I will have no peace until," and "my soul pines," Strecker's letters to the actor were charged with his *Sehnsucht*.[50] "Through his splendid correspondents," he sent Edwards as many exotics as he could get from Armenia, northern Persia, Siberia, Spain, and Senegal and, "as usual," wanted "anything and everything *in any quantity*" from Edwards. "The reason I have been so importunate in my letters," he explained, "is that I have five parties in Europe and West Asia with whom I exchange and they have things reduced to such a system that if I want to see the best of their season's accumulation, I must send my things, about the beginning or the farthest, middle of December. If I do not, their best things go to others. So please get going."[51] Edwards did all he could to satisfy Strecker's "desiderata," especially for his *personal* cabinet.

In 1871, Strecker opened the mail to find from Edwards his *very first* green-and-gold *Ornithoptera priamus,* the birdwing he had gushed over as a boy at the Philadelphia academy. Oh, this was something, a moment to weep over, a moment to stash in the deepest reserves of memory, as if some god had spoken to him. "There is no use trying to express my feelings at beholding the splendid ornithoptera. Only to think the dream of my childhood fulfilled for since I was five years old I coveted and fretted for the Green Ornithoptera."[52]

Over time, Strecker's attention to exotics and insatiable collecting would sweep away almost everything else. But in the early 1870s, he seemed determined to belong to the company of American naturalists who were exploring the fauna of North America. He bought and sold American species in preparation for his own study, trading, for instance, with fellow German-American Fred Tepper, a tailor in his early thirties, who worked at a skin-and-leather business in lower Manhattan. Like a few others, Tepper had an interest in exotics, although American insects held a primary place. He exchanged with Strecker specimens he had collected in Flatbush, Brooklyn, where he'd lived since 1872, a locale of a complex hybrid character, farmland but also wild, with tiny waterfalls, secluded wooded glens, and rare lepidoptera, a far cry from the asphalt metropolis it would become. On one occasion, Tepper also caught numerous *Hesperia massasoit,* a skipper given an Indian name by Scudder in 1864, "thirty or forty" of which Tepper shared with Strecker. One fall, he found many "different species [of moths]" in "one spot"; he noted, "I intend going to that place next season and to stay four or five days, just simply to catch Catocalas, so I expect to have several hundred duplicates of this beautiful family next year." He put aside "many good Catocalas" for Strecker, "comprising about twenty different species," which "I hope will please you."[53] In the same woods he came upon a caterpillar of a rare moth, *Smerinthus myops,* and "hoped" the beast "would get through all right," pupate, and emerge as an adult. Once, his brother brought him another moth larva, picked off a small swamp plant, "one of the handsomest caterpillars I ever saw. It is golden yellow covered with minute stripes of bright red and having large broad white stripes on the sides," he told Strecker.[54]

In 1872, Strecker began a catalog in twelve parts, each with its own lithographic plate, descriptions, and commentary, issuing one part every several months at fifty cents apiece. He published it under the ponderous title *Lepidoptera: Rhopaloceres et Heteroceres*, terms borrowed from Boisduval; the former refers to butterflies, the latter, moths. A friend of Strecker's, Arthur Fuller, the editor of a popular journal of the day, the *Rural New Yorker*, objected to the title on the grounds that "Lepidoptera is an unknown word to nine tenths of our people," suggesting, instead, "Day and Night Among the Butterflies," because it would "make Strecker more money." But Strecker stood his ground, faithful to the formality of the European tradition.[55] Each part was almost entirely the product of a single man, a distinction that set Strecker apart from all the other butterfly men of his day and, in the eyes of some, made him a superior figure.[56] Strecker wrote the text, set the type, and drew, colored, and lithographed the images—exhausting craftsmanship done mostly at night and all, doubtless, legacy of his exposure to artists and artisans on both sides of his family. He lugged a heavy lithographic stone by train from Philadelphia to Reading, drew his images on the stone in crayon, and then took the stone back to Philadelphia for final reproduction of the plates. He did this many times and at much cost to himself. "If it had ten times the circulation, it would never do more than cover expenses," he wrote a friend in 1874.[57] If every other leading lepidopterist had chosen to emulate Strecker in his mastery of every facet of conception and execution, the whole history of American butterfly science might have gone down a very different road.

The impact of the early parts of the catalog was impressive, especially in England. Many in London viewed Strecker as the best authority on lepidoptera America had yet produced. "Strecker stands high here, much higher than W. H. Edwards," wrote Richard Stretch, an English émigré and a butterfly naturalist, to Henry Edwards. Strecker's greatest fan there was William Hewitson, a British surveyor who used inherited wealth to become a preeminent amateur naturalist and an accomplished artist of many elegant plates of butterflies for notable volumes, including *The Genera of Diurnal Lepidoptera* (1852) by Edward Doubleday and John O. Westwood, the book that influenced the way William Henry Edwards grouped his butterflies.[58] A creationist ("each species is in itself perfect") and an anti-Darwinist, Hewitson despised naturalists who *analyzed* butterflies or who broke them down into parts or pieces, thereby destroying, he believed, their basic reason for being: the power of their beauty. For him, butterflies existed for the delight they gave to human beings, and he considered "any man as a personal enemy who abuses butterflies with unnecessary names and surrounds them with difficulties." "I wish Edwards and Scudder had never been born," he told Strecker. But he "thought the world of Strecker," sure he was a soul mate and fellow artist who longed for butterflies for the same reason Hewitson did: because, as he put it, "they were beautiful things."[59]

Hewitson showered Strecker with butterflies no one in America had ever seen, except as pictures. The *Antimachus* swallowtail, the biggest in Africa, "had never gladdened European eyes" until Hewitson himself got hold of it. Hewitson also sent Strecker a *Morpho menelaus,* one of the bluest of morphos; *Papilio antenor,* a lovely polka-dotted African swallowtail; and two sensational birdwings, discovered by Alfred Rüssel Wallace in the 1850s: *Ornithoptera croesus,* male and female, and *Ornithoptera brookiana* (now *Troides brookiana*). They made Strecker "the happiest man alive," partly because Hewitson had sold them to him for almost nothing; when Strecker offered to pay more, he objected: "Do not talk about recompensing me for trouble. The intense pleasure you express in the receipt of the things I send you gives me pleasure almost as great and I feel as if for the first time in my life I had met with one as *deeply inoculated* with these things as I am myself."[60] In a review of the early parts of *Lepidoptera,* Hewitson wrote that "the plates are all drawn by Strecker himself, after a hard day's work, and could only be done under such circumstances by an entomologist whose heart and soul are in his work."[61] Strecker later referred to Hewitson as "the greatest living authority on Diurnal Lepidoptera."[62]

More than anyone of his time, Strecker stood for or embodied the democratic American community of butterfly lovers, many of whom studied insects scientifically, but the majority of whom were merely engrossed in lepidoptera, grateful for their existence and for living in the same world with them. Strecker fostered a correspondence of many thousands of letters, written by individuals from around the country who were inspired—as one sheep farmer put it—by his "ardor and grand enthusiasm" and were willing to act as his hod carriers.[63] Once, on a walking tour of Williamsburg, the Brooklyn taxidermist John Akhurst stopped at a saloon, where he spotted a "case full of blue Morphos" above the bar, only a few of which he had ever seen. He begged the bartender for it, managed to get it, and then gave it gratis to Strecker, knowing how happy it would make him. "You may be assured there is no living man but your SELF," he wrote Strecker, "I would have taken so much trouble to please."[64] In the fall of 1871, James Angus, a farmer in West Farms, New York, parted with his only *Catocala relicta*, a gray moth with black underwings banded in white he "had labored hard to obtain for years," confessing to Strecker that "no hundred or more of my insects have cost me half so much trouble," and adding, "I know you will be pleased with it." Three years later, Henry Schonborn, a lawyer in Washington, D.C., thrashed about for hours in "thousands of lilac bushes ten miles square" to find moths for Strecker; and at the end of the decade, Adrian Latimer, a drugstore worker from Lumpkin, Georgia, sent Strecker a "splendid" selection of local "blue" butterflies, or Lycaenidae, after learning about Strecker's "needs" from one of his books.[65] Two of his most loyal patrons were George Hulst and Berthold Neumoegen. Hulst was a young pastor in Flatbush whose "Dutch ancestors had lived in Kings County, Long Island [later Brooklyn] for over two hundred years." His father ran a farm of corn, peas, lettuce, cabbage, and potatoes, one of the many farms that supplied Manhattan with its daily produce. He debated Darwinian theory and exchanged insects with Strecker. "I wish," he wrote, "I had money to back you up myself—all praise to a man who carries a 'hod' to such an end."[66]

Strecker formed a troubled friendship with Neumoegen, a German Jewish immigrant from Frankfurt in 1845 whose collection of butterflies would become, by 1890, the second or third largest to be assembled by an American.[67] In the late 1860s, Neumoegen became a stockbroker and arbitrageur on Wall Street and, at the same time, started serious butterfly collecting in the fields and meadows around his wife's family summer house in northern New Jersey. Some anti-Semitic butterfly men viewed Neumoegen with derision, but he was Strecker's most prolific correspondent, writing nearly three hundred letters to him over fifteen years. He celebrated their common German heritage, pursued butterflies for the same reasons, and, like Strecker, relied on the U.S. post office as his preferred collecting ground; he wanted "only fresh, intact specimens."[68] Early on, he sent Strecker the usual photograph butterfly men and other naturalists exchanged, his with its handsome, even Byronic, image, "the last one I have and flattering exceedingly," he had to say. "I am not such an Adonis as the picture makes me out to be, a small fellow of five feet, with eye glasses!"[69] Another German immigrant, Julius Meyer, a voice and piano teacher in Brooklyn who competed with Neumoegen over the richness of their collections, considered Neumoegen "a grasping Jew of the worst kind." "He doesn't love nature as we do," Meyer told Strecker. Yet ten years later, when Meyer and others were revitalizing the Brooklyn Entomological Society (created in 1873), he wrote Strecker that two significant members were "*Mr. Graef and Neumoegen . . . good examples of advancing Science.*"[70]

Old-time American naturalists Titian Peale, of Holmesburg, Pennsylvania, and John Morris, of Baltimore, both now in their seventies, fell under the spell of Strecker's overflowing butterfly devotion. Before Peale died, dirt poor, in 1888, he had hoped to begin a catalog of exotics; he had attempted an earlier version decades before and had failed, because no one then cared about foreign insects.[71] By the 1870s, when the competition was stiff, Peale lacked the strength to meet it, but he enjoyed the growing enthusiasm for foreign insects vicariously through Strecker, who reawakened in him "his old love for the beautiful among Butterflies." "I can readily appreciate your 'mania,'" he wrote Strecker, "being subject to fits of the same kind since infancy."[72] As for Morris, Strecker had known of his butterfly book for years and, in the early 1850s, soon after his return from Mexico, had visited him in Baltimore, making a lifelong friend. In Strecker's presence again in the 1870s, Morris felt his "old butterfly fever" return in its old ardor, and despite age and sickness, he sugared trees to snare moths for the first time in his life, bred his own butterflies, and did what he called "microscopic studies," enrolling as a student (!) in the "biology department" of the new Johns Hopkins University in Baltimore. Strecker invited him to visit Reading, and Morris wrote back, "I look forward with boyish exaltation to the times when I can revel over your collection."[73]

———

Augustus Grote saw little to fault in Strecker's work or in Strecker, at least at first. In fact, he wanted to stride forth with him, hand in hand, to transform natural science in America, a bond strengthened by a commitment to hard work and by the same Romantic tendencies, the same attraction to beauty. Grote was even more wide-ranging than Strecker (or than Scudder and Edwards, for that matter). In 1876, at age thirty-five, he felt confident speaking and writing on much besides entomology, at once a social critic, a historian of religion, a poet, a musician, and a composer. He appears to have gained knowledge of all these fields mostly on his own, and the ease with which he did so led some to ridicule him—a few called him "His Highness"—but he was always interesting and often bril-

Augustus Radcliffe Grote. Courtesy of the Charles Lee Remington Archives, Entomology Division, Yale Peabody Museum of Natural History. Originally published in The Lepidopterist News 2, *no. 2 (1948): 17.*

liant, incorporating much of what he knew into his writing on moths and butterflies, thus giving it a philosophic and idiosyncratic character.

Just as with Edwards, Scudder, and Strecker, natural history put Grote in touch with the complexity of the natural world, and reinforced the aesthetic side of his temperament. But Grote had an analytical side as well; he was inclined to look at butterflies and moths inside and out for their scientific value. Neither science nor art alone satisfied him. He wanted both, integrated into what he called "aesthetic entomology." Strecker shared this approach somewhat, although he never matched Grote in the sophistication of his systematics (and he was mostly indifferent to Darwin). William Henry Edwards and Samuel Scudder gravitated toward systematic analysis, although they never turned their backs on beauty. Hardships in Grote's life led to his repeated dislocation and

HANDBUCH

für

Schmetterlingsliebhaber,

besonders für

ANFÄNGER im SAMMELN

VON

J. W. MEIGEN,

Mitglied verschiedener naturforschen-
den Gesellschaften.

Mit 16 Steintafeln.

AACHEN,

1827.
Dübyen script.

Verlag von LaRuelle & Destez.

In the 1850s, as a young collector, Augustus Grote relied on this 1827 German guide by Johann Meigen, because no American guide then existed. Fortunately, however, some of the German species shared features similar to American ones, aiding in identification. This plate shows the title page, while the plate on the facing page displays German fritillary species resembling American ones (5a–6a). Nearly at the center of this plate, moreover, is a beautiful species common to

both countries (4), in Germany called the trauermantel *butterfly, and in America, the mourning cloak, which is what* trauermantel *means in English. The American name was doubtless the product of German emigration, but why "mourning cloak," when the butterfly is maroon, yellow, and light blue, without any black? Courtesy of Craig Chesek, © American Museum of Natural History.*

depression—a deadly moodiness that interrupted his work—but by the end of the century, he had created a remarkable phylogenetic analysis tracing certain butterfly families back to their ancestral roots. This achievement persuaded one gifted contemporary to praise Grote for his "keen powers of discernment" and "unusual perceptive abilities," used unsparingly to understand not only moths and nature generally but other people, and himself. Another champion—and one of the country's premier entomologists—would describe Grote in 1913 as "the best lepidopterist of America, living or dead."[74]

Augustus Grote was born in Aigburth, England, on February 7, 1841, the son of two transplanted parents: his father, Frederic, from Bremen, Germany, and his mother, Anna Radcliffe, from Wales. When he was seven, they moved to Staten Island, joining many German immigrants there. His father, who later helped found the Staten Island Railroad, bought a large farm in the southeast corner of the island, with a pond, ornamental trees, and a nearby woodland, and there, in a landscape similar to the one in nearby Brooklyn, Grote collected butterflies and moths in the mid-1850s, often with two other German-American boys, Fred Tepper of Flatbush and Edward Graef of downtown Brooklyn.[75]

The boys could not identify precisely a single bug they caught; they had no guide or catalog of American species because no such book existed. All they had to depend on was Johann Meigen's *Handbuch für Schmetterlingsliebhaber* (Guide for Butterfly Fanciers), one of the early fascinating German guides, published in 1827; it showed, of course, only German butterflies, but for beginning American collectors it was still an enticing window to the natural world.[76] Only three by six inches in size, Meigen's book explained nets, boards, pins, and mounting methods; it described butterflies through all their life stages, with tips on collecting and poetry enough to stir the imagination of the young. Meigen wrote that "with no class of the animal kingdom has the Creator produced so stunning a diversity of design, color, shade, and form, and in so great a number, as with the butterflies, equipped like colorful jewels, which cannot be emulated in human art, and seize us with wonder." An appealing example of German natural history, the handbook had pretty pictures of butterflies, some variants of which could actually be found in America. Meigen also portrayed in lurid detail the parasitic behavior of a female ichneumon wasp, a predatory stalker of butterflies that sticks its "stiff stinger" into larvae to lay its eggs, which hatch as worms to devour the prey from within; soon the caterpillar dies, but the wasps pupate, break out of the skin of the "dried-up larvae," and fly away.[77]

With Meigen in his pocket, Grote forged his way through Staten Island and, with Tepper and Graef, collected in Brooklyn. At first, they had two preferred destinations: a grazing meadow for sheep edged by woods, later converted into Prospect Park, and a large vegetable garden near the meadow on what became the junction of Flatbush Avenue and Fulton Street. Over time, they exhausted nearly all of Brooklyn.

In mid-nineteenth century America, both Brooklyn and Staten Island had an alluring mix of wild nature and rural topography, distinguished by fertile flatlands, handsome farmsteads, and "orchards abounding in fruit."[78] For a while in the 1840s, Henry Thoreau worked on Staten Island, as a tutor to the children of Judge William Emerson, a wealthy landowner. "The whole place," Thoreau recorded in his journal, "is like a garden and affords very fine scenery." He enjoyed especially "the sea-beach," so "solitary and remote" that it made one forget Manhattan, across the bay. "The distances, too, along the shore, and inland in sight of the beach, are unaccountably great and startling."[79] Young Grote hiked alone through southern Staten Island, going down to the beach, where he saw so many kinds of butterflies and moths that he would later write, "all good things naturally live in Staten Island."[80] He caught the handsome regal fritillary (*Speyeria idalia*), flying in unusual numbers in the wet meadowlands, and the harvester butterfly (*Feniseca tarquinius*), a lovely little orange-and-brown insect, among the rocks of riverbeds and in hedgerows, a butterfly with carnivorous habits, named in Latin after the vicious Roman emperor Tarquinius, its larvae dependent on aphids as food.[81]

Grote favored moths, or what he, Strecker, and others called the "night butterflies" ("*die Nachtschmetterlinge*") or "the night peacock butterflies" ("*die Nachtpfauennaugen*"). Although there are many exceptions to the rule, moths and butterflies both belong to the same order, Lepidoptera, each undergoing complete metamorphosis and each with wings covered by scales, shingled one upon another, and stamped with a single color that contributes to the total "tiled mosaic" of the wing.[82] Both have a proboscis, or a long, slender, coiled-up tube attached to the head, which the insects uncoil to suck nectar from many kinds of flowers, pollinating as they go; as caterpillars, however, they are much more choosy, some dependent on only one food plant, others on a few, and still others on many different species of plants. On the other hand, the differences between moths and butterflies abound. Again, in the most general terms: the majority of moths have feathery, tapered antennae; these, like radar, guide them through the dark, and the males rely on them to pick up the scent of females. Butterflies generally have clubbed or hooked antennae, used to smell and track down nectar and for sexual purposes. Moths have thick, commonly hairy bodies and large multifaceted, compound eyes and usually inhabit the night, while the majority of butterflies fly by day and have smaller eyes and thinner, relatively hairless bodies. (The classic exceptions for moths belong to the Uraniidae family; they look like butterflies in nearly every respect and are among the most stunning diurnal lepidoptera in the world.) Both are cold-blooded, requiring infusion from the heat of the ambient atmosphere. By the 1870s, Grote was an authority on many moth families, above all the Noctuidae, and for him, moths were no less beguiling than butterflies. In fact, he gloried in them as "among the most beautiful phenomena of nature" ("*gehören zu den schönsten Erscheinungen der Natur*").[83] He marveled at the "blue and green in the Wandering Hawk Moth" and at the "the pink and yellow of the Rosy *Dryocampa.*" He wrote, "The moths afford superb instances of the blending of neutral tints, unspeakable soft browns and grays, as in the Smerinthinae."[84]

In his search for specimens, the young Grote rose before dawn for long hikes along the edges of the woods of Staten Island and within them, "seduced," he said, "by mysterious silences and shadowy vistas," netting "hawk" moths of the Sphingidae family, robust insects that fly rapidly through the air like hawks, with thick bodies and often colorful, streamlined wings, each feeding on the white-and-purple flowers of the Jamestown weed, common in the region. A notable capture was of a pair of "great green vine hawks." Even more memorable were the migratory moths or "tropical wanderers" (as Grote called them), which dazzled him in the early fall, "coming up along the coast" on warm air currents from as far away as the West Indies and Surinam.[85]

Like the moths themselves, Grote came alive at night. So, too, did Strecker, for only then could he dwell on all his lepidoptera—butterflies and moths—freely and without interference. The night had a different magic for Grote, partly because of the moths but also because of the multiformity of life hidden in the darkness, as copious and various as the life of the day. Sometimes, on nocturnal forays, Grote forgot "when to stop and go home to bed," his quests taking him into other discoveries of what "the night was all about" and "how the world got along in the dark." But the halcyon days of his youth were short; in 1857, when he was sixteen, an economic panic, fueled by speculation in railroads on Wall Street, devastated his father, who had served as secretary and treasurer of the Staten Island Railroad he had helped found. Cyrus Vanderbilt bought him out and built a stately mansion in the heart of Staten Island, a grating reminder of Frederic's misfortune, which had dashed his son's hopes of attending Harvard College. For a time, apparently, Grote went to Lafayette College in Pennsylvania.

One thing was certain about Grote's life in the early 1860s: his good fortune in meeting Coleman Robinson of Brewster, New York, a successful stockbroker on Wall Street. Robinson, just three years older than Grote, was an enthusiastic amateur lepidopterist, like Grote self-taught about insects. He saw in Grote spiritual and intellectual kinship. Robinson had been buoyed by the capitalist vortex that had dragged down Grote's father, and by 1871, he would be worth $1.5 million and able to retire, at twenty-seven, and devote all his time to moths.[86] Even before this, Robinson had effectively subsidized Grote, pumping steam and money into his aspirations. In 1862 (the year his mother, Anna, died), Augustus, then twenty-one, became the curator of entomology at the new Buffalo Society of Natural Sciences, created that year partly from money donated by Robinson for discovery and assessment of "every animated being and every vegetable indigenous to Buffalo and its vicinity."[87] The two men collaborated there, describing new American moths in checklists and catalogs, and, at Robinson's expense, visited European museums together (to avoid naming what Europeans had already named accurately) and spent time with stellar entomologists, including Achille Guenée and Jean Baptiste Boisduval in Paris, Rudolf and Cajetan Felder in Vienna, and Gottlieb Herrich-Schaeffer in Berlin.[88]

But in 1868, Grote's father cajoled him to leave the museum and resettle in Alabama to join him in a new business. Frederic R. Grote and Sons, Cotton Factors and Commission Merchants, was set up to capitalize on the defeat of the Confederacy by distributing and selling cotton to northern buyers (Grote had managed—like Strecker—to escape the fighting). At the same time, Frederic took a new wife, a southern woman whose family owned a large plantation in Demopolis, Alabama, presumably a base for Frederic's operations. Augustus ran the business (or so he claimed) from 1869 to 1872.[89] But sick and morose, he disliked this "rebel" place. "Sometimes from my desk," he wrote Scudder in 1869, "I dream of my studies and envy you all."[90] He did manage to become an authority on the cotton worm, one of the scourges of the one-crop plantation system, and he married a southern belle, Julia Blair, the granddaughter of a judge and the relative of a general. The couple honeymooned in Paris, and the following year, in 1871, a daughter was born, adding to Grote's joy.[91] Then, in quick succession, two blows struck: Julia died while giving birth to a second daughter, and his friend and patron, Coleman Robinson, was killed at thirty-three, thrown from his carriage on his country estate in Brewster, New York.[92]

The outcome of the double tragedy was a plunge into despair, but it also yielded one of the most searching and gratifying periods in Grote's life, a product of some crucial shift in sympathies in his being, and a sea change from which he emerged a better naturalist. Robinson had left a large legacy of $10,000 to the Buffalo Museum of the Natural Sciences, and had apparently stipulated employment for Grote, should he want it. Eager now to leave Demopolis, Grote wrote George Clinton, the museum's director, that he did want the job—whatever it was—and, if necessary, would take it at low pay. A year later, he returned to Buffalo, leaving his two infant daughters behind in Demopolis, under the care of their step-grandmother.[93]

A whirlwind of activity followed for Grote, still in his thirties and now seemingly liberated. He wrote piano music in the classical and popular idiom, poured his heart out in love poems for his dead wife—poems so good that the *Atlantic Monthly* and other magazines published them—and took active part in the turbulent political and cultural debates of the 1870s, speaking out in Buffalo and elsewhere.[94] Buffalo, just like Boston, Philadelphia, and New York, had a circle of reform-minded people who believed (as one put it) that "every problem of the universe was in a melting pot." This circle anointed Grote one of its leaders.[95] His reputation reached New York City, and he became a leading contributor to a new journal, the *Evolution,* one of the many short-lived papers of that time, championing the rights of women and of labor and "of people who have outgrown their political and religious faiths." Its editors claimed to deal with problems almost entirely new to the United States—the rise of "our great moneyed corporations"; the spread of poverty ("begging from door to door"); the exploitation of women and children; and the status of religion in the new scientific era. In one piece, "The Laborer in Politics," Grote defended workers against corporations that "establish ruinous competition, and, when they fail to make money . . . cut down the wages of their employees" or "just the people that can least afford to suffer."[96]

Grote also explored in the *Evolution* ideas about comparative religion in America that would form the basis for a fascinating book, *The New Infidelity.* In a surprising turn—he was supposedly a secular naturalist who had no interest in churchly matters—he criticized the liberal Protestant churches for attempting to reconcile science and religion. He understood the historic role the Judeo-Christian tradition had played in shaping and preserving Western civilization and worried, therefore, that churches that had tried too hard to integrate modernity into their worldviews would move away from the simple religion of the heart and from spiritual appeals to self-sacrifice and the selfless caring for others. What mattered to him was not theology, which rested on myths and falsehoods, but the "morality of sacrifice" and emotional need. Hence his defense of Judaism and Catholicism (as well as evangelical Protestantism) on the grounds that they answered most profoundly to the heart, not to the mind (at least this was Grote's contention), and held fast against scientific thinking. The loss of the Judeo-Christian tradition, he insisted, would leave many human beings entirely isolated in the world, without the protection of others, a condition he found intolerable. He also faulted liberal Protestants for seeking "God *in* Nature," rather then *above* Nature. "God is not natural but supernatural," he insisted. Religion should serve "the Heart and emotions," not the head, giving succor and peace in a way nothing else could do. Liberal Protestants, the "new infidels," were abandoning the essence of the religious experience.[97]

All of these concerns demonstrated the wide-ranging character of Grote's mind, a way of thinking he shared with his forebears who drew no lines between art and science as well as with his American contemporaries—such as the merchant John Wanamaker, founder of the modern department store, and the philosopher William James, founder of modern psychology—who endorsed feeling over intellect as the foundation of religious life. But his *major* passion remained the moths of North America. Backed by his dead friend's money, he started out as a librarian and curator in 1872, at a modest yearly salary of $500, and by 1875 had become the director of the Buffalo Museum of Natural Sciences, remaining there for the next five years, at the same salary, despite his efforts to improve the place and despite repeated appeals for a raise.

The work was satisfying, however. Grote created a public lecture series and introduced microscopy; and he published more articles on insects and nature, in the 1870s, for the prestigious *Canadian Entomologist* than any other naturalist save one, William Henry Edwards—a key index of Grote's own popularity and influence among American naturalists.[98] He created two new journals: the museum's *Bulletin* in 1873, writing most of the articles himself, and, in 1879, the apparently privately funded *North American Entomologist,* which treated economic entomology and included extended and interesting life histories of moths and butterflies, written by both men and women.[99] He also translated from German, wrote an analysis in German, and drew on German scholarship on moths. Along with Strecker, Grote was among the first to bestow praise on Maria Sibylla Merian, a brave German naturalist and artist who in the 1690s, under Dutch auspices, sailed to Surinam, a Dutch colony in South America, to collect insects and paint their life histories. She painted butterflies (everything but the eggs) and made "a number of observations on the transformations of Insects that no one has since equaled in the same number of days or months," Grote wrote. "From the historical background of the Natural Sciences, it is a woman's face looking to us for well earned remembrance."[100]

Grote fundamentally changed his approach to the natural world. Throughout the 1860s he opposed Darwin's theory of evolution, to the chagrin and disgust of the brilliant English émigré Benjamin Walsh, with whom Grote briefly worked on the *Practical Entomologist,* a tiny monthly bulletin—it lasted barely two years—of the Entomological Society of Philadelphia, distributed free to farmers. Grote, at twenty-six, had been listed as one of the four editors, with Walsh, then at fifty-nine, as associate editor, but Walsh's effort proved to be of such high quality that he became the *only* editor, in control of nearly all the writing, "without any pecuniary benefit to myself" (as Walsh put it). The two men clashed by age and personal temperament and, especially, over Darwinism.[101] Walsh complained to William Henry Edwards that Grote was "excessively irritated with me because I have preached what [he is] pleased to call the 'Darwinian' doctrine, that our N.A. Fauna and Flora are of an old-fashioned type. The fool!"[102] In letters to Scudder, Grote bemoaned "Walsh's prejudicial style of reasoning," which "can never avail anything or aid the progress of aesthetic entomology," noting that Walsh "decried the assertions of others without denying the grounds and facts on which these assertions were made." And as late as 1872, in Alabama, Grote was unflinching about evolutionary doctrine. "I tell you," he wrote Scudder in that year, "Darwin has had his day already."[103]

But by the mid-1870s, Grote was penning pro-Darwinism salvos for the *Evolution* and defending evolutionary theory in Buffalo. "The question of the day," he wrote, "is evolutionary—how all these species came about."[104] In itself, Grote's new perspective was not a radical one. Even naturalists who believed in God had come to accept evolution as a key feature of the natural world, so long as God was considered the governing cause of the evolution.[105] But Grote—along with Edwards, Walsh, Charles Valentine Riley, and, over time, Scudder—*was* radical, since he believed that natural selection and the struggle for existence were the engines of evolution, not God.[106] After the sudden deaths of Grote's wife and his friend Robinson, any personal religious objections Grote may have had to evolutionary science collapsed. "The waste of life in Nature, the suffering and the wrong," he wrote, "refute the idea of any designing Mind which we can appreciate." He was now a disbelieving skeptic for whom, as he put it, "the butterflies, like flying flowers, only give pleasure for their beauty, and convey no lesson of immortality."[107]

Grote's conversion to Darwinism brought with it a gradual abandonment of conventional taxonomy, or what Grote called "descriptive Entomology," which he himself had practiced, and would continue to practice, entailing the systematic grouping of lepidoptera into species and genera. He remained loyal to the "aesthetic" features of an earlier German entomology but not to much of its taxonomic legacy. And, like Edwards, Scudder, Agassiz, Darwin, and all the inheritors of the natural history tradition, he turned to "living nature" as his subject, to how species emerged and where they came from, and to how similar species spread around the globe, as Grote put it, "once connected in geological time."[108]

Grote pondered the White Mountain butterfly, *Oeneis semidea,* the caterpillar he knew Scudder had found feeding on the sedges at the top of Mount Washington. How did such a little thing, with barely an inch wingspan, get to such an inhospitable place? Why, moreover, did the same insect exist on Longs Peak in the Colorado Rockies and in Hopedale, Labrador, to the north? "This is a strange distribution for a butterfly, and so the question comes up as to the manner in which it was brought about," he observed in an 1875 essay, "The Effect of the Glacial Epoch upon the Distribution of Insects in North America." For an explanation, he went back thousands of years, to a time when the glaciers surged down from Canada and gouged the topography of New England and Colorado, carrying with them thousands of species, most demolished along the way and a few, such as O. *semidea,* left behind as warmth forced a glacial retreat northward. "Far off in Labrador," Grote observed, "the descendants of their ancestral companions fly over wide stretches of country, while they appear to be in prison on top of the mountain" in New Hampshire. Grote's analysis increased understanding of the geographical distribution of species pioneered by Darwin and Wallace.[109] It encouraged lepidopterists to incorporate geographical distribution in their naming and grouping of butterflies and moths. Again, the older practice was exhausted; one had to show more—not geographical distribution alone but the character of evolution itself. "What is the question which at this time is *the* question among naturalists," Grote asked in 1875. "Is it not the question of how all the different species and genera came about, rather than a mere cataloguing of them for convenience sake? And will not, therefore, any system of classification which expresses clearly the interrelationships through slight modification of structure, be the classification thinking men will adopt?"[110]

Grote's entomology was elevated further in the 1870s when he got to know, through fieldwork as never before, the beauty of the natural world. Ordinarily, he did not go into the field himself, preferring his museum "closet," as he called it, a hideaway of sorts like Strecker's "butterfly room," where, like Strecker, too, he relied on the specimens of others to make his identifications. Nor was this necessarily a fault. "It has been objected," he wrote in his 1873 *Bulletin,* "that [museum studies] are of the Closet and not of the Field," but "I think the student at his books and dead specimens is the same we meet again, where grasses grow, collecting, and observing. So the Field is brought to the House with the Harvest, and can be rightly spoken of from the Closet. It is no excuse that we have been out of doors when we are called upon to speak. Perhaps the seed must be dry at first, to be properly green thereafter. This is the age of objective research. Let us, then, see what we can while we live."[111] This was surely an inclusive, defensible view: diversity of form and structure can be known in a personal cabinet of insects and in the museum room under a microscope as well as in the natural world. Nevertheless, Grote's memory of his youthful rambles through Staten Island, at dawn and in the dark, was etched powerfully into his being. The open fields beckoned, and Grote left the closet for the moths he knew abounded near Buffalo.

Sometime around April 1873 or '74, Grote set up a campsite on the southern border of Lake Erie, a favorite spot for him, staying until July and paying his expenses out of his meager salary. The camp stood on a wooded and sandy ridge, with the lake on one side and a farmhouse and level country on the other, meadows and fields of freshly planted corn farther out clearly visible from Grote's canvas tent. He arrived alone late in the afternoon, carrying gear, clothes, food, several bait traps, setting boards, and a bull's-eye lamp to lure in moths, "the air saturated with hazy light."[112] On his second night, he sugared according to the English style devised by the Doubleday brothers, Henry and Edward, in the 1830s, slathering a mix of beer and molasses over tree bark to seduce the moths, thereby increasing the number a collector could get. Then he sat patiently with his pipe, rewarded after a while by the arrival of an influx of American lepidoptera, rendered by him lovingly in Latin: "*Scopelosoma—walkeri, tristigmata,* and *morrisoni,*" all species of noctuid moths.

At night, asleep in his tent, he was awakened by a startling noise, and for a while he considered ways to escape, since "the tent is a sort of trap that the owner is caught in. You can even be prodded at through the canvas walls. You can see nothing. Through which end will you escape? After coming to this view of the case, it occurred to me that, instead of staying inside and frightening myself, it would be better to go outside and frighten somebody else." Grote left the tent, felt no danger, and was transported. "What a lovely night! There was no moon; but the radiant floor of heaven was trimmed with stars." The following morning all about him was "the spring, bringing out leaf and blossom." Day and night, animals of all kinds, not moths alone, came within reach of his tent—bluebirds, a kingfisher watching by the brook, a chipmunk "curious to see what manner of man had ventured into his realm," and two hapless flying squirrels, which, "blinded by his bull's eye" at night, had been trapped in his "butterfly net." In July, with warmer weather, the catocalas "swarmed like bats" around him, responding to his baits, hurling in from all points in such a morphological diversity as to amaze and exhilarate him.

Collecting on Lake Erie gave Grote insight into the extent of American species and an ecstatic and rejuvenating sense of nature, not unlike Humboldt's "soothing" experiences. "Skies are fairer in America," Grote observed, with shades of Doubleday as well as Humboldt, and "the collector can still find the possible known as well as the possible unknown, for I myself caught and first discovered specimens of the butterfly *Calephelis borealis* and many an undescribed moth has dropped into my collecting bottle." He later wrote up these species in detail for the *Canadian Entomologist,* under such euphonious Latin titles as *Hadena delicata, Mamestra vicina,* and *Dryobota stigmata grote.*[113] "It had been a happy time," he recounted in his final reflection, "stolen from Death and Bad Luck, full of Life itself strengthened by work. A time to realize the truth of Kepler's assertion that this world itself is heaven in which we live and move and are, we and all mundane bodies."

Grote shared these sentiments about nature with Herman Strecker, more perhaps than with any other contemporary, and each hoped to do great things with the other, a destiny that Edwards and Scudder seemed to share with each other as well. But fate had a different future in store for all four pioneer butterfly men. Scudder's and Edwards's friendship, so guardedly cultivated across the Atlantic, would wear thin and then break for a time, over matters profound and petty. For Strecker and Grote, the outcome would be far worse, fed by suspicions and doubts, and ending in disaster.

THREE

Beating Hearts

In early January 1859, Alfred Russel Wallace spotted a rare butterfly fluttering about a flower on the forested island of Batchian in Malaysia, a sighting since recounted many times over as symbolic of the collecting passion, hauntingly fictionalized by Joseph Conrad in *Lord Jim*. "On taking it out of my net and opening the glorious wings my heart began to beat violently," Wallace recalled, "the blood rushed to my head, and I felt more like fainting than I have done when in apprehension of death. I had a headache the rest of the day." "None but a naturalist can understand the intense excitement I experienced when at length I captured it." The insect was "fiery golden-orange, changing when viewed obliquely, to opaline-yellow and green," a birdwing later named by Wallace *Ornithoptera croesus*. "It is, I think, the finest of the Ornithoptera, and consequently the *finest butterfly in the world*."[1]

Natural history, as it spread throughout Europe and America, offered nearly unconditional communion with the beauty and power of nature. Its strongest expression came for many butterfly people when their hearts beat quickly in the presence of some extremely desirable thing, unleashing in their bodies the energy of the hunter; nothing seemed to rival collecting in the way it unlocked the most intense feelings, carrying or pushing the collector forward to gain possession in the face of often impossible odds. In 1874, Eugene Pilate, a physician and French immigrant in Dayton, Ohio, wrote to Herman Strecker about a moth he caught as a boy: "I can remember as if yesterday how my heart beat when I first caught a Lichenee blue (noctua catocala fraxini), like you when you got your first Catocala Amatrix." Around the same time, in a southern California desert, William Greenwood Wright chased after a black swallowtail "floating over low shrubs as only a papilio can" until the butterfly's "strong wings beat vainly against my net, sending thrills of pitying exultation through my fingers. . . . With hands trembling with excitement I stowed the prize away." James Fletcher, one of

Canada's most admired naturalists, recounted for Samuel Scudder the time he "caught" his first "Vanessa Antiopa with trembling hands and bated breath on fallen pears lying on the ground beneath a tree in our garden." Henry Edwards's "heart beat violently" when he came upon "a lovely black and orange moth, such as I had never before seen" in the Plenty Ranges just outside Melbourne. "I felt as if I should have gloried in making those primeval woods echo with my shouts," he wrote. In a mountainous region in India, Will Doherty, at twenty-seven soon to become America's greatest tropical collector, came close to fainting when he spied a butterfly feeding on the brilliant orange flowers of a Mussaenda bush, a seven-inch-wide "black and gold" *Troides minos,* one of the largest butterflies in India. Doherty took dozens of this butterfly and "thought the gold of their wings the most gorgeous yellow in nature or art."[2]

One hundred forty years later, Arthur Shapiro, a gifted California butterfly man and expert, remembered similar experiences that made his heart pound hard: "It began when I was ten or eleven and was especially true when I saw something 'special' and didn't have a net—which was, of course, most of the time. I got very good at catching stuff with my hand, and it was during the stalk that I got 'palpitations.' " "I occasionally have the experience today, [and] I certainly did in 1977, at Cambirumeina in the Sierra Nevada de Santa Marta in NE Colombia. I'd been on the trail for three days, [and] as I neared the top of the pass, the clouds parted, the sun shone brightly, and a female *Reliquia* flew right by. Of course, I could not get my net out in time, it was gone, the clouds closed back in—and I supposed that was the only female I would ever see! Fortunately, it wasn't. And within two days, I had eggs. But I'm not sure I ever had a higher adrenaline level in my life than I had at that moment—even in a fight-or-flight situation."[3]

In 1896, the philosopher George Santayana, then a young instructor at Harvard College, wrote a memorable book on aesthetics, *The Sense of Beauty;* in it, he tried to explain the beating hearts, drawing, perhaps, from his own personal experience and from his intensive reading of natural history and Darwinian evolutionary science at Harvard, most of which he did with the philosopher William James. His subject here was sex: "If any one were desirous to produce a being with a great susceptibility to beauty, he could not invent an instrument better designed for that object than sex." Every human being, he maintained, has a sexual desire to merge with "more and more definite objects," first "to one species and one sex, and ultimately to one individual." Many people fail in their attempts, "the differentiation is not complete," and "there is a great deal of groping and waste," especially for the young just coming to terms with desire. Santayana embraced the groping and waste. In a view not far from the secular and unsentimental one later expressed by Sigmund Freud in his 1930 *Civilization and Its Discontents,* he observed that "sex is not the only object of sexual passion. When love lacks its specific object, when it does not yet understand itself, or has been sacrificed to some other interest, [it] must be partially stimulated by other objects than its specific or ultimate one; especially in man, who, unlike some of the lower animals, has not his instincts clearly distinct and intermittent, but always partially active, and never active in isolation. We may say, then, that for man all nature is a secondary object of sexual passion, and that to this fact the beauty of nature is largely due." Without groping and waste, there was no culture or art, and "no ability to see the maximum of beauty" in "nature unadorned," in the "physical world, which must continually be about us." Nature requires human sexuality to expose "its deepest meaning and beauty."[4]

Santayana never mentions beating hearts or butterflies, and one might dismiss his views as reductive and propose other reasons for the arousal—and, yes, surely there *are* other reasons, such as a genetic disposition like "biophilia," as the Harvard biologist E. O. Wilson argues, or a psychological one like family conflict or abuse leading a child to escape into the otherness of other species, as Wilson also suggests in his autobiography, *Naturalist.*[5] Or was it the natural colors and forms, taking shape inside the mind even before the mind could see and then, in childhood, attaching itself to the colors and forms outside the self? Was it the sensuous wildness of the natural color, too, that once observed is never forgotten and is itself so generative?[6]

Yes, it was all these things, which, taken together with repressed sexuality radiating desire throughout the natural world, had the power to nurture in young and old alike a longing for nature and an ability to see it aesthetically. Collecting channeled and deepened this experience by bringing individuals purposely into physical contact with the world of butterflies and moths. It placed people *inside* the fullness of nature, where boundaries between species were weakest, and all was ebb and flow between them.[7] Collecting exposed a hidden generative realm, shared by both human beings and butterflies, that imparted to many an overpowering feeling of being alive and of knowing that *this is who you are and why you are.* James Tutt, one of the foremost butterfly men of England in the late nineteenth century, wrote that collecting connected the collector not only to the butterfly but to the context in which the butterfly existed, the flowers and carrion it fed upon, the surrounding insects and the predatory birds, the ambient air, the sunlight, the blended smells (and, he might have added, to the wildness of the natural color), climaxing in an "exquisite sense of enjoyment" nothing could replicate. Collecting brought individuals in touch with the forms, patterns, shapes, and colors of the natural world, which would all be meaningless or invisible were it not for the desire *inside* human beings to reach beyond the self. "If science grows out of collecting," Tutt concluded, "so much the better, but, with the feelings we possess even the charm of collecting cannot be altogether in vain."[8]

Collecting's cardinal motivation came from natural history, then a vital presence in the lives of Americans barely comprehensible today, when so many people avoid mingling with nature for fear of harming it. Natural history, on the other hand, allowed for the nearly full sway of the sensual drives of human beings. All the butterfly people in this book began their careers in this way, awash in the heat and smells of the meadows and forests, sensitive to something worth losing oneself in, worth knowing, worth a lifetime of vocational loyalty and reflection. "We unceasingly want to know," Grote observed. "Truth is one, and even a butterfly conceals it, though not, like man, intentionally."[9]

And yet this passion had the potential for harm, even sometimes sabotaging the entire enterprise butterfly people shared. The emotions that brought them to the fields and meadows of butterflies, and that kept them together as a community, often set them against one another, and this was assuredly the case throughout the 1870s, when Americans actually *created* the beginnings of a mature American science of entomology, one fully competitive with Europe. In this decade opposing camps formed, with opposing "authorities," journals, and associations in New York, Boston, and elsewhere. Americans got swept up in debates that had long marked the British scene. In the 1840s, members of the

British Association for the Advancement of Science (BAAS) attempted to end the flood of names that had followed in the wake of the discovery of a vast range of animal forms; it tried to impose a standardized Latin and Greek on all naming (no vernacular, no local naming) and to exclude all "déclassé naturalists" who lacked "refinement" and insisted upon naming species after "Peruvian princesses and Hottentots." Thirty years later, the American Association for the Advancement of Science (AAAS), founded in 1848, inspired by the British, and led from within by a phalanx of entomologists, attempted, too, to standardize systematic practices. Unlike the BAAS, however, the American society was more democratic, its membership open to amateurs and professionals alike; yet many Americans, too, opposed naturalists who acted in "unrefined" ways and refused to swallow the dictates of self-selected elites. It was in the AAAS that William Henry Edwards, Scudder, Grote (Strecker did not join the AAAS, nor did William Henry Edwards for that matter, preferring to operate through proxies), thrashed over matters of core significance to their science. The debate took place as well beyond the AAAS, among all kinds of butterfly people, in journals, letters, clubs, homes, and saloons.[10]

"What is there about our science," inquired William Henry Edwards in 1877 of his friend Joseph Lintner, "that makes one type of men so inflammable, and another rascals and thieves?"[11] "Fanatics, bigots, and dogmatists are more common in science than in religion (which is saying a good deal)," wrote George Hulst to Strecker a little later.[12] Feuds erupted over every aspect of systematics, with costly consequences. The butterfly people fought, above all, about collecting itself, where it should lead—to science, to beauty, to beauty *and* science, or to more collecting for the mere sake of amassing. All other issues, however fundamental, were secondary to collecting, because nothing could be gained without it.

Sometime in early 1875, Herman Strecker got in the post an unusual Brazilian moth from a friend who had bought it from a dealer in Rio de Janeiro. Gray and brown, with long, tapering tails, it seemed to be a member, according to Strecker, of the genus *Eudaemonia* (the word is from the classical Greek for happiness or contentment), first described under that name by the German naturalist Jacob Hübner sixty years before. Strecker considered it "the most remarkable Lepidopterous insect yet known," a species he was sure had yet to be noticed in America or Europe, having himself examined the existing literature (although not in *all* languages). He decided to publish a portrait in part 13 of his catalog, *Lepidoptera*. He called the insect *Eudaemonia jehovah*.

There was nothing new about this kind of naming. Ever since Linnaeus and Fabricius in the mid-1700s, naturalists had turned to Latin or Greek for names of *all* species (they still use Latin this way today). The royalty of the past lived on as butterflies or moths—from Linnaeus's *Morpho menelaus,* named for the king of Sparta and husband of Helen of Troy ("morpho" itself is a Greek word for form), to "the dusky red aborigines of America," the only nobility on the continent, according to Samuel Scudder, who, along with William Henry Edwards and Thaddeus Harris, attached Indian names to numerous skippers, butterflies with thick bodies and short, mostly dull wings (although many are richly colored). Fabricius, a student of Linnaeus's, named more than fifteen hundred "lepidopteran species," calling a skipper *Hesperia catullus* and a striking orange sulphur *Appias nero,* symbolic of the burning of Rome.[13] The Frenchman Pierre Latreille attached the Greek word "Parnassius," sacred home of the gods, to a lovely genus of alpine butterfly, the parnassians, which "fly around the summit of the sacred mountain of poetry, guarded especially by Apollo himself," and he coined *Colias philodice* (the "common sulphur"), the genus after one of Venus's many names, and the species after the golden maid of honor to Venus.[14] And so on and so forth, through the dense thickets of the classical past.

In choosing *Eudaemonia jehovah,* Strecker had simply taken the matter a step further, trespassing in foreign territory.[15] And he knew what he was doing. A toughened product of an artisan culture, run through the wringer in his youth, he detested most formalities, such as being addressed with "Esquire" after his name. (He told a friend that the term "belongs to loafers, to which paternity I hope I am no kin.")[16] Naming his Brazilian insect after Jehovah, he knew, "would be a fling in the face of the Christian world," as a friend of his put it.[17]

To John Morris, the aging Lutheran cleric who owed his revived butterfly fervor partly to Strecker, the name Jehovah was beyond the pale,

even for an ally. He promised to marshal the troops against its acceptance at the 1876 meeting of the AAAS. "You may rest assured," he
warned his friend, that we will "denounce it as irreverent, against all
refined taste as well as propriety." "Such names" should not be "applied
to such creatures."[18] George Hulst, also a minister and a Strecker loyal-
ist, pleaded with him to select another name, perhaps "of some warrior
female." "I don't see how you could help calling it Zenobia or something of the sort unless it was for the same reason a boy jumps into a
slough—just to prove that he dares to do it."[19] Charles Dury, a young
Cincinnati naturalist, lightly mocked Strecker's choice by proposing to
call one of his own recent discoveries "*Jesus christiana*," or "how would
satan do?"[20] Arthur Fuller, another friend and journalist, opposed Jehovah on wholly pragmatic grounds, because "it hurts your prospects in a
money point of view."[21]

Today, there is an almost incoherent general approach to naming,
either a lack of confidence in what is named or a demand for labels for
every group, nonhuman and human, while, at the same time, few people can identify more than two or three species of birds or butterflies.
In the 1870s, when the naming culture was still fresh, Henry Edwards
observed that "every schoolboy throughout the country knew that
the most common species in America was *Colias philodice* [the 'common sulphur']."[22] It mattered what butterflies were called. "Without
the groups of [insects] being named," wrote Hermann Burmeister in
his influential 1836 *A Manual of Entomology*, read by all the American butterfly people, "naturalists could not communicate together, and
without a distinction of the known and discovered all would speedily return to its former obscurity."[23] Sometimes naturalists thought up
unpronounceable and absurd names alien to most human beings.[24]
Many pompously behaved as though it were their right to name, or
to impose Latinate nomenclature on all natural forms, as if no one
before them had ever cared about nature.[25] They tended to magnify
human claims over nature: designate a thing so that you can find more
of it, translate it into wealth, or get rid of it. Yet names served generous and humane purposes as well. Along with the scientific identifying
and knowing that came with them, names helped to record and fix in
the mind otherwise shadowy beings, making visible the invisible. They
affirmed a bond between people and other organisms where none had

previously existed, raising the stature of a species, whether bird, butterfly, or any other natural thing. "You know when you don't know the names of things," observed Emily Morton, an heir of Linnaeus's and an amateur butterfly woman from Newburgh, New York, in 1879, "pages of names are no manner of use to you. I have had very rare [insects] for years and yet *I did not know to care anything about them* because *I did not know what they were.*" Three years earlier, another amateur, the tax collector William Holle, from Sheboygan, Wisconsin, told Strecker that "a species without a name and history and description is of no value." David Bruce, from Brockport, New York, wrote to a friend in 1888 that "I feel much more interested in the things if I know their names."[26]

At the same time, when people like Strecker took such liberties with it, naming caused an uproar, although Strecker professed not to understand the cause. Why all the carping against *Eudaemonia jehovah?* he asked. Is it because the name invokes the "sacred"? Well, he declared, "if such be the case, then indeed am I happy in my selection, for anything that would lead us to think of the Creator and would take our thoughts away from contemplation of the mimes [imitators] with which he has peopled the earth, cannot but be well. And what better than to reflect on sacred things—on evidences of the majesty and power of the Supreme Being?" Besides, people had used sacred names throughout history for heroes and saints as well as for animals or plants. "In Spanish countries almost every tenth person is surnamed Jesus, pronounced by them Hezoos. This may sound shockingly irreverent to the fastidious ears of Americans, but I doubt much if the Hidalgoes who bear the name of the second person of the Godhead would feel at all elated to know that, on account of their names, they were living offenses against decency and good taste." There could be no "reasonable objection to the bestowal of the Creator's name on one of the most interesting of His works," Strecker argued. Despising hypocrisy, unwilling to yield to anyone in the AAAS, he took shots at those who had named their discoveries after some politician or courtier: "To attach to scientific objects the names of political demagogues is without a doubt, the vilest of all [practices], especially in our own country where political eminence is now solely attained by the most corrupt means, and success ensured only by the sacrifice of every principle of honor and decency."[27] Strecker's *Eudaemonia jehovah* held up as a species and still survives, but the controversy he caused led many to suspect his judgment and character.

In the early 1870s, William Henry Edwards and Samuel Scudder, the innovative Yankees, were allies against the arrogant European and British entomologists. But beginning in summer of 1872, and continuing thereafter, Edwards, relying on members of the AAAS (he himself refused to pay what he thought an exorbitant membership fee), opposed Scudder's 1872 "Systematic Revision of American Butterflies." Edwards had read the document earlier in draft form; it had put the swallowtails at the bottom of the classificatory ladder, and it had also resurrected numerous older generic names and created a slew of new genera in accordance with the "law of priority." Edwards had disparaged these changes privately in letters to Scudder. Now he attacked them openly, persuading the AAAS to reject Scudder's views in favor of his own. In 1877, Hermann Hagen, a leading AAAS figure, invited Edwards to write a formal retort to Scudder, and Edwards did so, under the title "Catalogue of the Lepidoptera of America North of Mexico." Edwards reclaimed the system of the British naturalist Edward Doubleday, who in the 1840s had put the swallowtails at the top of the hierarchy of Lepidoptera and the skippers toward the bottom.[28]

Edwards had once told Scudder he didn't really care about classification, and, essentially, he didn't, just as Scudder, in his heart of hearts, didn't. What concerned them most was the *biology* of the butterflies, the way they lived and died, and why they existed at all. "I am indifferent to the matter of arrangement," Edwards insisted, just as he had when he wrote Scudder around 1870. "I have aimed at studying the stages of the insects and there I have done something that was needed."[29] Yet in his 1877 catalog, he sought to substitute for Scudder's "arrangement" a more Darwinian biological approach: "A great many systems of arrangement have had their rise and fall within the last half century on one character or other of the imago, and it is safe to say that none will be other than temporary which does not regard the egg, and the larva and chrysalis, as well as the butterfly. And it will be a long time before the knowledge of the Lepidoptera is so complete as to permit of any permanent arrangement." Moreover, "[i]n the preparatory stages, [the] two families [the swallowtails and the skippers] are as unlike as any of the series. And as to the butterflies themselves, they stand at the two poles."[30] Edwards flinched before what he considered Scudder's fiatlike imposed *laws,* conjured up, he thought, from nowhere and expecting universal acquiescence. He was not alone in his criticism. The Brooklyn tailor Fred Tepper wrote Strecker: "I read Mr. Edwards' criticism on Mr. Scudder's work with great interest—he treats him rather unmercifully, and with all rights, I think. Scudder's farfetched and most unreasonable ideas could hardly find favor with men of common sense and the sooner that work is pulled down and completely destroyed, the better for all concerned. We have confusion enough in our line of science, without having additions made by any Tom, Dick, or Harry that wishes to create for himself an immortal name."[31]

Scudder and Edwards looked quite differently at "the face of Nature"; that phrase, in common usage at the time, meant many things.[32] But Darwin singled it out in his 1859 *On the Origin of Species* to mean only one thing: "The face of Nature may be compared to a yielding surface, with ten thousand sharp wedges packed close together and driven inwards by incessant blows, sometimes one wedge being struck, and then another with greater force."[33] In time Scudder would come to see "the face of Nature" in a modified Darwinian form, but *not* in 1875, when he remained intellectually loyal to his teacher Louis Agassiz, who despised the idea that species "evolved." We have "different modes of seeing things," Scudder told Edwards that year. "I do not look upon species, genera, or any other groups, as the devices of man, but, just as far as they are truthfully, as the imprints of a creative mind. The common assent or dissent of naturalists has nothing whatever to do with them—agreement or disagreement of opinion does not alter *facts.*" He insisted that he did nothing to misrepresent nature. Trained in Agassiz's morphological methods, resting on comparative analysis of similar animal structures or an understanding of interrelationships, Scudder tried "to see the distinctions impressed on the animals by nature. No doubt I fail sometimes. I claim no shade of infallibility— All I claim is that I am endeavoring to discover the true relationships between the animals I study." "I am endeavoring to see things as they are and to express that view."[34]

Edwards perceived much of what Darwin saw in the face of nature and dismissed with more scorn and confidence than ever Scudder's stubborn clinging to Agassiz's "dogma." The notion that species and genera emerged "cast-iron" from "the thought of the Creator" or that one family of butterflies was "higher than any other" (as Scudder appeared to believe) "disgusted" Edwards, as he informed Henry Edwards. Today, "scientific men are looking for blood-relationships of species," not for separate creations. To Theodore Mead, Edwards observed: "If I understand Darwin's argument, many or all families would radiate from a common center, and be of equal rank to the last." "Species are forming all the time," he told his old friend Joseph Lintner in February 1873.[35] "The proposition is enunciated by Mr. Darwin," Edwards wrote in his 1877 catalog, "that 'distinct species present analogous variations, and a variety of one species often assumes some of the characters of an early progenitor.' And what is true of species is true of genera and families." But what of Scudder's search for structural or morphological "relationships" among animals? Surely it could be carried on—and *was* carried on by Edwards and other butterfly people—without adopting the evolutionary argument as a necessity.[36] Edwards was not in a conciliatory mood, and he could not tolerate what he considered Scudder's many other generalizations based on limited data, such as "*all* caterpillars, after hatching out, eat their shells," or "*all* swallowtails and lycaenids should be classified together because their caterpillars have retractable heads." What upset him most was the word "all." How could Scudder speak of *all* swallowtails or *all* retractable heads? Why not "perhaps all" or "probably all" or "many" or "most"? Scudder "is a well-trained naturalist with a kink in his head," Edwards wrote Henry Edwards.[37]

The depth of Edwards's opposition came partly from a man who had spent hours, days, even months in the field or around his property, looking at the slow, grudging molting of thousands of larvae and at the wiggles and twitches of countless numbers of pupae, tedious beyond imagining, making Edwards intolerant of those who, like Scudder, spent too much time in a room peering through a compound microscope.[38] Ironically, of course, Scudder had watched and struggled in the same way (although he never knew the field as Edwards did); he'd attached a heated laboratory to his home for purposes of constant viewing. But he did not go at the throats of his competitors in quite the lecturing way Edwards did.

Edwards had an explosive temper, to the chagrin of many of his friends, and by the late 1870s, his tone had ceased to be friendly. He did whatever he could to get others to disparage Scudder's ideas in print, while he stayed behind the curtain, enjoying the spectacle. Theodore Mead performed this service willingly, as did others such as Selim Peabody, a New England naturalist and insect generalist who would become nationally renowned in the early 1890s, for co-organizing the Chicago's World's Fair of 1893. Peabody wrote children's nature books after the Civil War, under the series title of Cecil's Books of Natural History, named after his own son, Cecil, and in *Cecil's Book of Insects,* he celebrated butterfly caterpillars as the source of "wings of beautiful form, exquisite coloring, and most delicate plumage." He had a collection of American butterflies big enough to put on display at an "exposition" in the Chicago Academy of Sciences, where he was a member and later president.[39] Edwards respected Peabody and backed up his dismissal of Scudder's vernacular names in the *Canadian Entomologist.* "Scudder," Peabody argued, "is seized with a certain Adamic afflatus, and begins the work of naming afresh. [He has, for instance,] dubbed Danais Archippus the Monarch, [because it lives supposedly longer than any other butterfly, but if that were really the case,] and its longevity were proven, then the insect might be called Patriarch, as Mr. Edwards suggests."[40]

To be sure, Scudder was a highly respected naturalist, and certainly no pushover for Edwards. In 1875, despite rejection of his theories, the AAAS honored Scudder for his work on fossil butterflies with a prestigious grant from a fund created by Elizabeth Thompson, a wealthy Boston widow with a devotion to scientific causes; it was the first such award given by the AAAS to any naturalist.[41] Nevertheless, Scudder's status had been weakened, though he was unwilling to defend himself at public meetings. But in his formal articles, he did not back down, offering other versions of his revision of genera without making any concessions. He "never retracts an error," Edwards wrote in high dudgeon to his friend Joseph Lintner.[42] In a private letter to Edwards, Scudder insisted that it was Edwards, not he, who had distorted the "face of nature," a claim that infuriated his opponent, who, in turn, charged that Scudder "attributed the distorting characteristics to the wrong individual. I could run down the page with your efforts at improving on Nature. It is not I who's in that business, not by any manner of means." Edwards went on to fault Scudder's entire systematics as based on speculation, not on the "facts." Scudder had "learning and ability and industry" but had invented a system that only Alice in Wonderland would understand. "A few years ago," said Edwards,

> we had a fairly settled and well working system of arrangement and nomenclature, as regards the fauna of this country. You first set the fashion of glorifying the upside-down, and taught how to plant tops in the ground and set the roots in the airs, and see what has come of it! . . . If I can lead any of my lepidopterological friends and others interested back from the bogs and pitfalls into which you have been sedulously beguiling them I shall be able to sleep the sleep of the just. We will then hope to travel on Nature's highways instead of spending our days in a bedeviled maze. There will then be some study of things, instead of harping on words. In one year, in my plan of action, we shall learn of what it befits us to know, as students of Nature, than in ten decades spent in evolving truths from the depths of one's consciousness, or by spinning themes as do the spiders their webs, from their own entrails.[43]

Harsh words. Scudder stopped the correspondence. "Scudder does not write me anymore," Edwards told Henry Edwards, rather smugly. "I suppose he's taking it hard. But he will have to come back some day for what I know he can't learn elsewhere."[44]

The most heated contests occurred over collecting itself, the activity crucial for all butterfly people, rousing them out of bed and into contact with the life around them. The Americans had a lot in common as collectors. They all used a net, an indispensable talisman of sorts capable of conjuring up a butterfly person out of the welter of childhood yearnings. ("Do you know what a butterfly net is?" my own father asked me when I was nine years old. "Yes." "Well, good, I've made a net for you from cheesecloth, a coat hanger, and an old pole. Go out and catch some. Here's a cigar box for you, too." So that's all it took—no guidebooks, no other collectors beyond my own family, since there were none I knew of, nothing but my father's unexpected proposal, the fragmented nature around me, and a net.)

Probably invented in the early 1700s, in both England and Germany, the net later took many incarnations.[45] Young Theodore Mead purchased twelve-inch net-rings and nets made on demand from Akhurst.[46] For fancier things, he shopped by mail at Deyrolles in Paris or Janson's in London.[47] So did George Crotch, an English immigrant and specialist in moths at the Museum of Comparative Zoology in Cambridge, Massachusetts. "What a thing it is to have a net you can catch things with," Crotch wrote a friend in 1873, of his Janson's purchase. "My net you never saw, it is a buster, two feet across and two feet and six inches long. It just catches things itself."[48]

Americans also had few or no reservations about killing insects or using poison to kill them, unless they had read too much Romantic poetry (especially by the English) or spent time in London long enough to absorb a climate hostile to killing.[49] Augustus Grote, having lived in London for several months in 1882, announced that he had "abandoned collecting" on the grounds that moths and butterflies "were part of the Universe of Stars and Sun. I could not understand the life I was taking; and then I felt the grief that arises when we become conscious of the role played by Destruction. I hope that the enthusiasm of the student will not cause him to forget that these little creatures suffer and feel pain."[50] For Grote, however, Romanticism worked both for and against collecting; in London, he opposed it, but a few years afterward, he reversed himself, observing that "poison bottles are indispensable to the collector, and can be recommended on account of the probably painless death they inflict. When we recollect that insects are the main store of food to numberless birds and animals, besides falling prey to each other, so that the greater proportion meet a violent death in any case, the comparatively small number which fall a sacrifice to the pleasure of the collector, or supply the studies of scientists, cannot be in reason objected."[51] Later, he wrote simply that "the poison bottle should be kept handy" when "taking Moths with the net."[52]

In America, a seeming superfluity of nature, along with conditions that condoned or depended on the routine slaughter of animals for food on family farms, reinforced an outlook easily tolerant of the killing of insects (and again, of birds as well, as shown by the behavior of Theodore Mead, William Henry Edwards, Samuel Scudder, and Herman Strecker). In this context, most Americans agreed with Grote (the later Grote, that is). Strecker explained, in one of his books, how to kill smaller moths with chloroform. Use heavy doses, he suggested with

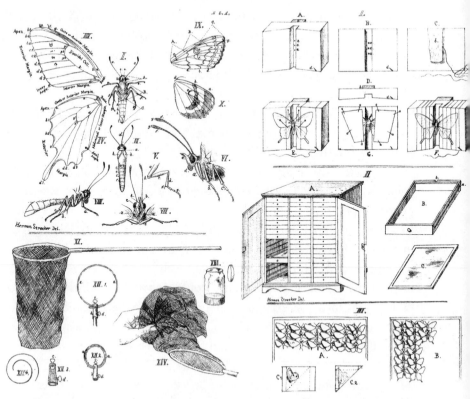

Observe, at bottom left, a "bag net made of fine strong gauze—a mosquito netting from which the stiffening has been well washed," according to Strecker. He wrote further of nets that they "are made in various ways; in some the rim folds up in sections, in others it is made of steel and can be coiled up like a watch-spring, all with the object that they may be put in some big pocket to be put out of sight until we are in the fields, for in this enlightened land a man can easily earn a reputation for lunacy if he lets it once be known that he is a butterfly hunter or any kind of hunter except a money hunter" (Butterflies and Moths of North America [Reading, PA, 1878], 7, 32–33). Also shown, at top left, are the different parts of butterfly morphology. On the right are shown standard mounting boards at the top; display cabinet and drawers in the middle; and storage trays and storage envelopes at the bottom.

irony, since the insects "may recover from the effects of the drug—more tenacious are they of their worthless lives than are we greater human beings."[53] Theodore Mead sugared trees to catch moths in a style characteristic of his cat-killing methods and one that cleverly had no need of nets. One August night in 1872, on a visit to the Catskill summer home of William Henry Edwards, he painted surrounding trees with a "wonderful mess of molasses, sugar, and water flavored with vanilla and cologne." At the same time, he learned by experience that moths do not fly up when disturbed but head down; this discovery allowed him to enlist "a large cyanide poison bottle with a mouth as wide as possible," rather than only nets, to capture his prey. Stepping gingerly about with a lantern from lathered tree to lathered tree, he poked one moth after another, inducing each to dart downward like clockwork into the jar, until many hundreds "piled together, making a stratum an inch or two thick in the bottle." Another American, Andrew Foulks, a cattle herdsman from upstate New York, also dispensed with the net by converting his daughter's bedroom (she was living elsewhere) into an insect trap. He attached a muslin funnel to the window, and behind it plate glass and "a large [oil] lamp with a powerful reflector." "I left a long narrow opening at the bottom of the glass," he told a friend, so that "the flies [moths] come in, strike the glass, slip down in the room, and in the morning I go around and pick what I want."[54] Even John Muir, one of America's most Romantic and tenderhearted naturalists, accepted the merit of the butterfly net, and presumably of the poison as well.

The enthusiasm for nets, poisons, and other strategies underlined what Americans shared most of all: the love of collecting itself, although such a love came in diverse forms. Many collected merely to fill their cabinets with specimens, having no other aim than to beat out their friends next door or at the local tavern. More serious naturalists knew about these people, called them "stamp collectors," and expected little else from them.[55] On the other hand, some naturalists, like the remarkable Eliza Fales Bridgham of New York City, collected selflessly, as it were, for the pleasure of sharing their insects with others and for advancing the cause of science. Born in Boston to a rich Yankee family in 1813, Bridgham enjoyed the usual sheltered upbringing of upper-class girls, until on a holiday trip, she noticed spiders crawling about her family's summer place in St. Augustine, Florida. "I am trying to get some Spiders to bring home with me," she told her sister. "The Spiders here are very beautiful."[56] When she married Samuel Bridgham, a prosperous New Yorker, she branched out from spiders to moths, collecting many near their summer home in Newport, Rhode Island. She got her son, Joseph, who was later a respected illustrator of scientific books, to go with her, and together, the two of them, with her husband's blessing, often stayed up until two or three o'clock in the morning, wondering over, arranging, mounting, and chatting about their quarry.[57]

Bridgham's "extensive collection is a model of useful collecting for scientific purposes," Augustus Grote observed. "She has allowed me" again and again "to examine most of the species occurring near the seaboard of the Eastern States."[58] Strecker and William Henry Edwards both studied the moths in her cabinet in Manhattan, and Alpheus Packard spent hours in 1862 pondering the whole array.[59] "I felt rebuked by her industry in collecting," Packard told his father. On one visit, he spotted a bizarre specimen of one of America's most lovely large moths, a blond-and-pink emperor moth with blue eyespots or ocelli, of the family Saturniidae (from the Latin name of the goddess Saturnia, or Juno, the daughter of Saturn), bred by Bridgham herself and discussed by Packard in his *Guide to the Study of Insects*. "It is," he wrote, "a curious instance of an imperfect hermaphrodite. The left antenna and left primary [top wing] are male; the right antenna and left secondary [bottom wing] are female; the right primary is also female, but the right secondary is something between the two, neither male nor female. In this hermaphrodite, the confusion of the sexes is conspicuous."[60] Eliza Bridgham often gave away specimens to those eager to have them, such as to her equally generous neighbor Berthold Neumoegen; in 1877, Bridgham surprised him with a gift of a "splendid pair" of American lappet moths ("lappet" meaning, in Middle English, a small flap or looking like a folded garment), male and female, "as a present out of her own collection."[61]

The differences between selfless and selfish collecting seemed unbreachable, although they often overlapped in the same person. They rarely caused serious trouble among collectors (at least not at first), and they never interfered with what Americans seemed to share most of all: the thrill of knowing an immense number of natural forms on the American scene—the various beauty of the natural world—wherever and whenever they looked. "When I go anywheres," wrote Fred Tepper to Berthold Neumoegen, "I want to take in all I can, everything gives me pleasure, from the grandest Papilio down to the interesting little Noctuids and Geometrids—But I suppose we were not all born alike."[62]

In 1873, a feud broke out between Grote and Strecker more extreme than any between Edwards and Scudder, and with repercussions deadly enough to alienate Strecker from many of his contemporaries, sully Grote's reputation, and change the course of American work on butterflies. In the early 1870s the most critical thing William Henry Edwards said about Strecker was that he was a "queer genius," and Edwards happily lent him specimens. In 1874 he informed Henry Edwards that Strecker "is a rascal" and "too dirty a fellow to handle." An amateur naturalist himself of the greatest distinction who depended on other amateurs, Edwards now pounced on Strecker as an "amateur" collector whose "work is too cheap and ill-done for consideration," "an illiterate, uneducated man who could not possibly identify all the insects he collects. He is a maniac." Edwards was delighted to see "Grote punch his head" in the *Canadian Entomologist*.[63] In the summer of 1873, Samuel Scudder had traveled to Reading, eagerly studied Strecker's collection, and borrowed specimens from him (something Strecker rarely allowed for anyone), but after 1873 Scudder came to see Strecker as a mere compulsive collector and, despite Strecker's later overtures, refused to deal with him. Earlier, Grote, too, had treated Strecker in the most amiable way, making him an honorary member of the Buffalo Society of Natural Sciences, borrowing his insects, publicly praising his submissions, and inviting him to write articles for the museum journal.[64] "Mr. Strecker kindly sends me" or "I owe to the obliging disposition of Mr. Strecker" or "both sexes of this species have been obligingly communicated to me by Mr. Strecker" frequently marked the pages of Grote's journal.[65] Both men dreamed of collaborating; Grote wrote Strecker that "you and I must be good friends," and Strecker told him, "As we are about the hardest workers on Lepidoptera in America, we might as well go it hand in hand as not."[66] One year later, however, a friend told Strecker, "There is *bitter* feeling against you in high quarters."[67]

The shift began on a day in April 1873, when Strecker visited the new American Museum of Natural History to inspect the butterflies and moths that Theodore Mead, the temporary curator of insects, had just put on display.[68] Strecker wanted to see, in particular, the specimens Augustus Grote had recently given the museum, and two insects caught his eye: a yellow-and-black swallowtail and a sphinx moth. He told Mead that both specimens *belonged to him* and that Mead should give them to him. Dismayed, Mead asked Strecker to have a friend—preferably, Grote himself—verify that the insects were Strecker's. "If you want to do me an everlasting favor," Strecker requested of Grote, "write me a line saying that they are mine, or write a line to Mead, telling him to give them to me so that he don't make me out a humbug or worse."[69] Instead, Grote wrote everyone—including Strecker's friends—that Strecker was a "thief" and a "liar."[70] In March 1874, he reported to Philipp Zeller, one of Europe's most admired authorities on moths, that he had "caught Strecker stealing specimens from the Museum of Central Park." "He is not respected here" and "has been refused admittance to the Academy of Natural Sciences."[71] Grote sent Zeller copies of Strecker's letters to him that, Grote implied, confirmed Strecker's guilt. By the mid-1870s probably all the principal butterfly people in England, Europe, and America had received the news through the Grote grapevine. Even William Hewitson, whom Strecker worshipped as the finest lepidopterist in the world, heard the gossip, but he refused to take it seriously. "The American entomologists are I hear greatly incensed with you," he wrote Strecker in 1875, "because they say that they cannot trust any butterflies in your hands . . . , which I told my informant is I am sure an infernal lie."[72]

Grote embarked on a public smearing that lasted throughout the decade and into the next. He went out of his way to target Strecker as a member of a "certain class of collectors who 'covet' diverse and pretty specimens without any higher philosophical value" and as the kind of person who takes advantage of "uninformed young collectors, whose rarities are speedily transferred out of their keeping by false statements and industrious letter-writing." There were further implications: "As indulged in by such persons, Entomology loses much of its refining influences and educational value, and becomes merely an opportunity for the display of human passions and idiosyncracies."[73] In the entomological press, many people read Grote and trusted him, and Scudder and Edwards respected him. In the mid-1870s, Ezra Townshend Cresson, the head of Philadelphia's Academy of Natural Sciences, ordered all the specimen cabinets to be locked up whenever Strecker entered the building. Strecker never defended himself in any periodical or newspaper; nor did Grote present persuasive proof of a crime, beyond the letters

Pap.ᵗ Antimachus
(S. Leone)

9. Another tropical species that fascinated Strecker was the giant African swallowtail, *Papilio antimachus,* the largest butterfly in Africa, with a seven- to nine-inch wingspan. The image here, doubtless seen by Strecker as a boy, was figured by the English naturalist Dru Drury in his *Illustrations of Natural History* (London, 1770). Drury named the butterfly "antimachus," after an ancient Macedonian king. Latin and Greek were (and are still today) used as the standard languages of naming, to give all naturalists throughout the world a common way of discussing the same organisms, which often exist in many places at the same time, in defiance of national or local boundaries. Swallowtail species, for instance, can be found nearly everywhere in the world. *Courtesy of Craig Chesek, © American Museum of Natural History.*

XII.

Herman Strecker del.

10. This plate of moths was created by Herman Strecker. The insect with the long tails is a moth from Brazil, *Eudaemonia jehovah,* named by Strecker in 1875. *Eudaemonia* is Greek for happiness or contentment, and *jehovah* is the proper name of the God of Israel in the Hebrew Bible. A symptom of the risks many naturalists took with their nomenclature, the name outraged many of Strecker's friends, especially the religious ones, who thought Strecker had breached conventional standards of decency. Published in Strecker's *Lepidoptera* (1878).

PHILODICE 1.2 ♂, 3.4 ♀, 5 ♀ var., 6 ♀ albino.

a a²	Eggs	....	magnified	e	Larva after 3ʳᵈ moult.	
b	Larva (young)		„	e²	„ „ „ „ 'magnif.ᵈ.	
c	„	after 1ˢᵗ moult	d	f	„ „ 4ᵗʰ „	
d	„	„ 2ⁿᵈ „		g	„ mature „	
		h	Chrysalis.			

Drawn by Mary Peart.   L. Bowen, Col.

11. The plate featuring the common sulphur is from William Henry Edwards's *Butterflies of North America*. A Frenchman, Pierre Latreille, first named it *Colias philodice,* the genus after one of Venus's many names and the species after the golden maid of honor to Venus. In the 1870s, Henry Edwards observed that "every school-boy throughout the country knew that the most common species in America was *Colias philodice.*" William Henry Edwards said of it that "where *Philodice* is no one can fail to notice it, as it gently flits from flower to flower, or courses along the road or across the meadow, with sustained and wavy flight. It is sociable and inquisitive, and may often be seen to stop in mid-career as it overtakes or meets its fellow, the two fluttering about each other for a moment, then speeding on their way."

A.H. Searle, del. et lith.

Mintern Bro's imp.

39.Sabulosa. 40.Abbreviatella. 41.Chelidonia. 42.Beaniana.
43.Mira. 44.Frederici. 45.Lunilinea.

12. Augustus Grote correctly identified more American moth species than anyone in history. This plate contains several underwing moths described by him in his *An Illustrated Essay on the Noctuidae of North America* (London, 1882).

Herman Strecker Del.

1 PAPILIO EURYMEDON ♂. 2 P. MARCHANDII ♂. 3 COLIAS DIMERA ♀.
4 C. ab. SEMPERI ♀. 5 CHIONOBAS UHLERI ♂. 6 SATYRUS RIDINGSII ♀.
7 S. STHENELE ♂. 8 S. var. HOFFMANI ♀.

13. Depicted by Strecker himself in his 1878 *Lepidoptera,* and perhaps an homage to the time he spent as a very young man alone in Guatemala, *Papilio marchandii,* the bright, orange-colored swallowtail at the center of the plate, was originally identified and described by the French butterfly man Jean Baptiste Boisduval in 1836.

XV.

1 *Melitaea Alma*, Streck. ♂. 2 *Ægiale Cofaqui*, Streck. ♀. 3 *Macroglossa Ulalume*, Streck. ♂. 4 *Sph. Vashti* Streck. ♂. 5 *Hepialus Sangaris*, Streck.
6 *Gloveria Arizonensis* Pack. ♀. 7 *Coloradia Pandora* Blake ♂. 8 *Pseudohazis Hera* Harris ♀. 9 *P. Hera* ♂ aberr. 10 *P. Pica* Wlk. ♀. 11 *P.Pica* ♂ var.

Herman Strecker del

14. This plate by Strecker contains several moth species known as buck moths (figures 8–14), all belonging to the genus *Pseudohazis*, in Latin meaning "false" (*pseudo*) "Syrian god of war" (*hazis*). The logic of the name is perplexing, although not to Strecker, who described, drew, and colored many of the species. They "may be found," observed Strecker's contemporary William Holland, "in vast numbers in the morning hours on bright days in their favorite haunts in the region of the Rocky Mountains. They frequent flowers in company with diurnal lepidoptera, and they may be easily taken. They are characteristic of the country of the sage-brush, and the ranges of the western sheepherder." Holland, *Moth Book* (New York: Doubleday, Page, 1903), 93.

15. The butterflies in this wonderful plate by Strecker, from his *Lepidopetera: Rhopaloceres et Heteroceres* (1878), belong to the family of Lycaenidae, boasting today more than six thousand species in the world, including such subfamilies as the coppers, blues, and hairstreaks. They are all small, often brilliantly colored insects, otherwise known as the "gossamer winged" due to the delicate appearance of their wings.

16. Strecker's portrait, in his 1878 *Lepidoptera,* of a multifaceted species belonging to the Sphingidae, a family of robust flying moths, with large bodies, streamlined wings, and tongues usually long enough to reach far into flowers for nectar. Also called sphinx or hawk moths.

Strecker had written to him that he had copied out for Zeller, and these letters contained no admission of theft.

Being shut out from the cabinets in the place where he, as a boy, had first plunged irretrievably into his paradisiacal pool would go far to explain why Strecker hit back just as hard with so much invective in his writings. If, in 1872, he had good relationships with William Henry Edwards, Scudder, and Grote, none of whom found fault with him, he now despised those whose status had risen as his had fallen and partly at his expense. He loathed Edwards for being the "demigod of North American Lepidopterology," "a charlatan," and a "Coalburgh Ass," he wrote Theodore Mead. With penetrating sarcasm, he stereotyped Scudder as "a tight-lipped Yankee" who "treated" his competitors "with silent contempt." "Grote flares out if you touch his vanity, but Scudder. Well, if he and I traveled in the dark together, I would want a shirt of chain-mail under my coat, if I had done ought to offend him, though in the sunlight he would smile, oh! so blythely."[74] Strecker never lacked bite before in what he wrote, but after that summer, and especially by 1875, at the time of the *Eudaemonia* controversy, he heaped abuse even-handedly on all his major competitors, leaving only Henry Edwards, Theodore Mead, Berthold Neumoegen, John Morris, and a few others unscathed or even praised.

In 1878, two major studies by Strecker were published: his completed catalog, *Lepidoptera: Rhopaloceres et Heteroceres,* and his instructional guide and catalog, *Butterflies and Moths of North America,* books of much merit, each proof of his fighting spirit. Strecker used both for retribution, striking at issues of systematics and occasionally scoring points. He observed, for instance, that Edwards sometimes mixed up his genders and sometimes "imagined" differences in species when there were only "varieties" (Strecker did the same thing, of course). His most cutting charge was that Edwards overemphasized the early stages of the butterflies as a way to establish species identity, especially in the fritillaries (Edwards's favorite butterfly group and his favorite method of establishing species). "Too much stress by far," Strecker argued, "is laid on the circumstance of whether the larva differs or not from that of the ordinary form.[75] Strecker pointed to Edwards's descriptions of species to show how "worthless" modern taxonomy had become, reduced "always to same monotonous thing, the same stereotyped greyish under

side, the same tedious 'sinuous rows of spots,' and the same everlasting this shaped or that shaped discal bar, spot, or mark." He ridiculed
Edwards's contention that two forms of a genus of blue butterfly were
separate species requiring separate names. "The most critical eye will fail
to detect the slightest difference between them." And, in a concluding
exposé in *Lepidoptera,* he noted how Edwards had botched a description of another blue. "Since Mr. W. H. Edwards described this species, it
very nearly had the misfortune of losing its birthright: the author having
through accident, lost his types; and what was equally unfortunate, his
memory."[76]

As for Scudder, Strecker called his generic names the "Phantasies" of
a "mad man" and claimed that "it would take volumes" to "recapitulate
all" his "entomological vagaries." "Language fails us, our hand refuses
to go further—even the ink on our pen pales—must we record" the
character of "this puerile affair? . . . Scudder's lists, theories, etc, seem to
be gotten up to show what amount of time and labor one human being
is capable of wasting." With Grote, Strecker was pitiless, mocking his
"fancy" language and his understanding of moths as "so replete with
errors and inaccuracies that to eliminate them all would leave it in much
the same condition as the result of that arithmetical problem where
'nothing from nothing and nothing remains.' " He berated Grote's habit
of hitching his own name to the end of every species he described, with
hundreds of "groteiis" populating the literature, a "habit" that exposed
Grote as someone who cared nothing for science or nature, only for
himself, "ever ready and mad for any means that might bring his name
into notice," seeking to "exalt himself above others," even "above all
creation."[77]

After visiting Strecker in 1878, young Theodore Mead returned
home, certain, to the utter bemusement of William Henry Edwards, that
Strecker had been misunderstood and "deserved credit for much that
he has done." Mead excused him "for much that is objectionable from
his imperfect education." In a letter to Joseph Lintner, William Henry
observed that Mead "says that Strecker has not the slightest idea that he
ever said anything that the most sensitive could find fault with." Taking
into account that Edwards here is paraphrasing in his own fashion, we
might wonder who was hoodwinking whom.[78] Edwards, at one point,
had been angered not by Strecker's supposed thievery but by what he

considered an attempt by Strecker to upstage him in identifying a new species (by stealing the glory, in other words, a crime equal, in Edwards's mind, to stealing the real thing). Was all this warfare a matter of social class? Did Edwards and Scudder et al. look down on Strecker because he had little formal education and worked with his hands as a stone carver?

Ironically, even some of Strecker's supporters took Grote's side, unaware of the cause of the feud between the two men. "Every candid reader will conclude that you and he have had an undecided quarrel," John Morris wrote Strecker, "and that you are striking him from a distance. I regret this very much on your account" since "Grote is nothing to me."[79] Fred Tepper and Edward Graef, Grote's boyhood collecting buddies but friends of Strecker's too, denounced him for "degrading science" with "personal" politics. "I see with sorrow," Graef wrote to Strecker in 1878, "that you do not avoid your gross personal remarks, but are getting worse then ever. . . . No matter how Grote may have wronged you, he always, as far as I could find out, treated you as a gentleman . . . which is certainly more than you have done to him."[80] In 1873, Joseph Lintner, the state entomologist of New York, wrote to Strecker, "You deserve enormous credit for what you are doing, under so many difficulties and impossible circumstances." But, after reading Strecker's catalog in 1877, he urged Strecker to drop the personal attacks. Strecker refused, and Lintner wrote back, "I am sorry to hear you say that you need have no care whom you offend. It is dangerous ground to take. If one outlaws the whole world, then the whole world may outlaw him, and I cannot conceive how any person can willingly place himself in that position."[81] George Hulst considered Strecker the most original of all American lepidopterists but, fearing that his angry outbursts would discredit everything he did as a naturalist, pleaded with him to think about "Science" before venting his anger in the press again. "Whatever our personal feelings, whatsoever our grievances, we should obey the voice of Science. We are nothing; the truths she utters are everything. As in your work, you well say, one new fact of science is worth a score of lives such as most we live, so apply your own sermons, and for science sake, put yourself on the altar."[82]

Strecker, at first, tried to get out of Reading, Pennsylvania, hoping to find work as a naturalist, a change encouraged by a new friend, Arthur Fuller, an amateur moth specialist and well-known journalist.

Fuller visited Reading in 1873 to see Strecker's collection and returned home, certain of Strecker's destiny—that is, if he'd only quit "cutting stone," a dreary trade for so gifted a man. "A man of your talents must eventually find his place."[83] He praised Strecker at length in his newspaper, the *Rural New Yorker,* and led efforts to find him employment in leading museums. When Strecker inquired of Spencer Baird, at the Smithsonian, about possible openings there, Baird insisted that if he wanted a job he would have to stop collecting for himself, collect only for the government, and donate his existing collection to the Smithsonian. Strecker also believed he might get a position in Iowa, at the Davenport Museum of Natural Sciences, a flourishing institution founded in 1873 by a resourceful woman, Mary Putnam, under the directorship of her delicate, tubercular son, Duncan, an ardent butterfly man and a proponent of Strecker's. Strecker did prepare insect descriptions for the museum journal and, in 1878, a few plates, and Duncan paid Strecker to paint butterfly tiles for the butterfly room. But the Putnams finally discouraged his job search.[84] Meanwhile, Strecker found Spencer Baird's demand that he trade in his collection for a job at the Smithsonian beneath contempt. Leaving Reading ceased to be an option.

One other escape route remained for Strecker: block out your enemies by immersing yourself in your butterflies. He had always preferred collecting, anyway, to purely scientific work. Now he was free—by default, as it were—to allow his *Sehnsucht,* his spiritual guide, to possess him fully. It did, and to such a degree that he would build the greatest monument to the spirit of collecting by an American naturalist, a magnificent treasury so packed with specimens, many of which were rare and endemic, as to transcend the confines of the man who assembled it. And here he found a partner in his pursuit, his new friend Berthold Neumoegen, whose heart beat for butterflies nearly as strongly as Strecker's own and who didn't care a fig what Strecker wrote about Grote, Edwards, or Scudder or they about him. Neumoegen had behind his collecting a great deal of capital (at least for the moment). Briefly blindsided by the business turmoil of the 1870s, he had managed to "establish" himself, late in the decade, "as a Broker" for "German millionaires," he told Strecker.[85] He bought many "exotics" from Otto Staudinger and Staudinger's protégé Heinrich Ribbe, who, in 1877, left Staudinger to create his own business in rare species, mailing many to Neumoegen.[86]

Neumoegen hired his own collectors and his own curator, Jacob Doll, a German immigrant with extensive knowledge of lepidoptera who had fought on the Union side in the Civil War and made his living as a baker in Brooklyn. In 1879, Neumoegen had a large butterfly room built in his house; Doll tended it twice a week.[87] Neumoegen proposed making Strecker his "entomological secretary." One can only ponder what Strecker thought of that.[88]

The two wrote to each other nearly daily, exchanged photographs of their children, vacationed and spent holidays together, and visited each other's homes. Sometimes Strecker stayed for days at a time at Neumoegen's place on East Forty-seventh Street, the men sharing butterfly lore for hours over bottles of wine.[89] They bought insects together, from Staudinger and others, quibbling over who should get what and occasionally withholding information about particular specimens. "You never told me about the grand Saturnia you got from the South," Neumoegen complained. "Is there a chance for me, poor fellow?"[90] When Strecker seemed remote, or when both men were troubled by problems interfering with their common endeavor, Neumoegen struggled to get the butterfly exchanges flowing between them again. "Trouble should not estrange you from me," he told Strecker, "as I have a full load of it myself. Come, come, be a little more lively and send something and give a *signium vitae* of your locked up self! I am always thinking of you, but you are—forgetting me."[91] Strecker enjoyed Neumoegen's company immensely, because of his wealth and the comfort and the access to butterflies it promised but also because of his warmth, his accepting graciousness, and his easy drinking style. Neumoegen, for his part, hoped some of Strecker's expertise on butterflies and moths would rub off on him, and he appreciated their common German heritage.

Still, from the start of their friendship, Strecker may have harbored resentment toward his friend, its depth and character perhaps beyond Neumoegen's understanding. Strecker had begun, at unsustainable cost to himself, to shell out hundreds of dollars for "exotische Schmetterlinge" from Staudinger, but Neumoegen's appetite for exotics pushed Strecker to find makeshift ways of matching his competitive friend, even in the face of serious family illness and of an economic downturn that led to a drop in his wages as a stonecutter.[92] He reached beyond his means, relying on credit extended by Staudinger and others. Already poor, he made himself even poorer, "purchasing Lepidoptera at such a rate," he told Duncan Putnam in June 1877, "that ere I knew it I was head and (not last) heels in debt."[93] He borrowed money from Neumoegen, who, in turn, when "the market" sometimes turned against him, had to borrow money himself to lend to Strecker, "not an easy thing" for him "to do," he wrote Strecker. "Yet in order to help you in a dark hour I will do it."[94] When he and Strecker began visiting each other's homes, Neumoegen shared his "finds" unexpectedly, even divvying up a shipment from Utah of rare cocoons of *Samia gloveri* (a big maroon-colored moth described and figured by Strecker himself): three for him, four for Strecker. "Nobody but you would ever receive them from me," he wrote, "and may it serve again as a sign of how much I am attached to you."[95]

Strecker had passed something of a threshold, refusing to lend specimens to anyone for any purpose, including scientific ends. (Grote, for one, had always "generously placed all his material" in hands of any creditable naturalist who requested it.)[96] Prouder than ever of the size of his collection, for the range of his American acquisitions spanning the continent but also, with ever-increasing fervor, for the extent of the species from far-off lands, exotic and tropical, he became a partner not only with Neumoegen but with much younger and naive collectors who, out of respect for him, went off to foreign places to get insects, promising to mail them to Strecker so he could identify and name them and then return them. Sometimes, without his eager young acolytes knowing it, he kept the best specimens for himself. Titian Peale admired his "mania," Arthur Fuller made fun of his "insatiable maw," but, in 1879, John Morris, then seventy-five, warned him about yielding so unconditionally to his *Sehnsucht*: "Why, man, you are too covetous—you must get over that 'heartsickness' and 'take things cooly.' "[97] To make up for the grinding deficits, Strecker turned more and more to *selling* butterflies and moths; in five years, he would be publicized in *The Naturalist's Directory*, a widely circulated record of naturalists, as "a man who buys and sells Lepidoptera from all parts of the world."[98]

The butterfly people who fought over naming, identification, the face of nature, and, especially, over collecting suffered a good deal because of their contending beating hearts, although hardly in equal measure. After nine months of estrangement, the Yankees Scudder and Edwards reconciled, not by apology but by resuming a correspondence that, all things considered, they could not do without. Scudder buckled down to work on American butterflies, research that would go on until the end of the 1880s and the publication of his magesterial three volumes. Edwards, hobbled substantially by the collapse of his coal business, in March 1873, after some fierce bargaining, signed a contract with the Chesapeake and Ohio Railway, which, he noted, "changed our whole mode of doing business"; instead of sending his coal from West Virginia by barge and boat, he'd decided to move all of it by railroad. The transaction "drove butterflies nearly out of his head," he told Henry Edwards, and did nothing to enrich him. Two years later, his company went bankrupt, making him poor despite owning "a great deal of property." It was a condition from which he would never really recover.[99] He lost his salary, could not raise enough money to pay his artists, and even planned to sell part or all of his butterflies for cash to pay creditors. Among his buyers was Scudder, who, in an act of unusual generosity despite their falling out, mailed him hundreds of dollars for some specimens. "I ought not to make you pay for these things," Edwards wrote Scudder. "And yet I am compelled to utilize what I can."[100] He sold some of his "other" property "for a good sum," and soon was back to "normal," although no longer managing coal mines. Sometime in 1878, a neighbor bought the remains of Edwards's coal business, leaving him with seven acres for his house and family, as well as several other scattered properties.[101] Nothing now of consequence impeded Edwards's work on butterflies, and he took the reins with an admirable ferocity. A great work would be the result.

Grote and Strecker fared less well. By the end of the 1870s, Grote's family life was a mess, his children, whom he'd left with their step-grandmother, perhaps still in Alabama (it is unclear exactly when he got them back) and his father impoverished and extremely ill on Staten Island. Once at the center of cultural life in Buffalo, Grote now had few friends and little money, and in 1879 he lost his job as museum director, pushed out by the trustees, according to him, because of his espousal of evolution but also, apparently, because of his shifting moods gave offense to many people. To make matters worse, when he left his job, the Buffalo Society of Natural Sciences kept his moth collection, on the ground that he owed the society money; the collection was his only valuable possession, with many rare type specimens (type specimens are those first described and named by the discoverer and play an indispensable role in the identification of an organism, against which all future revisions of the name are measured).[102] In time, he retrieved his precious insects, but only after a volley of recriminatory letters charging that what he owed the society was "nothing compared" to what it owed him—his "collecting expenses," his books, his assistant's salary, "his seventy-four glass boxes of moths," all out of his own pocket. He wanted them back. "The Society has no more claim on them, than it has on my clothes. . . . It has stripped me of everything, turned me out . . . I am quite a ruined man."[103]

Circumstances worsened for Grote in the aftermath of the Strecker "Central Park Affair." Despite the fact that the leadership had accepted his accusations, many criticized him not only for his assault on Strecker but also for emphasizing *himself* in his writings, a criticism that must have stung him, since, in his writings on religion, he had made such an issue about placing the "Self" at the service of some larger moral purpose.[104] In November 1875, Grote, in a startling turn, wrote to Strecker, "I hope that now there can be a possible truce between us. I desire to be good friends with everyone. My health is bad and I am afraid I have overworked myself." In one year, however, he was at Strecker again, calling him, in a checklist of moths, "an incompetent writer" with "undoubted capacity for misunderstanding the simplest structure in insects."[105] Three years later—in the midst of writing *New Infidelity* and of his personal distress—Grote tried again to bury the hatchet, this time coming close to expressing doubts that Strecker had committed any crime at all in the American Museum of Natural History. "I have written to you several times endeavoring to heal the difference between us. You must know how you misrepresent me and I spare myself any written defense in consequence. If I have ever done you wrong," he wrote, "I am willing to make you honorable amends. I desire for the sake of public decency that the warfare between us should cease and I offer you my services in any way in which I can assist you."[106] On one occasion, Strecker was tempted to give in, but after a friend reminded him that Grote had told too many people over too long a time that he was a liar, a

light-fingered amateur, and a "forger" who had violated the ninth com-
mandment ("Thou shalt not bear false witness") and the tenth ("Thou
shalt not covet"), he stopped short.[107] If many people blamed Strecker
for all the ugliness, Grote carried the burden as well, and he knew it; his
feud with Strecker increased his instability and his isolation in Buffalo
and, later, on Staten Island.

Of all the pioneers, Strecker perhaps suffered the most, no ordinary
collector, a true Romantic about butterflies and moths, with a *Sehnsucht* for them that, once aroused, never waned and was never grati-
fied. Arguably, it was his years sculpting stone and his lack of education
and privilege that set him apart from other butterfly people, leading
him to violate customary courtesies and sometimes to risk breaching
ethical boundaries. But many other butterfly people were from meager
backgrounds and worked under bosses, and Strecker, of course, had
talented artists in his lineage. It was common gossip in the late years
of the century that he wore a stovepipe hat lined with cork on visits
to other people's collections and, when no one was looking, poached
a butterfly or two and pinned them on the cork. But the only actual
evidence to confirm this gossip came long after his death. Thus, for
example, in the 1910s, Preston Clark, a rich Boston manager of silver
mines with a huge moth collection, "thought" he found in Strecker's
collection (by then housed in the Field Museum in Chicago) a specimen
Strecker "probably" lifted from Philadelphia's Academy of Natural Sciences. "My guess," he wrote a friend, "is that Strecker stole it from the
Philadelphia Academy." A guess? No word on how or when Strecker
took the insect? Before or after Ezra Cresson locked the cabinets?[108]
Another collector, an affluent physician with a huge butterfly collection,
William Barnes, had the same suspicions, in 1929 remembering Strecker
as "a very peculiar man when visiting you and it was better to stay close
by his side as a protection to valuable specimens!" Perhaps. But Barnes
observed, at the same time, that Strecker had "a most peculiar habit of
crawling in between the sheets with his boots and clothes on." Barnes's
and Clark's anxieties may have run along class lines. How, anyway, did
Barnes know about Strecker's sleeping habits?[109] And how trustworthy
was he, a man who resented Strecker for humiliating him twenty-five
years earlier over some butterfly matter? "The fool," Strecker wrote a
friend, "I only laugh at him (and he knows it)."[110]

Whether Strecker walked off with a butterfly or two, we may never know for sure. What we do know is that the image fashioned by Grote, and supported by Scudder, William Henry Edwards, and others, dispatched him to the ash bin of "déclassé naturalists." "It is his own fault that he is not at the head of America's lepidopterists," George Hulst told Henry Edwards. "As it is with his unquestioned ability and work he has no standing whatever."[111] Not everyone put him down or fled his company, certainly not Neumoegen, or even allies of William Henry Edwards's, such as Theodore Mead, Henry Edwards, and John Akhurst, who remained friendly long after William Henry Edwards himself shut the door. The banker Joseph Drexel, who bought Strecker's art pieces in the 1860s, remained loyal to him all his life. He helped Strecker pay for many of the rare butterflies he got from Otto Staudinger and, later in the 1880s, made it possible for him to move his family into a four-story home Drexel owned in Reading, asking in return only that Strecker make a down payment, at 20 percent of its worth, and freeing Strecker from paying for the other 80 percent. Strecker converted the entire upper floor into a butterfly room.[112]

The friendship of these people must have done much to salve the injury inflicted by Augustus Grote, who, despite whatever doubts he may have had about Strecker's behavior at the American Museum of Natural History, never stopped vilifying him. "That entomological *spider*!" he wrote to the young naturalist Harrison Dyar in 1898, underlining "spider" twice.[113]

The disputes among the butterfly people had more consequences than the alienation of Strecker and the weakening of Grote; they also harmed the commonalities that tied the butterfly people to one another as a group and to the butterflies and moths of America. If, before 1870, most naturalists had little trouble accepting a full range of purposes for collecting, from collecting for beauty to collecting for science, by 1880 they were beginning to move in contrary directions. Scudder and Edwards, who were much affected by the beauty of butterflies—color drew them into nature—began to fault those who collected for beauty as bad for science. The "very charms which often attract men to the study of butterflies," Scudder protested, had "grievously checked" advances in the natural sciences. "There is such a rage for their collection by amateurs," he charged, "enchanted only by their exquisite beauty"—as *he* had been at Williams—"that the study of butterflies has been largely abandoned by those who are best fitted for this work by specific scientific training."[114] Was Scudder saying that scientists were staying away from butterflies because others—amateurs, children, illiterate people—found them too "pretty"? Did Edwards, who had little "scientific training," agree? Must people sacrifice such behavior in order *to be* scientists? If so, this broke with the Romantic vision espoused by Humboldt, who saw art and science as integral to each other.

Butterfly naturalists found it more and more difficult to work together, a state of things illustrated best by the short life of what may have been the first magazine in the world devoted entirely to the study of butterflies and moths. In 1881 Henry Edwards, Neumoegen, Mead, and Grote met at Neumoegen's house in Manhattan to establish a new society and a new journal, similar to those created by Scudder a decade earlier, in Cambridge. Neumoegen hoped to use the new journal to advance his reputation as a butterfly expert, and it was his money that launched it. Grote and Edwards had just arrived in New York, Grote having recently been fired from the Buffalo museum and returning after his father's death (joined, this time, by his children and his stepmother) to live in a cheap cottage on the northeast side of Staten Island, and Edwards to take a choice position, with time to "entomologize," offered by Wallach's Theater in New York, one of the most fashionable theaters in America. Edwards had left San Francisco in 1878 for Boston, attracted mostly to the exciting climate of the natural sciences, but soon hating the "straight-laced, puritanical audiences," which made acting in theaters around Boston, he believed, akin to "being in a graveyard." "I should have never have left the 'Hub' [Boston]," he wrote Hermann Hagen, "[but] my head and cheese depend on my coming to New York. I must look out for the main chance."[115] With his wife, Polly, and his sprawling bug collection, he set up house on 116th Street in East Harlem, a short walk away from a new railway station at 125th. All the men envisioned a naturalist core "worthy of the great City of New York." They called the society the New York Entomological Club and the journal *Papilio,* its name picked by Grote; with etymology in Ovid and Pliny, it meant literally, a tent flap, chosen for the way butterflies hold their wings at rest.

At first everything was "on the high road to success," as Henry

Edwards, who was editing *Papilio,* told Hagen. "Grote—Neumoegen— and myself intend to see it safely through its first year's journey, and by that time I think our subscription list will make it self-supporting."[116] *Papilio* paid most attention to American species but also treated exotic lepidoptera, presenting the first article ever published in America by a Japanese butterfly specialist, Charles Ishikawa of the University of Tokyo.[117] The essays of William Henry Edwards gave the magazine particular distinction, one describing a yellow-and-black swallowtail, *Papilio machaon,* that, over millions of years, had arisen in various forms throughout the northern latitudes, all looking very much alike.[118] Another of Edwards's articles presented a rather salty account of the "courting" behavior of male sulphur butterflies (genus *Colias*) in a dry desert habitat in the West, where the males far outnumbered the females. "From what I know of the frenzied eagerness with which certain male butterflies watch the coming of the females from chrysalis," Edwards observed, "I am confident that they would seize upon the females of any allied species just as readily, if one of their own were not at hand. If such things occur in the mild climate of the Mississippi Valley, where the females are as common as the males, what may not occur in a sage-brush desert? The panting male cannot fly over hill or valley, under these conditions, seeking its mate, as we often see male butterflies doing in a Christian country. Nature impels him, and he captures the first female he meets."[119]

Papilio lasted only two years under Henry Edwards's editorship, done in partly by unbusinesslike practices (Edwards, for instance, gave out free copies) but mostly by the continuing struggle of Strecker, Grote, and their proxies, in a context of worsening economic crisis in the country.[120] Theodore Mead, who'd never engaged with the magazine anyway, abandoned ship in 1881 for a new life in Florida, buying an orange grove with his father's money, taking Willie Edwards's advice to chase something more practical than lepidoptera. A few years later he married William Henry Edwards's daughter and, settled in Eustis, Florida, tried to make a living from the orange business, also commencing what would be a lifelong study of orchids and ornamentals. (William Henry was happy to have "Ted" as a son-in-law but much saddened to lose "Mr. Mead" as *the* man who would carry on his achievements in the study of butterflies. His son Willie must have felt justified, having won his little war for Mead's soul.)

Berthold Neumoegen lost nearly all his money on Wall Street, in turmoil caused by speculation in the railroads; he was forced to move into his father-in-law's house and find ways of selling his collection.[121] ("Neumoegen is really poor," William Henry Edwards informed a friend, "and *Papilio* is poor.")[122] But it was Grote's and Strecker's behavior that perhaps most damaged the magazine. Strecker, feeling excluded, struck from the outside, complaining to friends of "the Club composed of 4 whole persons" and "that New York thing *Papilio*." "They meet in each other's houses," he sneered. "No respectable man" should "pay for their rubbish."[123] Strecker hated the efforts of the club's committee on nomenclature, led briefly by Mead, to diminish his credibility as a describer and namer of new species, and, understandably, he turned against Neumoegen for joining with Grote. Neumoegen was mystified.[124] "What the promptings on your part, for such a step are, I do not know," he wrote Strecker, "but I regret its existence, for I always shall and will bear friendly feelings toward you and will never attempt crossing your path in an inimical way. Nobody shall ever bias me in that direction."[125] Neumoegen's longing to be respected as a naturalist had got in his way. In any case, he assured Strecker later, when this storm was about to pass, that he was "no satellite of Grote, nor of Edwards."[126]

Grote let his prejudice against Strecker spill onto *Papilio*'s pages. His belated apologies to Strecker notwithstanding, and all Henry Edwards's efforts to stop him to no avail, Grote seemed trapped in the conviction, expressed best by the poet Heinrich Heine, that "one must, it is true, forgive one's enemies—but not before they have been hanged."[127] In one essay in *Papilio*, without mentioning Strecker's name (though no informed naturalist could have missed his target, least of all Strecker), Grote rampaged against Strecker for "slovenly descriptions and confessed unacquaintance with structure" that "places him on a level with the worst amateur who has 'coined' a 'species.' In vulgarity and misrepresentation he is, fortunately, without a rival."[128] Angry at the diatribe in their magazine, Henry Edwards and Neumoegen called a private "crisis" meeting in November 1881 (with Grote absent) and in Neumoegen's parlor denounced all references to "personality" in *Papilio*.

Meanwhile, mercurial Grote, desperate to escape his own poverty, was trying to sell his moth collection—first, fruitlessly, to the American Museum of Natural History, in New York, or the Museum of Comparative Zoology, in Cambridge, and then to the British Museum, which took it after haggling him down from £1,000 to £800. (At least the British were interested!)[129] Then, telling no one in New York, he sailed off to England with seventy-four glass boxes of priceless moths, packed with nearly eleven thousand lepidoptera and more than twelve hundred valuable type specimens, to deliver them himself.[130] After returning, he fell into a depression, stayed in bed for days, suffering from "insomnia and worry," and then, feeling betrayed by many people, accepted an invitation from a European friend willing to pay his passage, so that he could "work in Germany." Grote sailed alone again across the Atlantic, this time to Bremen, where one of his sisters took him in, as well as, later, his children and stepmother.[131]

Henry Edwards was sickened—Grote had not prepared him (or anyone else, for that matter) for his flight, and he now had the full editorial burden of *Papilio* on his shoulders.[132] He was disturbed, too, that the war of "personalities," so visible in *Papilio,* had given the wrong impression abroad—that American naturalists did more to tear each other down than to build each other up. Worst of all, subscriptions to *Papilio* had begun to fall; at the end of 1883, Edwards quit as editor. Eugene M. Aaron, a young naturalist from Philadelphia, took over, only to hasten its decline by siding with Strecker against Grote and by publishing an article (written by his brother, Samuel, a callow youth in his early twenties) that argued against William Henry Edwards's strongly held belief that the study of caterpillars helped significantly to establish valid species.[133] William Henry's "preposterous" positions, Samuel insisted, "do not work for the good of science. . . . Larval characters will not do.")[134] In November 1884, *Papilio* was shut down. Three years before, William Henry Edwards had warned Henry Edwards that *Papilio* would "overtax" him and that he "would have to give it up," but when it finally happened, William was saddened, despite the silly attack on his views that had meant to validate Strecker.[135] He thought of "continuing Papilio myself," but had hardly enough funds to pay for volume 2 of his *Butterflies of North America,* let alone anything as expensive as a journal.

By the end of the 1870s, the hearts of many Americans were beating hard in response to the beauty of the world. Thousands were in pursuit of lepidoptera, a manifestation of that beauty: some collecting specimens for their own sake, others to get insight into natural diversity, or to achieve scientific understanding, or to help others find their way. All adopted the tools of collecting—above all, the net. Collecting was the first step in assembling insects into a coherent whole, in preparation for study, reflection, and understanding. It was an indispensable entry point into experiencing the many-faceted forms, shapes, designs, and colors of nature. At the same time, like a thief in the night, collecting and everything tied to it (naming, identifying, and classification) became occasions for rivalry and conflict, even greed and villainy. Some American butterfly people entertained serious doubts about collecting for beauty rather than for science, thereby forfeiting the satisfying wholeness of an older approach. In the process, two apostles of aesthetic pleasure, men joined at the hip by the same impulses, were in effect lost: Herman Strecker, who would never write another significant book on butterflies, and Augustus Grote, who would spend the rest of his life in Germany.

FOUR

Word Power

The emotional toll inflicted by the conflicts of the butterfly people proved costly, but it should not obscure or detract from the character of what they achieved as naturalists. On the most basic level, that achievement took shape as lists of words, lists of American butterflies and moths tracked and recorded by diligent butterfly detectives, who drew out of the shadows for the first time an account of the wonderful species life of the North American continent. At the next stage were the first guidebooks, butterfly manuals, and textbooks, for young and old alike, at once instructive and entertaining. Then followed a steady outpouring of periodicals and journals, popular and scientific, presenting a new world of knowledge about butterflies. At the summit were the grandest word-carriers of all, the catalogs by the primary butterfly people. Creativity squandered in quarrels and backbiting was redeemed in books of beauty and distinction, in the excellence of the writing and illustration, the very best demonstrating an integration of art with science.

All the butterfly people understood the power of words and books, with each of their catalogs stamped by an identifiable character as unique as the men themselves. Thus, Herman Strecker produced a feisty, combative book, every feature conceived and executed by himself, and Samuel Scudder an imposing, huge systematics leavened with poetry and lucid reflection. Augustus Grote mixed the poet's voice with that of the philosopher and naturalist; William Henry Edwards constructed an Apollonian interplay of word and image. Taxonomic dullness sometimes marred their work, but, more often, their volumes burst with energy and life. At home, in Cambridge, Scudder had unusual access to books from around the world, and he incorporated what he absorbed into a full vision of nature. He loved big canvases, and his were among the biggest ever realized by a naturalist, reflecting an intimidating quantity of hard work. At least until the railroad offered regular passage northeast from

Coalburgh, West Virginia, to Philadelphia and Washington, Edwards had a much smaller library than Scudder. "I work under great disadvantages here," he complained in 1871 to Hermann Hagen, "in the absence of libraries."[1] But less, for Edwards, may have meant more, freeing him to read wisely and comprehend thoroughly, and he did—much of Darwin read three times over, near equal amounts of Thomas Huxley, and, nonscientifically, all of Jane Austen's novels, six times over; he preferred her, as a writer, even to Darwin or Huxley, and his writing style may have been partly shaped by hers, testimony, perhaps, to the interrelationships of novel writing and modern natural history, each subtly influencing the other.[2] Grote and Strecker read more unconventionally than both Yankees, Grote in philosophy, sociology, poetry, and religion, Strecker in myth, poetry, art, and history, each injecting what he learned into his writing on butterflies and moths.

The masters of butterfly word power knew that if their work was to mean anything, it had to be housed between covers, whether by established publishers who cared about good writing and good science or by the butterfly people themselves, as in the case of both Strecker and Scudder. In 1878, after the American Entomological Society, which had published Edwards's first volume, refused to publish Strecker's catalog, rather than seek another publisher, he printed and issued his best book under his own name. Scudder, too, would break from his publisher to print his three-volume study at his own expense. Grote, Edwards, Scudder, and Strecker, along with a legion of others in the United States and across the Atlantic, were confident that what they did had purpose and made sense and that words—especially words with pictures—had power.

Making lists of American butterflies, as of other beasts, had wide currency among naturalists in the 1880s and beyond, perhaps best represented by one of William Henry Edwards's dearest friends, a tireless list maker, Joseph Lintner. The son of German immigrants, he was born in Schoharie, New York, in 1822, the same year as Edwards and very near Edwards's own birthplace in Hunter. Lintner began his insect work later in life than most others, after years in the woolen business.[3] At thirty-nine, he met Edwards, who saw at once that Lintner was "too much absorbed" in butterflies to stay much longer in wool. By 1880, at fifty-eight, Lintner was a seasoned, full-time naturalist at the newly reopened New York State Museum of Natural History, in Albany, and the state entomologist for New York; in close touch with farming, he became an exponent of economic entomology intent on wiping away insects as pests and villains. Yet he remained all the while far more engaged with insects for their own sake.[4] His painstaking reports, produced entirely by himself in a small, stuffy office, displayed a wealth of fresh insight into butterflies and moths, with eloquent encouragements to children to treat all "creeping and crawling" things with respect. "Ignorance of everything that creeps must be avoided," he wrote, because it "tends to the development of cruelty towards lower forms of nature."[5]

In the 1860s, Lintner and another German-American and Albany butterfly man, Otto Meske, hunted together in a wet and boggy place called Center, forty square miles of "a butterfly good time" (to quote Walt Whitman) midway between Albany and Schenectady, along the line of the New York Central Railroad. Samuel Scudder had walked his "blue roads" in the White Mountains of New Hampshire; Lintner waded through the "blue air" at Center, concocted, he observed, out of the swarming azure blues "driven up from the damp sands by our approach." By the late 1870s, Center was a "famous collecting ground," sheltering a rare blue species Vladimir Nabokov would later name *Lycaeides melissa samuelis* for Scudder himself, today known as the Karner blue (for the town of Karner, as Center is now known; it is a protected site in the Albany Pine Bush preserve).[6] In 1872, Lintner penned the earliest list of butterflies and moths indigenous to New York State, documenting more than one hundred species—fifteen more, Lintner boasted, than Scudder had found four years earlier for all of New England.[7]

As the New York State entomologist, Lintner urged collectors everywhere, "in nearly every State in Union," to furnish for publication "authenticated lists of Lepidoptera" known to exist in their neighborhood. He appealed to them to reveal their findings, proposing as a model a list by Roland Thatcher of Massachusetts in *Psyche,* with data on more than 300 species of butterflies and moths found in Newton and its vicinity—a fantastic tally, perhaps, but many were moths.[8] Possibly in response to Lintner's call, in 1882, Eugene Pilate of Ohio, a man with a large collection, delivered to the journal *Papilio* a list of 450 species of Lepidoptera (72 butterflies) from the Dayton area; and Charles Fernald, a moth fanatic and an educator at the University of Maine, published three lists on all the known moths and butterflies of Maine.[9] Another compiler, a young Philadelphia physician, Henry Skinner, cut back on medicine for butterflies and soon emerged as a commanding presence in the butterfly world. By the early 1880s, he had raised "nearly all the diurnals" (day-flying butterflies) in his area, studying "them from egg to imago in three summers," most caught in or near Fairmount Park, the spacious pastoral site in Philadelphia where the 1876 Centennial Exposition had been held. "We feel certain," he wrote, "that this list of eighty-seven species is a remarkable one for so restricted a locality," and "we do not suppose that it is entirely complete."[10] In Wisconsin, Philo Romayne Hoy erected in his backyard a "little house expressly for his cabinet," with glass-enclosed cases for his butterflies and moths, in splendid systematic order, and all designed so that people could comfortably see his bugs. He prepared the first lists of Wisconsin butterflies, one for the *Geology of the Wisconsin Survey* (1883), another for the *Canadian Entomologist* (1884). Scudder turned to Hoy for data.[11] Lintner, meanwhile, continued with checklists of his own, above all for the vast Adirondack region of New York, "perhaps second only to the White Mountains in point of interest," though as late as 1879, its "entire entomological wealth" was still untapped and unstudied. In his state report that year, Lintner began to erase this neglect with an account by W. W. Hill of the more than 250 species of butterflies and moths inhabiting the area, scratching the surface of what Lintner suspected was there. In 1880, Lintner himself finished a more complete summary, "The Lepidoptera of the Adirondack Region," but this list was also incomplete, due to the area's impenetrability as a wilderness. Even in 2010, Arthur Shaprio, a well-known California butterfly man, observed that "the butterfly fauna, let alone the moths, remains only partly known!"[12]

Instructional literature for young and old alike, from the first textbooks and guidebooks to the earliest children's books, reached the public at the same time as the lists. As far back as 1870, Henry Edwards had planned to write a "textbook of our butterflies, so that the young people springing up may be able to recognize their captures." He'd hoped, also, to finish a monograph on a family of lovely clear-winged moths, but neither book appeared, the latter because he could not afford color plates. In 1889, Edwards did complete the *Catalogue of the Described Transformations of North American Lepidoptera,* written for the Smithsonian, but it, too, lacked images. Edwards came closest to getting his desired plates in 1884, when he consented to write a chapter on butterflies for *The Standard Natural History*, a multivolume study published by Samuel Cassino, an entrepreneurial naturalist with several other nature books to his credit. Besides Edwards's, there were five chapters on other insects in the Cassino *History* by American experts, all testifying to the extraordinary strides Americans had made in entomology since the Civil War. Edwards's chapter, with its black-and-white drawings and plates, had the earmarks of his special delight, the words "very beautiful" or "most beautiful" appearing repeatedly. He had much to say about American butterflies ("our fauna"), but his main interest was in the world's lepidoptera, especially those that "have no *special* home but are scattered over the whole surface of the globe"—like Edwards himself, an emigrant from England to Australia to San Francisco to New York City, the tension between home and homelessness mirrored in his favorite butterfly families: the fritillaries, the blues, the skippers, and the swallowtails, all with worldwide distribution. His account ended with a tribute to the "glorious Ornithoptera" or the luxuriant birdwing butterflies of southeast Asia, coupled with a long quotation from Alfred Russel Wallace's reminiscence of the way his heart "beat violently" in Batchian when he captured the golden birdwing, *Ornithoptera croesus*.[13]

In 1886, Cassino recruited Edwards to write a "popular entomology book for the large number of students of this subject," to be issued first "in parts at a dollar each," then as an inexpensive book. He entreated William Henry Edwards to do the same, but William had come to mistrust Cassino, ever since the late 1870s when Cassino had promised to sell "a large lot of Natural History books" for Edwards that Edwards had wanted to get off his hands. Both men were to share the profits. But the Boston businessman kept most of the books for himself, never giving Edwards a penny, something Edwards learned only too late. "I fancy, if I pressed him at law now," William wrote Henry, "he would plead the Statute of Limitations. He is a great rascal."[14] This letter effectively erased any hope that either man would ever publish a "popular book on entomology" for Cassino. But two years later, John Comstock, a creative young professor at Cornell University, managed to marshal private funds to bring just such a book to fruition; called *Introduction to Entomology* (1888), it covered the "elementary principles" of insect life. Comstock, a Wisconsin native, had grown up poor and orphaned and, as a young man, had made his living as a cook on merchant vessels on the Great Lakes, but at Cornell, he formed one of America's earliest departments of entomology and became one of its most original naturalists, with a talent and passion for Darwinian science to match that of William Henry Edwards. His book remained standard for one hundred years, passing through ten editions, with many woodcuts done by his brilliant wife, Anna Botsford Comstock, who acquired new engraving methods at New York's Cooper Union in the mid-1880s. The 1888 edition had nothing in it about butterflies, but Comstock would remove this deficit five years later with "Evolution and Taxonomy," a long and bold essay all about butterflies, especially about the wings of butterflies; this, too, was destined to become standard.[15]

Contemporary with these texts were the earliest compact guidebooks intended for young people, which, like so much else entomologically speaking in the United States, piggybacked on an already mature tradition in France, Germany, Britain, and elsewhere, begun, perhaps, when Sweden's Linnaeus printed his own botanic handbooks for his students in the 1740s.[16] In Germany alone, well over fifty books on German butterflies had been published in various towns, many for the beginning collector, offering instruction on collecting and on gear (nets of different sizes, mounting boards, pins, cork) and on rearing and breeding. Calendars and diaries indicated where and when to find particular species throughout the year.[17]

England pioneered the first guide to reach a mass audience, Henry Stainton's *Manual of British Butterflies and Moths* (1857), complete with biographies of species and collecting places, written in lucid prose, but unillustrated.[18] After 1880, *pictorial* guides in England dealing with many entomological fields "grew in number astonishingly," according to Theodore Cocherell, an English immigrant to America in the 1880s, a skilled observer and butterfly man, and an expert on bees and on much other natural life, as well. Cocherell considered guides on butterflies and other organisms actually "*superior* to class instruction" because they made young people think for themselves and required that they go out into the fields to collect. "I believe a strong argument could be presented," Cocherell said, "for the abandonment of formal instruction in science as a means of education, except in relation to manifest utilities and technical trades, and the substitution of something more like the apparently haphazard method of the English amateur."[19]

In 1880, demand was still insufficient to support a genuine butterfly guide. "It is too bad," William Walters, a butterfly man and book dealer in New York City, wrote Strecker that year, "that America, the greatest Butterfly Country in the world, has no book to encourage its people to study its beauty and wonders." "What is wanted is the *names* of our Butterflies, so that our children can know them." For a while, Walters thought he might do it. "Like yourself," he told Strecker, "I have to work hard for my corned-beef and cabbage, but I *make time* for little recreation such as this." Still, the task was too great, and he urged Strecker to write the book instead. "You are at the pinnacle, and I am at the foot of the ladder," but "if I had *your* knowledge and collection, I could do a book that would make the children rise up and call me blessed. What a book you could compile, if you would."[20] Snowed in or over by work, family, and collecting, and perhaps dispirited by months of entomological sniping, Strecker snubbed the proposal.

Others acted more decisively, among them Helen Conant and Julia Ballard, who wrote books for the children's market: *The Butterfly Hunters* (1881) and *Moths and Butterflies* (1880, revised in 1889), respectively. Conant crafted each of her chapters around a single popular American butterfly, such as the mourning cloak or tiger swallowtail, that the children could find in the fields and byways. She explained taxonomy, noting how "moths are divided into two great classes—Hawk-moths, or Sphinxes, and Moths, or Phalaenae," these, in turn, "subdivided into many smaller classes in regard to which nearly all naturalists differ in opinion."[21] Ballard, the wife of an old college chum of William Henry Edwards's, took a bit more humorous approach. She had a butterfly narrate its own life history: one chapter on the pupa, one on metamorphosis, and so forth. "I am only a day old," the butterfly says in chapter 1. "I was born in a prison," and though "I can see right through my walls, I can't find any door."[22] Both women, unblinking about killing, readily described how to end insects' lives, Ballard, with a little more clinical precision, as if following a recipe for making pudding: "Take a glass jar with a large mouth and closed lid, perhaps your own candy jar, into which put four or five lumps of cyanide of potassium. Dissolve enough plaster of Paris in water to cover the cyanide evenly over, forming a hard, smooth surface. Put the moth (or butterfly) into the jar, close the lid, and let it remain five or six hours, after which it can be taken out and mounted."[23]

George French, an Illinois professor and a faithful follower of the entomology of William Henry Edwards, produced America's first full-scale reliable guidebook, *Butterflies of the Eastern United States,* a far more substantial work than Conant's or Ballard's, although devoid of their lightness of touch and, unfortunately, nowhere near as successful as its European predecessors. Designed mainly for students in zoology— who had "clamored" for it, according to French—it had the virtue of being small (one could stuff it into a knapsack), with the latest knowledge compactly delivered, arousing one butterfly man to recommend it to a friend as "another gift of the period for young entomologists."[24] French borrowed from William Henry Edwards in every way that mattered, from the formal ordering of butterflies (swallowtails first, skippers last) to the life histories. Not surprisingly, Edwards told French in a letter, "The more I have read the book, the better I like it. It is exactly what beginners and many who are not beginners want."[25] Scudder, on the other hand, in a review of French's book in *Science,* faulted "the book's arrangement" as "unnatural, holding its ground only as a legacy from the less-informed authors of fifty years ago. . . . It is but the rehabilitation of the dry husks of a past generation," he insisted. "We fail to see how the work can be of any pedagogical service, although this is claimed as its chief end."[26] Holding his fire on Edwards, Scudder had decided to pummel one of his proxies instead.

The periodical press fed this growing outpouring of words, with frequent forays by the broad literary magazines, such as *Harper's* and the *Atlantic Monthly,* into the butterfly world. The naturalists themselves, of course, had more to say, beginning just before the Civil War with the *American Naturalist* in 1867, edited by Alpheus Packard, the author of the first general insect guidebook, and Scudder's *Psyche* in 1868, followed by *Papilio*, the *Canadian Entomologist, Popular Science Monthly, Science Magazine, Insect Life, Entomologica Americana*, and *Entomological News*. Altogether, these periodicals surveyed every facet of the natural world, seeking to "popularize the best results of scientific study." All had a national and, most, a transnational character ("for science is cosmopolitan," Packard observed).[27]

Brilliant in many ways but spoiled by feuds, by weak subscriptions, and by a policy of sending out free copies, *Papilio* lasted only three years. The *Canadian Entomologist*, subsidized early on by the Canadian government and still published today (as are *Psyche, Science*, and the *American Naturalist*), was edited by William Saunders and Charles Bethune, both Canadians, and remained the bible of Americans right up to 1900. "More than any other similar undertaking," Grote observed in 1886, "the publication of the *Canadian Entomologist* has assisted the progress of Entomology in America."[28]

Popular Science addressed the whole panoply of science, issuing news on butterflies and moths, if only in spurts, while *Science Magazine*, one of the most venerated such journals ever published, was more generous with up-to-date information on butterfly work, largely because, between 1882 and 1885, Samuel Scudder served as its editor in chief. Thomas Edison had funded an earlier, short-lived version of *Science*, and Scudder revived it, after getting backing from another rich inventor (and relative by marriage), Alexander Graham Bell.[29] Free to proceed in any way he wished, with no one meddling, Scudder imposed a new agenda on *Science*, departing from Edison's focus on electricity and totally in line with the vision of Alexander Humboldt surveying the map of science to a degree unimaginable today, from physics and chemistry to meteorology and entomology (reviews of the work of Grote and Edwards, and of other butterfly people, regularly appeared there). He wrote weekly summaries himself in every area, "rejoicing the hearts of scientific men of the day," as one of them put it, but *Science*, too, had a short life, lasting only until 1885, so financially vulnerable that, in the winter of 1885, Scudder complained to Bell that Bell owed him $500 and he (Scudder) was thinking about jumping ship.[30] Offended, Bell wrote "Surely—my dear Mr. Scudder—you cannot mean to convey the idea that, so far as *Science* is concerned, the main thing to you is *money—regularly paid!*

I am sure that at heart you really wish to see the child of your creation grow to manhood and prove a power in the land." Very likely Scudder did not care about the money so much as about Bell's plans to centralize the magazine in Washington, meaning Scudder would have to leave Cambridge, the heart of his life in countless ways, and so Scudder quit, and *Science* went down again, to be resurrected in 1890, this time with stable backing (and an established subscription list) from the American Association for the Advancement of Science.

Insect Life, a government journal popular well beyond Washington, was created by Charles Valentine Riley, the Darwinian head of the U.S. Bureau of Entomology inside the Department of Agriculture, in 1888 and published until his death in 1895. As an instrument of federal government policy, it was, from the outset, economically driven, and in time would help take the country aggressively away from natural history. In *Insect Life*'s early years, however, Riley had invited naturalists to write pieces on butterflies for their own sake; in response, Lord Walsingham, an English aristocrat and butterfly man, discussed his favorite microlepidoptera; A. H. Swinton (also English) explained the "stridulations," or grating sounds, made by the mourning cloak butterfly, a lovely maroon-colored species first on the wing in the spring; and Henry Edwards recalled hearing, as a youth in England, the nearly imperceptible "rasping sound" emitted "by the beautiful *Vanessa io,* a large moth with striking eyespots, when several flew together, or when a male was in hot pursuit after the opposite sex." One had to listen carefully, Edwards explained, "when all around was still," lest one miss "the insect's expression of love."[31]

Entomologica Americana, emerging phoenixlike from the ashes of *Papilio,* was published by the Brooklyn Entomological Society throughout the 1880s and edited by a hulking, young, tendentious German-American lepidopterist from Brooklyn, John Smith. The society's members—Smith, George Hulst, Edward Graef, Franz Schaupp, and others—had long felt estranged from the Manhattan naturalists led by Henry Edwards, Neumoegen, Grote, and Mead.[32] The Brooklyn journal resembled both the *American Naturalist* and *Psyche* in its breadth, while yielding prime turf to butterflies and moths. Upset by the way *Papilio* met its demise, William Henry Edwards disparaged the new journal as a "mongrel" creation of "ill-informed entomologists" and refused to publish in it. "No one connected with the Brooklyn Society is able to run such a magazine," he decided. "I will send my papers henceforth to the *Canadian Entomologist*."[33] But, unlike *Papilio, Entomologica Americana* survived for several more decades.

Backed by funds from the American Entomological Society and founded in 1889, *Entomological News* soon prospered in the butterfly world, especially under its long-lasting editor, the medically trained Henry Skinner, who only a few years before had counted the butterfly species in Philadelphia's Fairmount Park. Skinner, appointed curator of butterflies at Philadelphia's Academy of Natural Sciences, lived part-time on a sprawling farm just outside the city, where he planted corn, melons, and peaches.[34] He was not the first editor of *Entomological News;* Eugene Aaron was, after serving as the last editor of *Papilio*. But Aaron proved unreliable and, later in the 1890s, abruptly quit the *News* for a career as a "quack cancer specialist in New York City," Skinner wrote Strecker.[35] Skinner then "took full charge," although he had long been at the helm de facto, making the *News* the best written, best illustrated, and best printed periodical of its kind. "We now give more for the money than any journal [on insects] in the world."[36] Skinner's journal cast light on several insect groups but on butterflies abundantly, due in large measure to Skinner himself; the *News* navigated the globe from Calcutta to Paris, Berlin to the Dutch Indies (as did Scudder's *Psyche,* but not to the same degree), and did more than any previous serial to cultivate collecting and to instruct thousands in the elementary basics of entomology. Skinner made reading it "a necessity to every student of insect life," "guided entirely," he wrote in one of his later editorials, "by an unselfish love for our interesting study."[37] He kept it in touch with the Romantic tradition, making only grudging concession in the beginning to the merely clinical approach to butterflies, or for what he called "the dry details of descriptive and classificatory work."[38]

Skinner had a kind and sympathetic nature, befriending the sometimes friendless stonecutter, whose 1878 *Lepidoptera* he greatly admired. He and Strecker exchanged hundreds of letters, as well as hundreds of butterflies and moths, both native and exotic. While Skinner sent Strecker spectacular moth cocoons (most obtained from nearby farms in New Jersey), which Strecker immediately recycled throughout Europe, Strecker sent Skinner the most stunning exotics from southeast Asia. The men regularly visited each other's homes, and Skinner welcomed his friend back to the Academy of Natural Sciences, although Strecker seemed never inclined to visit that place again, having earlier been blackballed as a thief.

On its face, a catalog may seem dreary and lifeless, and many of those on natural history subjects did seem so, at least from the late eighteenth century on, in their presentation of name, synonymy, taxonomic description, classification, and so forth. Yet catalogs were also often fluid and open-ended, ideally suited to depict the staggering number of new species and genera of butterflies from throughout the world that Western naturalists came to know. Over a brief twenty-year period, from 1870 to 1890, Americans wrote fascinating catalogs in all sizes, many even more complex than their European predecessors. It was a culminating moment in the history of the engagement of Americans with their own fauna, a "Renascence," Grote called it, "during which a great deal of work was performed with good humor and at considerable self-sacrifice."[39]

The most impressive volumes were compiled by the main butterfly men, although one middle-aged New Englander, Charles Maynard, who sold birds' eggs, stuffed birds, minerals, and butterflies in a small natural history shop in Boston, tried to compete with them. In 1886 he issued a quarto-sized catalog, *The Butterflies of New England,* with 108 descriptions and eight lithographic colored plates, all his own, "for the student of whatever age and sex, to aid in identification of the beautiful, aerial, and almost evanescent forms, which haunt woodlands, fields, and meadows, during the brief summer months of our vigorous climate."[40] The book seemed to one-up Scudder's 1889 study, but it actually derived in large part from Scudder's own system of classification and adopted many of Scudder's vernacular names. Maynard coined his own names, too, such as "Quaker Butterflies" for the wood nymphs, loping low-fliers usually with eyespots on the wings, because "all in this genus are grayish in color."[41] He ignored, for the most part, the early butterfly stages; opposed giving names to polymorphic forms (something Edwards had insisted upon); and cared nothing about priority in naming, "frankly confessing" his impatience with nomenclature, since "we have arrived at a point where we must choose between wasting our energy upon what is merely secondary to the object in view or advancing scientific knowledge in the study of objects themselves." He defended readable natural history, hoping to cultivate interest in the beautiful things of nature.

Scudder met Maynard in the mid-1860s, when both were hired as young men at the Boston Society of Natural History. Maynard stayed only a year, to help curate a bird collection, and Scudder held various executive posts over a period of twenty years.[42] He never reviewed Maynard, although he was well disposed toward him, and quoted freely from his catalog later in the decade in his own big book on butterflies. William Henry Edward, on the other hand, picked the book to pieces in a review for the *Canadian Entomologist*, despite Maynard's very Edwardsian reservation about Scudder's nomenclature and his frequent and favorable citations of many of Edwards's descriptions. The book got everything wrong, Edwards said: the colors, the classification (which was Scudder's), the numbers of molts and broods, the food plants. "The text, so far as it is correct, is worse than nothing." Edwards also disparaged the vernacular names. "In Europe before the binomial nomenclature was invented, it was natural that there should be local names for such striking objects as butterflies. But these names have nowhere come to be commonly used here. No one but contrivers use them; they do not stick to the insect." "The field is open for a well-illustrated book on the same butterflies," Edwards granted, but it must be "written by one who is well-acquainted with his subject."[43] Edwards could have been more temperate, having nothing to fear from Maynard, and the same might have been said of Scudder regarding French. It is a wonder, given the readiness of these extraordinary Yankees to plunge like hawks on prey (one from above, the other from the side) that anyone else would risk writing a new book at all on what Scudder later called "the frail children of the air."

In 1886 Augustus Grote, residing in Germany, told Scudder that he had nearly finished a "big popular book on North American Butterflies and Moths, and I really think it might be called the Butterflies and Moths of North America." "I think everybody in the line will buy it and want to read it."[44] Grote had been living in Bremen, suspended between two worlds: on the one hand, marrying the daughter of a wealthy Dutch tobacco merchant and taking part in the city's cultural life, and on the other hand, "homesick nigh to crying."[45] "Sometimes I smell the Staten Island woods [where his parents were buried] and then I see Lake Erie in my dreams! Those two bits of water, the New York Bay and Lake Erie, seem to draw me to them, especially the Lake."[46] His big book never surfaced, perhaps because his "exile," as he called it, had sapped his energies or because his multiple selves—poet, musician, sociologist of religion, entomologist—pulled him in too many directions. The fact is, the only coherent book he ever wrote—*New Infidelity*—had nothing to do with lepidoptera. A large-scale venture was beyond him. Nonetheless, he did complete two slender catalogs, or extended essays, the first, in 1882, *An Illustrated Essay on the Noctuidae of North America,* with colored plates of forty-five moths, and the second, in 1886, *The Hawk Moths of North America,* the skeleton of the never-to-be study, with an unexpected portrait of "the man of science," who "observes the small changes which underlie the endless succession of life. It is clear to him *how* we are drifting if, with the rest of humanity, he does not know *where*. Within certain limits he believes that the will of man counts for something and that, in the perpetual struggle, that which is useful, good and beautiful will prevail."[47]

Grote's constant stream of articles was more imaginative than his catalogs, including his engaging "Moths and Moth-Catchers," in two installments in *Popular Science Monthly,* in 1885.[48] He became so identified with the *Canadian Entomologist* that he was called on to write the memorial poem marking its "quarter century."[49] Unlike Edwards, whose latest fieldwork appeared in those pages, Grote presented mostly theoretical or philosophical analysis. He knew, of course, how crucial fieldwork was to his science, and he criticized "a tendency in Europe" to emphasize "a sort of book working of nature" over "living material," but after leaving Buffalo, he did little fieldwork himself and depended on others to harvest insect material for him.[50] In the *Canadian Entomologist,* he defended the vernacular in the naming of insects that Edwards always faulted and Scudder celebrated, and he considered at length the structural basis for reclassification of the principal moth families. He reflected on the geographical distribution of insects, the forces that induced migration, the transforming power of climate, the ways in which the American species of moths and butterflies generally were fashioned by the world around them and by evolutionary change—all themes basic to a bioecology of insects. He may not have produced the big popular book, but his output matched on many small canvases the catalogs of the others.[51]

Herman Strecker, on the other hand, did complete two significant catalogs in the late 1870s. The first, *Butterflies and Moths of North America: A Complete Synonymical Guide of the Macrolepidoptera,* was dedicated to William Hewitson in "loving remembrance" and listed all the names given to American butterflies since Linnaeus, along with their localities. Though it lacked satisfactory descriptions and had no color plates, it did contain black-and-white illustrations of butterfly morphology by Strecker himself and an informative bibliography of writings on butterflies by both foreigners and Americans.[52] It showed amateurs how to collect, mount, preserve, and store lepidoptera, where best to find them; and how to rear them from eggs and larvae, protect them against pests, pack and mail them, and label specimens correctly (including date of capture, and synonyms, if necessary).

Strecker's second publication was his 1878 catalog, *Lepidoptera: Rhopaloceres et Heteroceres,* which his subscribers combined into a single volume, integrating all the parts he had issued since 1873. The original plates were the volume's most remarkable feature, in the tradition of European craft and rivaling the more scientifically oriented, precise art of Mary Peart and Lydia Bowen in Edwards's volume. Nearly all the species were American, except for a few tropical ones, such as *Papilio marchandii,* from Guatemala, which stood out, amid several other butterflies in the same plate, in an almost radiant orange. His first plate was of an American silk moth, a big, handsome reddish-brown or purplish moth, *Samia gloveri,* named after Townsend Glover, a naturalist and the pioneer architect of the fledgling U.S. Department of Agriculture, who had given Strecker his type specimen. Strecker named several new underwing, or *Catocala,* moths for the first time, including *Catocala agrippina, Catocala sappho,* or *Catocala amestris,* honoring three famous women of classical ancient Greece and Rome, and arranged the images into ensembles on three separate plates to emphasize the brightness of the colors, the yellow, pink, orange, jet black, white, and crimson of the underwings. Strecker illustrated American swallowtails, hawk moths, coppers, and blues, but despite promising in his "advertisement" to depict larvae, pupae, and eggs "whenever possible," all he could manage was one caterpillar at the bottom of one plate (his idea of "habitat" was "New York" or "Texas"). He cared most for "the picture" of the adult form because it carried, in his mind, the principal imaginative value, and he faulted naturalists for countless words in descriptions "no one ever reads." Words had power but only in relation to the image he felt—an anthem struck as well by William Henry Edwards, Scudder, and Grote. "Oh! that we could throw out every description that is unaccompanied by a figure," Strecker exclaimed, "how our labour would be lightened, how we would be spared the maledictions of after generations for all time to come. With what boundless veneration do we look on the tomes of Cramer, Seba, Drury, Huebner, Hewitson, and Herrich-Schaeffer, not winding into countless useless descriptions in all sorts of scattered periodicals, but a great massive work—grand, compact, solid, every description accompanied by coloured figures. I never open these mighty volumes but feel my soul expand in Hallelujahs to the Almighty that through his great goodness such intellects were allowed to sojourn here and to bequeath to us the result of their vast labors."[53]

Lepidoptera conceived of butterflies and moths as parts of a unified field, not separate from the rest of nature and man. Strecker was not a naturalist as Scudder, Grote, and Edwards were; he had a limited interest in life histories and did most of his work in a butterfly room, not in the field. But he did have an ecological sensibility, like the other trailblazers, placing his butterflies in a larger context, if not in evolutionary or Darwinian terms. Strecker worshipped Hewitson, as well as Humboldt, and, like them, believed that the connections among natural forms came from some mystical energy that bound all together. In a section of *Lepidoptera* entitled "Entolomological Notes," he recounted a visit to the Smithsonian Institution of Natural History in Washington, where he saw many "pre-Adamite animals," octopuses in alcohol, the wood carvings of Indians, and the "mass of native copper" in the Geology Department. Now how, he asked, do these things relate to butterflies? In every way, he responded, for "each page of God's great book is connected with the other, bound in its mighty cover the Universe, and we cannot admire one without admiring the other; we do not love our mistress' hands alone, but also her brow, hair and eyes, her whole beautiful form, the entire faultless work."[54]

Strecker was thinking of other relationships, in a metaphorical sense. When he described the orange, yellow, and black Guatemalan exotic *Papilio marchandii,* he placed that butterfly—or, more exactly, its appearance and *its name*—inside a stream of dreamlike associations. "I have never looked at this lovely thing, with its delicate form and brilliant hue," he wrote,

> without my thoughts reverting to the long past builders of the temples and altars of Palenque and Copan, the butterfly flitting through the tropical groves of their day, as now, but the inhabitants of the old dead cities have passed away, their names, their history unknown! birds, reptiles, and insects now alone tenant the forest where once stood the populous cities, the kings and priests of which, with their slaves and sycophants, long ages ago have gone to rest; naught remains of their past greatness but the moss-coated and timeworn ruins of altar and idol, and the frail golden butterfly hovers, suspended in mid air, over the monster face of some fallen Dagon [an ancient fertility god] which far back beyond even the "night of time," received its meed of human sacrifice.

Other butterflies conjured Xanadu or "visions in a dream." "What a flood of thought suggests itself when we gaze on the gorgeous Ornithoptera Priamus! The court of the old Trojan King arises and is 'followed fast and faster' by each varied scene of the Iliad." And how the "Ornithoptera Croesus reminds in an instant of the magnificence of the Lydian monarch and the death of the hapless Atys."[55] Butterflies could be understood not only in natural ecological terms but in epic aesthetic-poetic ones as well.

For the depth and variety of his life-history studies, William Henry Edwards had no rival, and Samuel Scudder created the crowning achievement of the era on American lepidoptera, touching every aspect of butterfly existence. Both men built monuments to American butterflies as well as to the American naturalist amateurs, or the hod carriers who helped them research and write their books.

Many Americans, outside the naturalist community disliked the Linnaean approach to nature. Walt Whitman, on one of his many walks down the farm lanes in Brooklyn or in New Jersey, imagined how good it would be if nature had no names (which is peculiar, given how he loved to catalog the American experience). "Many birds, I cannot name," he wrote in *Specimen Days and Collect,* "but I do not very particularly seek information. . . . You must not know too much," he said, "or be too precise or scientific about birds and trees and flowers and water craft; a certain free margin, and even vagueness—perhaps ignorance, credulity—helps your enjoyment of these things. I repeat it—don't want to know too exactly or the reasons why." Even Thoreau, writing in his 1860 journal, disparaged the "science" that reduced animals to dry shells with recondite Latinate names. What he found "most requisite in describing an animal" was to capture "its vital spirit," or "anima." Science, however, tended to ignore the "living creature." "A history of animated nature must itself be animated."[56]

Scudder and Edwards (as well as Strecker and Grote) never went as far in their criticism as Whitman and Thoreau; nor did they ever cast aside Linnaeus. What they did do was to combine taxonomy with natural history, the dead with the "living creature," as Buffon, Humboldt, and Linnaeus himself championed it. Often, they got tangled up in the nitpicking of pure naming and classifying, the lumpers (those who preferred to combine various forms into one species) and the splitters (those who tended to declare each form a species) fighting it out over fractions and millimeters. But essentially they had the same deep interest in living things as Whitman and Thoreau.

After the collapse of his West Virginia coal business in 1874, Edwards had worked like a demon on his insects, supporting his family partly with revenues from selling parcels of his land. By 1875 he had arranged with the Chesapeake and Ohio Railway to get his own depot near his house (he could see it from his living room), giving him ready access to the big eastern cities and a speedier mail route for his larvae, eggs, and imagoes. Ten years later, the post took only two days to arrive from New York or Washington, D.C., three days in the other direction; and in an entire twenty-year period, 1868 to 1888, he did not lose a single express package and barely a bug (although insects hatched from eggs in transit and often died because of the lack of fresh food).[57] With rails continuing to penetrate to the whole country west of the Rockies, down to southern California, skilled collectors were able to travel regularly over the lines, trunk and branch, and if the ecological costs of such change might prove incalculable, scientists and naturalists reaped riches not only in experiences at the end of the line but in the safe transport of new material (dead and alive) for study.[58]

When the *Canadian Entomologist* asked Edwards to explain his entomological methods, he proposed breeding as a therapy against "despair" or "fear" that one might "go mad" with nothing to do. "Any one can take it up and follow it with a pleasure that becomes rapidly absorbing."[59] "I am deep in larvae. Never so much," he wrote his friend Joseph Lintner in July 1880. He would go nowhere between each June and September, lest he miss a clue to resolving some tiny mystery.[60] He criticized Europeans for looking only at "dried butterflies in cases" to establish species, and, to be sure, even in Britain, supposedly *the* nation of naturalists, life-history studies based on regular breeding had barely gotten under way.[61]

Edwards enshrined much of his fieldwork of the late 1870s and early 1880s in volume 2 of *The Butterflies of North America* (1884), by all accounts his best book and by some accounts the best book ever written on American butterflies. If volume 1 presented only a few life histories, or "stories," of butterflies, volume 2 glowed with the most complete accounts of well-known American butterflies, among them, the common sulphur, the wood nymphs, the alfalfa butterfly, the zebra heliconian, the baltimore checkerspot, the pearl crescent, and the tiger swallowtail. His systematics superceded anything done earlier, marked, above all, by a sophisticated Darwinian analysis that tied the history of butterflies to a larger evolutionary story, thereby making his book one of the best applications of evolutionary theory of the age. Scudder wrote that it "showed what one man, remote from associates, libraries, and even from much of his own field work, may accomplish."[62] There was a fullness to his life portraits, a keen awareness of connections to living places, a feeling for organic relationships, with each portrait benefiting, as never before, from the labors of many selfless people.

In the early twentieth century, the naturalist Fordyce Grinnell Jr. remembered Edwards as "a great *teacher*," able to arouse unusual "cooperation" in the interest of a common aim. For the *Lepidopterist* of 1917, he wrote, "We must all recognize W. H. Edwards as the greatest butterfly student which this country has ever produced or probably ever will. He described a good majority of our species; but his work on the life-histories was greater yet. The key to his great success in these two lines was his numerous correspondents in every part of the country to which he exhibited the greatest unselfishness in help and encouragement."[63] Edwards owed a great deal to knowledgeable women, sometimes chronically ill women, who worked out of their homes. Annie Wittfield, daughter of William Wittfield, a physician from Indian River, Florida, suffered from a heart ailment that would kill her at age twenty-three, but she found for Edwards caterpillars, eggs, and pupae of previously unknown species, which he described and Mary Peart figured. In the summer of 1887 Wittfield captured a dimorphic Mimic butterfly (*Hypolimnas misippus*), a tawny brown female first identified from a single specimen by Linnaeus and perhaps introduced into the Caribbean during the slave trade (the male was black with white spots on the forewings, edged in purple).[64] Her discovery of the insect's food plant (mallows and morning glory) and of its life stages stunned Edwards. "It is mostly owing to her zealous, friendly, and intelligent assistance," he wrote in a plaintive obituary of her in 1888, "that I have been able to learn the history of so many Florida species. Her death is a loss to science and to me. For several years she had been a correspondent of mine and gave me intelligent aid in obtaining eggs and in making observations on the habits of butterflies."[65]

A greater boon by far to Edwards was Mary Peart, his premier artist, who had begun working for (or *with*) him out of the love of it and without regard to payment, partly because Edwards, through many letters written over many years, treated her as an equal, and partly because she had matured as both an artist and naturalist, proud of her work, able not only to draw insect imagoes with great skill but the entire life histories of the butterflies as well. She *saw* things that even Edwards failed to see. After her marriage in 1876, when she was twenty-nine, to an Englishman, John Peart (same surname, but unrelated), she and her husband lived in a "nice, little house, such as you only find in Philadelphia," Edwards observed. He feared she would quit working for him and depend entirely on her husband, but she didn't. She couldn't. She "hated to resign her Plates to any one else," a relieved Edwards wrote to Henry Edwards. "I shall not find the like of this lady as an artist."[66] Of course, he paid her well, as he did his colorist, Lydia Bowen, although Bowen never ranked as high as Peart in Edwards's mind. By 1880, Peart and Edwards were nearly collaborators, became friends, visited, and even attended cat shows together in Philadelphia, seeing "no end of pretty kittens in all colors, from Angora to Siamese."[67] Edwards would tell Scudder in 1885 that his "discovery of Miss Peart" had proved "as important a find as a new planet almost."[68] In order to figure well, she had to breed well, and, just as Edwards did, she took hours to wait and watch as one butterfly after another passed through all its stages. Edwards himself often carried a hand lens on his walks, but was inept at using anything more advanced in his home (although he did borrow a microscope from Theodore Mead); Peart, on the other hand, had begun to achieve an expertise with the microscope, familiar with every new invention on the market. The most precise drawings of the tiny pupae of some of Edwards's rare insects were the result.[69]

Besides depending on the expertise of women naturalists, Edwards continued to rely, whenever possible, on Henry Edwards and Theodore Mead, well into the 1880s, but three other men—two émigrés from Britain, Thomas Bean and David Bruce, and one American, William Greenwood Wright—emerged as crucial to the success of Edwards's labors. The men formed a triad of stalwart Victorians willing to do nearly anything to assist Edwards in his work. A lumberman from San Bernardino, California, where he lived with his wife, Wright was past fifty when he began collecting. Gingerly he walked the uneven California coastlines looking for insects, scrambled over mountains and through canyons, and crossed the scorching Mohave Desert, just east of San Bernardino, or Death Valley along the Nevada border (the "mouth of hell"), often going alone with a covered wagon. He withstood the worst in nature, certain that however "desolate and comfortless" it seemed, it harbored as many secrets as "rustling leaves or babbling waters."[70]

Thomas Bean and David Bruce were like night and day, Bean single and shy, Bruce sociable and twice married, with six children.[71] When Bean was seven years old, in 1851, he came to America from England with his family to live in Galena, Illinois. His father died six years later, leaving Thomas's mother to support the family and Thomas himself, at thirteen, to clerk in a bookstore. He fought in the Civil War and returned home to a string of dreary jobs made bearable only by the insects that dressed the Galena meadows, especially his favorites—the fritillaries, the wood nymphs and satyrs, and the sulphurs. Sometime in 1883 he found work as the lone telegraph operator in a remote railroad outpost called Laggan, in Alberta, Canada, at the end of the line of the Canadian Pacific Railway Company. Laggan rested, he told a friend, "right on top of the divide within six miles of the actual summit" of a major mountain pass in the heart of the Canadian Rockies. Bean was soon picking butterflies off flowers or snatching them from the air by hand, until he bought a proper net. When he could, he camped out at extreme elevations, usually at the timberline and often in open air, but always with a fire burning, to collect the rarest specimens of the same butterflies he had known in Galena, but this time they were all *alpine* species.

Isolated and mostly alone, the monkish, religiously agnostic Bean matured into a brilliant alpine collector who acquired more knowledge "about the high altitude Rocky Mountains than anyone else."[72] He treated his butterflies with a tender humor. In a letter to Edwards, he described the female Alberta alpine fritillary as "habitually un-vexed, her flight deliberate." "She has a certain dignity of manner that commands respect. An air of speculation marks her, denoting a mind preoccupied with problems." The male arctic ridge fritillary, on the other hand, cruises the high "upper slopes of the mountains stealthily" and "seems always on the lookout for an entomologist, whose advent is carefully noted." "The female has no apparent preference for these extreme heights. She does not devote her valuable time to racing madly across the windy summits for the mere nonsense of the thing."[73]

David Bruce, too, knew the Rocky Mountains, although these were the Rockies of Colorado, the eastern part of which held a fair share of undescribed fauna. Born in Perth, Scotland, in 1833, he worked as a milliner, kept entomological diaries, and caught insects for Charles Darwin and Edward Doubleday. Around 1870 he sailed with his family to New York, moving to Brockport, an upstate village, there to paint houses out of necessity and to collect butterflies when he could. He inspired many "lads from around here" (as he put it) to take up "butterflying."[74] By the mid-1880s, with all his children grown, Bruce set off for Colorado, partly because it was healthier (he had asthma and rheumatoid arthritis) but mostly because of the lepidoptera hidden in the hills.

William Greenwood Wright curtailed his lumber business soon after his wife died of a long illness and occupied the rest of his life with butterflies. Bruce solicited subscriptions for his expeditions to the Rockies (otherwise he could not have gone), but he had no desire to make a profit from his butterflies. "It is just as necessary as food and drink that I observe and collect insects," he wrote another naturalist, "and while I am able to crawl, I shall do it. I don't make half a cent by it." Bruce was willing to buy insects from other collectors, as was Thomas Bean, but neither man would sell his own, not even to one another.[75]

Bean, Wright, and Bruce did what they did for science, for fame, "*con amore,*" in Edwards's words, and, above all, for Edwards himself, who made them feel at the center of modern natural science, just as Agassiz had done for Scudder. In letters delivered almost daily over many years, he taught them how to see and raise insects and about food plants and breeding boxes (his favorite were powder kegs). He urged them to plant butterfly gardens, and sent seeds to get the gardens under way. "Nothing is so good as single Zinnias," he explained to Bean, when he first met him, in the spring of 1875, but "they are hard to get, as the fashion runs to double ones, which are worthless for the purpose."[76] When they were down emotionally, he picked them up. ("You are a *real* lepidopterist," he told Bean, who had briefly lost faith in himself.) He published their thoughts and descriptions, in generous chunks of their prose, in his books, everything they could send him on "the habits and localities" of the butterflies they specialized in. He commiserated with them when their caterpillars died after being tended with the utmost care for several days, even weeks. "Sorry to hear of the demise of [your little alpine]," Edwards wrote Bean. "But in this transitory world, life is uncertain, caterpillar or mammal. Take the more care of those who are left." Edwards promised these men a place in history, and even more. "Can't you immortalize yourself," he inquired of Bean. "I do exceedingly wish to give you credit for your hard work and in your own language, with quotation marks. Magazines are ephemeral, but the Butterflies of North America goes down to the ages."[77]

In exchange for immortality, these men fell down shafts in gold mines, stumbled through matted bogs, and tumbled over cliffs, breaking arms and legs. Bruce was "the greatest aid to myself," wrote Edwards, "always ready to do everything to get eggs. He goes out every year and gets rare butterflies."[78] Though hobbled by arthritis, he scaled the peak of Mount Bullion, in the Rockies, in 1887, many thousands of feet above sea level, seeking the eggs of a satin-colored insect, the rockslide alpine, "the most difficult to capture of all our native butterflies from the nature of its habitat," so named because it flew only above the timberline in places known for crashing, precipitous rock slides.[79] "The collector cannot follow it," Bruce reported, "and when it is at rest on the black rocks it is almost invisible." Inching along the darkened mountain terrain on his hands and knees, he managed to secure only two eggs, nearly buried within rock crevices, and these he jubilantly mailed to Edwards.[80] "No one but Bruce would have stuck it out and accomplished the result wanted," Edwards recounted to Scudder in 1895. "He has borne in mind the need of the Butt. of N.A. and generally without payment, or other rewards than my thanks and his."[81] As for Bean, he delivered "the eggs of so many butterflies of exceptional rarity" to Edwards that his knowledge of them exceeded that of "anyone else in Canada."[82] Wright took "a vast deal of trouble expressly to aid me in making known the history of *Papilio rutulus*" (the western tiger swallowtail), Edwards recounted, consuming "days and weeks of experiment, and many disheartening failures." Wright himself remembered that it was because of Edwards's "instigation and encouragement" that he remained a butterfly collector.[83] What Edwards always wanted, what he "yearned for," as he put it, more than anything else in the world, were the eggs, so that he might follow the butterflies through their life histories. And here Bruce, Bean, and Wright surely satisfied him by sending thousands of eggs over the years, inserted into pen quills, sheltered in cork, squeezed gently down small glass tubes, or, in Wright's case, carefully dropped into the leftover morphine bottles he had used to care for his wife in her last illness.[84]

Even before the ink had dried on volume 2, Edwards began volume 3, in October 1883, hoping its completion would make him "go out, if go I must, in a blaze of Entomological Glory."[85] But the work was hard ("if I had known what I had to go thro' when I began Butt. N.A.," he wrote Wright, "I never should have made the beginning"), particularly, because of continuing money problems.[86] Although his wife, Catherine, had a legacy of around $3,000, and his son, Willie, a recent graduate of law school, had begun investing in the gas and oil industry in West Virginia, Edwards himself, throughout the 1880s, was strapped for cash, forced to resort to a patchwork of means to pay his expenses.[87] He beseeched his publisher, Houghton, Mifflin and Company of Boston, to extend "advances," even without a clear prospect of completion, and the company did so, frequently and generously.[88] He sold much that was dear to him—land in Paint Creek to the Chesapeake and Ohio Railway; his "ancestral home," in Hunter, New York; many of his books on butterflies (including rare ones by John Abbot and Dru Drury); and, most heart-wrenching, *his own butterflies.* His buyer was William Holland, a wealthy young Presbyterian minister from Pittsburgh who had begun a massive buying spree of insects, both domestic and tropical, and who would, in time, enter the bloodstream of American butterfly science in a major, even malign, way.[89] "He is buying left and right," Edwards informed Henry Edwards.[90] It was one of the ironies of the times that, at the very moment Edwards had reached a vocational pinnacle, he should have had to sell off the source of his achievement to an avaricious butterfly man.

"It was painful to part with the collection or division of it," he told Henry Edwards in May 1886. "It was pulling eye teeth. Still I greatly desire to publish."[91] Holland offered him $2,000 to cover costs of the artists and of the printing, in exchange for ownership of "the most complete collection in existence of North American insects," as Edwards expressed it. "No other approached it in completeness," he explained to Holland. "You yourself, after using no end of money, had written me that there were many species you despaired of ever getting." Well, here they all were, "rich in varieties and localities," and the "whole correctly determined."[92] Holland proposed to take possession in installments, so that Edwards might continue to turn to them for scientific study, but by summer of 1886 Edwards had already exhausted the first half of Holland's money ($1,000) and told Scudder that "soon the rest will go." "It is therefore of great importance to me and to the issue of this work to get what help I can. Please aid me in this matter."[93] Scudder contacted the trustees of the Elizabeth Thompson fund of the Association for the Advancement of Science, which, years before, had awarded him a grant for research on fossil insects. They gave Edwards $200, enough to defray some of the costs of his plates.[94] A few years after that, Scudder intervened again, winning a $500 grant for his archcompetitor.[95]

He stepped in on Edwards's behalf in a nonfinancial way as well, after learning that neither the Academy of Natural Sciences, in Philadelphia, nor the American Entomological Society subscribed to *The Butterflies of North America*. This was a terrible sign since writers like Edwards depended on the endorsements of such institutions. "How does it happen," Scudder wrote the director of the academy, that it is not "down as a subscriber" for Edwards's volume 2? "Unless the list can be increased," he explained, "there is danger that its appearance may be checked or even discontinued. It would be an immense pity if this should happen, as the *Butterflies of North America* is a great credit to American science and art, far superior in matter and expression to anything done abroad."[96] Scudder complained that Edwards's "great work" had to be "carried on by him at more or less a dead loss, under very adverse circumstances," and "at so heavy an expense" to the author himself.[97] No one realized more than Scudder how much money and sweat Edwards had poured into his book. His own massive three-volume study, overall, took an even greater effort than Edwards's.

Between 1879 and 1889, Scudder was a neo-Calvinist tornado of productivity, his only escape vacations with his much-loved son, Gardiner, a student at the time at both Harvard and the Harvard Medical School. The two summered together in a remote wild enclave on the coast of Maine, there seeing their first migration of monarch butterflies; in the winter, with paths blanketed by whiteness, they climbed Mount Washington, in the White Mountains, Scudder's old haunt full of unusual fauna.[98] Mountain climbing was a metaphor for Scudder's life as a naturalist; he reached many peaks. As assistant librarian of Harvard College, he completed two comprehensive scientific catalogs, the first commissioned by Harvard, *Catalogue of Scientific Serials,* on *all* the scientific writing in periodicals since 1633, and the second for the Smithsonian in Washington, *Nomenclator Zoologicus,* on the generic names for *all* animals, fossil or otherwise, used by naturalists from ancient times to 1879. Each book cost him tremendous pains, "dreary," as he put it, beyond imagining. But together they yielded a vast reservoir of data for naturalists, who no longer had to track such information down for themselves; Scudder had done the work for them.[99] From 1882 to 1885, he edited *Science Magazine,* eating up time, energy, and creativity. Its collapse liberated him, however, for what became his next extraordinary work, on fossil butterflies.

Scudder had seen his first fossil butterfly and the first ever found, in France, when caring for his wife at the end of her life; it had been unearthed in Marseilles in 1873, and was so curious a find that he left her bedside to see it for himself.[100] He soon became one of the most advanced authorities on fossil butterflies in the world, worked the fossil beds in Canada and Colorado, and, in 1886, got appointed paleontologist to the United States Geological Survey, a post he would hold for the next six years.[101] Today, we know of only forty-four fossil butterflies, reaching back more than sixty million years, a fragile residue of the time when flowering plants and other animals first appeared.[102] Scudder exhumed more fossil lepidoptera than anyone else, before or since, and in 1886, he described them for the first time in an essay, "The Fossil Butterflies of Florissant" (a fossil bed in Colorado) for the *Eighth Annual Report of the United States Geological Survey to the Secretary of the Interior.* He named a number of species, among them the River Styx nymphalid and Charon's nymphalid, and two snout butterflies, including a tropical vagabond snout, with relatives in Africa, "separated from its ancient home by the wide ocean."[103] All extinct, they dated from a time when the continents were fused and the earth was wetter and warmer. This publication alone would have established Scudder as a leading paleontologist, but at the end of the decade, he completed a greater task, a 750-page book on *all* the fossil insects ever known and named, of which James Fletcher, a close friend, asked, "How on Earth do you manage to do so much work, and keep your senses?"[104] Joseph Lintner "felt a just pride that such a work has been accomplished in our country, and by one of *our entomologists.*"[105] William Henry Edwards, in a letter to Scudder, called the book "a stupendous amount of hard work," high praise from one so stingy. "It is wonderful that such a number of species managed to inter themselves so that you could find them, and do honor to them."[106]

Any one of these contributions, then, would have been enough to satisfy any ambitious naturalist. But Scudder was grappling also with *living* lepidoptera, for research on his three-volume *Butterflies of the Eastern U.S. and Canada* (henceforth, *Butterflies*).[107] Compared to all previous works on butterflies in America, his volumes were gargantuan, published so rapidly in parts as to seem to come out *all at once,* in one sweeping moment, like a great burst of light at the end of a long corridor. Scudder defrayed the cost of the production of *Butterflies* with fifty-dollar subscriptions, but they met only a small part of the expense. The rest he paid for himself, having dropped or been dropped by his earlier publisher, Hurd and Houghton, after it became Houghton Mifflin. He proofread every page, consuming four hours daily on it for days ("Above all, take care of your eyes," urged his friend James Fletcher), and wrote, as well, many letters, almost always *by himself,* to get much of the data about the lives of butterflies with which to construct his life histories.[108]

By any measure, *Butterflies* is awe-inspiring, unlike any work ever published. One feature stands out: like Strecker, Edwards, and Grote, Scudder tied his science to art, but doubly—pictorially and with poetry. His fervor for pictures derived from Agassiz, who'd derived his from the great European pictorial tradition. Agassiz had already conveyed how critical pictures were to the transmission of knowledge about the natural world.[109] In 1862, eager to apply photographic techniques to scientific illustration, he supervised revision of Thaddeus Harris's 1841 *Treatise on Some of the Insects Injurious to Vegetation,* with help from Scudder and others, adding to it drawings done with "the utmost accuracy and perfection."[110] In 1869, Scudder released his own first plate for *Butterflies,* done as a chromolithograph. It elicited great praise from Henry Bates in England: "Your specimen of chromolithography as applied to the representation of butterflies is the most beautiful I have yet seen."[111] Twenty years later, in his volume 3, Scudder published pictures on nearly every feature of butterfly existence, each plate reproduced photographically by a gelatin process developed in Boston. In several lithographs and drawings, the eyes, antennae, genitalia, legs, and palpi (jointed sensory organs on each side of the mouth, bedecked with hairs and scales) of the imagoes of several adult butterflies from different families appeared in orderly rows or arrangements, the body parts exaggerated in size, delicately depicted, and dramatic in their repetitive progression. Other plates of the eggs of butterflies, often in color, magnified and drawn through a microscope, had a character never before seen in this way, each as lovely as any snowflake. For the majority of imagoes, Thomas Sinclair of Philadelphia prepared the chromolithographs, relying on eight to fifteen stones; others appeared as black-and-white woodcuts or engravings.[112] If Scudder's pictures lacked the artisanal beauty of

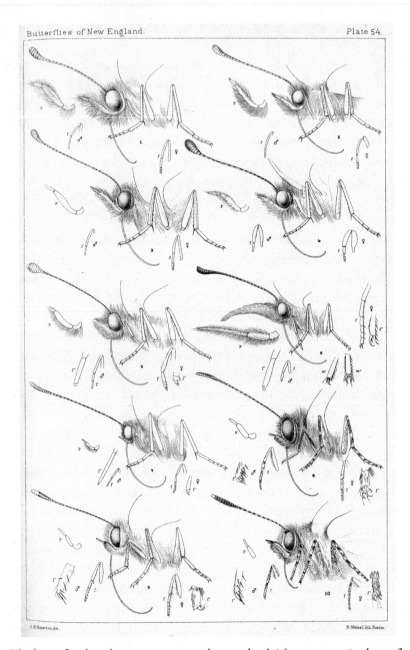

J.H.Emerton,del.   B.Meisel.lith.Boston.

The butterflies here boast eyes, tongue, legs, and palpi from two major butterfly families, the Nymphalidae and the Lycaenidae. Palpi are small body parts that protrude from the head of the butterfly and that contain—according to modern science—numerous hairy sensors able to detect scents. See Butterflies of the Eastern United States and Canada, *vol. 3, plate 54.*

Strecker's and the lucid precision of Peart's (Peart also depended on the microscope, and used Lydia Bowen's exquisite colors), they were unsurpassed for depth of knowledge, drawing on the latest technical processes, just short of relying on photography itself, though still in touch with natural history traditions.

Scudder relied on more than visual art to enhance his science, employing the art of *real, unashamed* poetry, often long excerpts taken from some of the greatest poems in the Western tradition, occupying nearly every page of his massive fifteen-hundred-page tour de force. Poetry was enlisted to introduce *every* life history and *every* excursus, Scudder's digressions on ecological themes. He opened the work with a remarkable excerpt from *The Spanish Gypsy,* a book-length poem by George Eliot; spoken by Don Juan, the aristocratic hero of the text, the excerpt reads:

Repent? Not I.
Repentance is the weight
Of undigested meals ta'en yesterday.
'Tis for large animals that gorge on prey,
Not for a honey-dipping butterfly.
I am a thing of rhyme and redondillas,
The momentary rainbow on the spray
Made by the thundering torrent of men's lives:
No matter whether I am here or there;
I still catch sunbeams.

What a curious way for a middle-aged Congregationalist to begin his book. Did those who inhabit the realm of sunbeams have no need to repent? Are people themselves free of sin in such a world?

Some excerpts came from Seneca and Horace, but since the early poets showed little affinity for butterflies, Scudder quickly exhausted them, preferring later authors on both sides of the Atlantic, above all the Romantics, with their incredible number of references to butterfly life. Some poems were by Scudder's American friends Augustus Grote and Thomas Wentworth Higginson. A respected American reformer who studied with Thaddeus Harris at Harvard, Higginson had a passion for butterflies and championed the work of Emily Dickinson. "Thou winged blossom! liberated thing!" Higginson observed in "Ode to a Butterfly." "What secret tie binds thee to other flowers / Still held within the garden's fostering?" Grote, in an untitled poem wrote, "Pretty flower that June remembers / Blossom that July forgets."[113] More often, Scudder quoted from such figures as Blake, Wordsworth, Coleridge, Christina Rossetti, and especially the late Romantic Robert Browning. Thus, from Browning's *Dramatis Personae:*

> On the rock, they scorch
> Like a drop of fire
> From a brandished torch,
> Fell two red fans of a butterfly;
> No turf, no rock, in their ugly stead,
> See, wonderful blue and red.

And from *The Ring and the Book:*

> Some finished butterfly,
> Some breathing diamond-flake with leaf-gold fans,
> That takes the air, no trace of worm it was.[114]

Naturalists had drawn on poetry before, but never to the same degree. In the end, Scudder tallied more than 150 poets.

At first, his purpose was quite simple: to use a few poems "here and there" to keep the interest of the reader, who, he thought, would want "relief from the dry synonymy." But another motive took over, and he began to select poems of "some special significance." But what this meant Scudder never made clear or even addressed. Perhaps the meaning can be found in the overriding Romantic character of the poetry, for nearly all of it encouraged the reader to live in a sin-free land of beauty and butterflies, and to do them no harm. His reliance on poetry as a whole, however, Romantic as well as non-Romantic, had an even stronger cause: a respect for the word and for the educational power of the written word. This distinguished nearly all of his writings, as, indeed, it did those of Grote, Edwards, and Strecker, and of the natural history tradition as a whole. Just as he observed how butterflies lived and died in relation to the whole environment around them, so he showed how poetry and art related to butterflies, and how the study of butterflies belonged to the Western tradition. Through an imaginative work of natural science, he exposed Americans to the greater creative culture.

Scudder's *Butterflies* had many other distinctions, of course, well beyond its poetry, beginning with a ninety-page introduction on the butterfly stages (egg, caterpillar, chrysalis, and imago) and backed by plate illustrations in volume 3 of all the stages, often in microscopic detail. Also, his informative survey of the history of butterfly classification (from Linnaeus to William Henry Edwards) reflected great knowledge of and respect for his precursors. The plates in volume 3 on different ways to classify species—four pages on genitalia, five on wing venation, sixteen on external morphology, and so forth—were exemplary, as were the two long and instructive essays commissioned by Scudder and written by two Darwinian naturalists, William Morris Davis, a professor of physical geography at Harvard, and Charles William Woodworth, a young man from Illinois who had studied insects under Hermann Hagen at the Museum of Comparative Zoology, in Cambridge. Davis discussed the biogeography of New England, especially the glacial carving out of the landscape; Woodworth, the evolution of the egg of the mourning cloak butterfly from a mass of similar cells (the blastoderm) to a more complex differentiation. Both essays revealed great advances in glaciology and embryology since the 1820s and Scudder's own conversion to Darwinism.[115] Volume 3 of *Butterflies* boasted the first plates ever published showing the distribution of species in America, and it concluded with a map of Scudder's grandest butterfly spot: the alpine districts of the Great Range in the White Mountains of New Hampshire.

Scudder never wavered on many of his older positions, those that William Henry Edwards had reviled in the early 1870s. Following Henry Bates and others, he continued to place the Nymphalidae (the wood nymphs, satyrs, fritillaries, the "purples," and so forth) at the top of the butterfly chain, while kicking the Papilionidae (the swallowtails) toward the bottom, near the Hesperiidae (the lowly skippers). Almost shamelessly, he described the Nymphalidae as a "higher grade of life," because, among other things, they "are the most sprightly and vivacious" insects, "the most audacious" and "the fondest of propinquity to man and his cultivations," besides being "endowed with the most varied psychological traits." Scudder never budged from his enthusiasm for Hübner's often bizarre generic names. Nor did he—nor would he ever—renounce vernacular names, despite what he called the earlier "violent opposition" to his approach (he even added some new ones, especially for the skippers).[116] It helped, of course, that other naturalists—including Grote—had come around to agreeing with him.

The meat of the book was the life history of hundreds of American butterflies and the essays called "excursuses" on butterfly existence. In the life histories Scudder freely acknowledged the help he'd gotten, above all from William Henry Edwards, who, more than any butterfly man of the age, had inspired entomologists to give up their emphasis on classification and "dead, dried" specimens, and to embrace *life,* the living, breathing butterfly in all its glory, as the subject of interest. A great beneficiary of Edwards's example, Scudder quoted him profusely on nearly every page.[117] "Mr. Edwards thinks" and "Mr. Edwards states" echo so often in the life histories that any ordinary reader might have thought them the work of two men, not one. Scudder incorporated Edwards's prose into his own, fulfilling a promise he had made to Edwards shortly before resigning his editorship of *Science:* to "re-write the life histories" (as they had appeared in his 1879 book *Butterflies*) "and take from you largely!" ("He is right there," Edwards wrote Lintner, "and I trust then we shall have no more romances.")[118]

Scudder began each history with the butterfly's name in both its Latin and its vernacular forms, with a synonymy or record of all the names imparted to it over time, followed by a taxonomic description of the butterfly—almost invariably, in wearying detail—but then, almost as invariably, by a lively and engrossing account of the *real* life history of the butterfly, its various stages, its food plants, its enemies and defenses, and its relationship to other species and genera. He finished each portrait with a constructive reflection of what remained to be done, which often amounted to a great deal: more on eggs and larvae, more on the mysteries of metamorphosis, more on variable forms, more on parasites, more on everything. As his opening surveyed the history of the butterfly systematics, so his concluding remarks placed *Butterflies* in a stream of investigation, alive with the present, tied to the past, and preparing the way for work to come. The reader feels the amazement Scudder felt in the face of all America's butterflies, their shapes, forms, patterns, and colors.

The excursuses in *Butterflies,* interspersed among the life histories, gave Scudder himself his strongest sense of pride, encompassing a vast range of subjects, from the "lethargy of caterpillars" to biographies of early naturalists. No American—no writer of any kind—had ever before dwelled so thoughtfully on butterfly life.[119] The results led the respected British butterfly experts Karl Jordan and Walter Rothschild to conclude in 1906 that "no other work on Butterflies can be compared to it," and Nabokov, seventy years after that, to declare of Scudder's "stupendous work" that it "inaugurated a new era in lepidopterology."[120] Even William Henry Edwards—despite his critical take on everything Scudder wrote—conceded, with ever-increasing certainty as he mulled them over, the value of Scudder's volumes. "I differ with Mr. Scudder radically about many things," he explained (he never got beyond "Mr. Scudder"), "but in other important and essential points this work of his is and will forever remain unapproachable." He praised Scudder's "wealth of illustration" as "amazing, not only of the butterflies themselves, but of every part and organ of them, and what has never been attempted before except on a limited scale, the eggs and larvae are shown in greatly magnified and admirably executed figures." Furthermore, on matters "of anatomical details, worked out with wonderful ability, and the life histories and distribution worked out with exceeding care, the *Butterflies of the Eastern U.S. and Canada* will be a standard work, and no student can possibly get along without it. Therefore, I say to my friends, subscribe without delay."[121]

To finish *Butterflies,* which he had decided, in the end, to publish himself, Scudder saddled himself with a debt that took him three years to erase. In 1892, when the U.S. Senate stripped away appropriations, he lost his position as chief paleontologist for the U.S. Geological Survey. The experience caused an insomnia that lasted ten months, until he went off for three weeks into the woods and mountains with his beloved son, Gardiner, "living completely in the out-of-doors away from all entanglements."[122] He struggled to find other work, including, once again, as librarian of Harvard College, but failed and turned to journalism to make a living, writing articles freelance for the popular press and, over time, mining his three-volume study. Within five years, he had added brilliantly to both guidebook and children's literature, including two short books, *Brief Guide to the Commoner Butterflies of the Northern U.S. and Canada,* the first designed explicitly for boys and girls "who had caught a common butterfly," and *The Life of a Butterfly,* a biography of the monarch that synthesized all existing knowledge buried in hard-to-find places; it proved extremely popular, especially with countless grade school teachers of natural science throughout the country, who adopted it for their classes, and transformed *Anosia plexippus,* as Scudder called the monarch in Latin, into the "typical American butterfly."[123]

At the end of the century, he completed two other books, *Frail Children of the Air: Excursions into the World of Butterflies,* using many of the best excursuses in *Butterflies,* and another popular guide, *Every-day Butterflies: A Group of Biographies,* a lovingly written series of portraits of more than sixty of America's most common species, with woodcuts by Anna Botsworth Comstock. *Every-day Butterflies* was a special book, the last of its kind by Scudder, with most of the species found "in open fields, meadows, roadsides anywhere in the open country," on "paths in dry pastures," or down "overgrown pasture tracks in the vicinity of woods," "on grassy meadows," "wet meadows," or "low meadows," "in open woods and orchards as well as along roadsides and stone walls," or amid "shrubbery bordering on cultivated fields," a few species hibernating "in cellars of old buildings, barns, or outhouses," or pupating "under the sides of fence rails."

This was the landscape of the American family farm, familiar to most people of the time, home of butterflies seeable by anyone, prodigal gifts of nature in numbers unimaginable today: "The tiger swallowtail collecting literally by the thousands, and when startled, filling the air with a yellow cloud"; the clouded sulphur or *Colias philodice,* bursting forth "on meadow-bordered highways after rain, coloring the ground as they sit by the thousands with erect wing"; the pipe-vine swallowtail, a splendid iridescent blue butterfly with tropical ancestors, "particularly fond of flowers and sometimes clustering on them in vast numbers"; and the regal fritillary, "with its rich orange-red and blue-black coloring above, marked with black, orange, and white," and, like all fritillaries, bearing "large gleaming silvery spots on the under surface," "frequenting open breezy meadows or pastures in close proximity to marshy lands or ponds," and "most abundant in the middle pastures of Nantucket."[124]

Every-day Butterflies, however, did not break new ground, nor did the other books; all lifted whole passages verbatim from *Butter-*

flies in ways that seem shameless today. Scudder knew how costly his three-volume work was, that it was—as one impoverished butterfly lover called it—"grapes too high for me," so Scudder pulled the grapes down. All his derivative books sold cheaply, were highly readable, and so small (three by six inches in size) that they could have been easily slipped into the back pocket of his three-volume tome. He democratized the word power of his *Butterflies,* "opening the door of Nature" to some special American boy or girl who might want to "wander into the byways for more eager personal search."[125] Edwards never took a similar path, the popularization of his knowledge left to Scudder and others who had no trouble or reservations about spreading it around in the public domain. By the end of the century, Americans had produced a foundational literature on butterflies, everything from lists to massive catalogs, dealing especially with native American insects. But we have barely begun to tap into the ideas and insights this word power carried. These butterfly people had a great deal to say about the creatures they or their disciples collected in the mountain meadows, marshes, deserts, semitropical forests, and backyards of America. The portrait they gave was beyond anything achieved before, capturing how the butterflies survived and persevered against all odds, fought and cooperated with one another and with other animals, and contributed to the making of the beauty of the world.

FIVE

The Life and Death of Butterflies

The writings of the butterfly people made nature seem fascinating and magical, just as Romantic Alexander Humboldt had insisted such science should do. In the introduction to his *Cosmos*, Humboldt rejected the notion held by Edmund Burke, an eminent conservative English thinker of the late eighteenth century, that nature loses its "magic and charm" as people "learn more and more how to unveil her secrets." Quite to the contrary, Humboldt argued, "the excitement produced by discovery, the vague intuition of mysteries to be unfolded, and the multiplicity of paths before us, all tend to stimulate the exercise of thought in every state of knowledge." In a way never known before, nature invokes a "sense of the sublime" and a "feeling for infinite things."[1]

In only a few decades, the American encounter with American butterflies yielded new insights into "Nature's shifting panorama of form and color," to quote Scudder. It took the art of seeing to a new level, casting light on the changing forms of insects; on the complex ways in which they emerged, developed, and died; and on how they existed in relation to other forms, either as partners with them or as their enemies and predators. The American investigation also led to a more sophisticated understanding of natural beauty. Americans, generally speaking, were not much known for their aesthetics, a situation due to an overriding utilitarianism or, perhaps just as likely, because beauty was (and is) a difficult thing to fathom in the first place. Yet there were some individuals, such as George Santayana, a man trained in both aesthetics and natural science, who in the 1890s attempted to understand beauty. And there were some American butterfly people, too, who knew much about its purpose both as adaptation in the preservation of life and "as a balm for all worldly ills," in the words of Herman Strecker.

By the 1880s, all the leading thinkers about butterflies in the United States had adopted the Darwinian evolutionary outlook, one that viewed natural selection as the guiding impulse behind evolution, that emphasized fertility of form over the confinement of form, and that seemed to exclude nonscientific ways of understanding nature; nothing organized nature except by chance, through a chain of life in relation to surrounding forces and reaching back to the beginning of time. For the fully converted modern evolutionary biologist, beauty in nature lacked interest except as adaptation to change. Among America's butterfly people, William Henry Edwards was the earliest and most complete exponent of this position. Herman Strecker never seriously entertained it, pro or con, although he did say, in 1879, "I am no believer in the special creation of each species." "Moreover, a careful study of the moths and butterflies of different parts of the world will show how wondrously today's species are all linked together, not only in a continuous line, but interlinking, as the rings in a coat of chain armor, and that, too, by species and genera in countries as widely remote from each other as Buenos Ayres and Australia."[2]

Like Edwards, Augustus Grote, too, was a spokesman of Darwinian evolution, and by the late 1880s had adopted a similar phylogenetic approach to butterflies, or that approach, in other words, that traced the identity of species back to shared ancestral roots. Following Darwin, Grote understood nature not as a ladder of ascending species, one species higher or lower than the other, but as a web with branching lines, each adapting in its own way to change. "The evolution of Lepidoptera," Grote observed, can be "represented as an inverted and spreading bell of net work, in hanging threads of unequal lengths, branching variously and in different directions. The depending tips of the threads represent the existing species, all connected to the past, and the task before us is the tracing of the threads, always running here and there together, grouping themselves around thicker strands, converging in the hand of time." "As nature did not produce these creatures in a linear series, one after the other, we can only approximately exhibit their relations in our catalogues and collections."[3]

Samuel Scudder appears never to have abandoned high/low distinctions, yet he did, along with Grote and Edwards, adopt a phylogenetic approach as indispensable to classification if nearly impossible to achieve. The war of all against all in nature's realm was also part of his outlook. "Nature," he wrote, "always seems on her guard." "The struggle for existence is the perpetual inheritance of the individual."[4]

He, as much as the other main butterfly people, became engaged with how butterflies evolved or adapted to conditions around them, defending themselves against an army of enemies, escaping concealment, above all, by mimicry. A dark picture emerged in Scudder's work. Like the others, he had become an ecological Darwinist, no longer loyal to Louis Agassiz (who died in 1873). Scudder not only accepted evolution as a fact of life, he also viewed natural selection as the major animating pivot of evolution, acting through such pressures as competition and predation to "select out" the most fit individuals. It is not clear exactly why or when Scudder took this new route. In late December 1879, Charles Fernald, a Christian naturalist from Maine, visited him in his home in Cambridge and was impressed that Scudder still maintained the "observance of the good old custom of saying grace at the table." "There are so few of our scientific men at the present day," Fernald lamented, "who observe or tolerate any form of Christian worship that when we meet one who does it seems like finding an oasis in a desert."[5] Scudder probably never lost a strong faith in God, but he did come close to believing that something else besides God dictated change in the universe.

American butterfly people, by the 1880s, had done as much as any other group of naturalists in the world to advance the understanding of the life and death of species in Darwinian terms. At the same time, they illuminated each stage of the insect life cycle, "adding to the accumulation of anatomical detail," as the historian Jakub Novak has written, and presiding over "one of the greatest accomplishments of nineteenth-century life science."[6] Americans had discovered much with the microscope, which they used in an unprecedented way. Charles Valentine Riley, then the Missouri state entomologist, explored pupation accurately for the first time; and Edward Burgess, Scudder's colleague at the Boston Society of Natural History, established—again, for the first time—how the monarch butterfly's proboscis worked to suck nectar from flowers, as well as showing in impressive detail the insect's internal anatomy based on Burgess's original dissections in the 1860s.[7] Scudder described, often with microscopic care, the surfaces, shapes, and colors of the eggs, larvae, and pupae, and their diversity.[8] His drawings in volume 3 of the wing veins of butterflies and of adult male genitalia showed how these

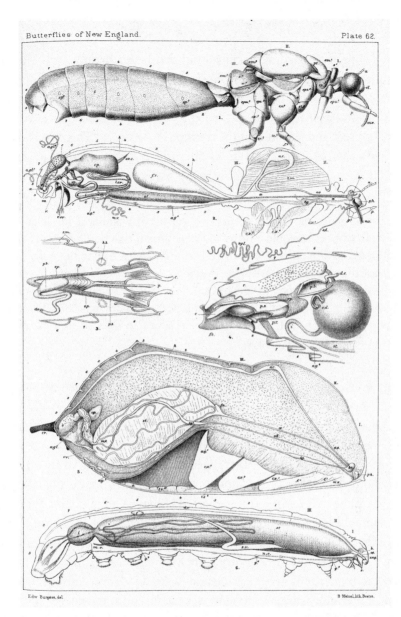

This plate offers an unprecedented glimpse into the insides and outsides of the monarch butterfly, based on the brilliant dissections of Edward Burgess, Scudder's colleague at the Boston Society of Natural History, who later gave up entomology for the glories of building and racing yachts. From the top down: (1) the external anatomy of the female monarch, side view; (2) the internal anatomy of same; (3) horizontal section of the end of the male abdomen, with emphasis on the genitalia (see p. for penis); (4) lateral view of the same male abdomen, showing the genitalia in position (see t. for testis); (5) the internal anatomy of the female pupa, three days old, side view; and (6) the internal anatomy of the male caterpillar. See Butterflies of Eastern United States and Canada, *vol. 3, plate 62.*

features might be used in classification, and better than anyone else at the time, he explained metamorphosis.

Scudder posed and answered a basic question: How did the butterfly escape from "its iron prison, hardened by months of exposure to wintry cold and sleet and sun in rapid succession?" Answer: "There is a weak point in every structure, and in the chrysalis it lies next to the point of greatest strength in the captive butterfly." After winter ends, and

> the more genial showers of spring or damp air of a summer's night have softened the texture of its prison-walls, they are further weakened by the moisture now exuded by the twice-bound prisoner, feeling the hour of release draw near. A suture along the crest of the thorax gives way, often with a perceptible click, to the force of the great muscular mass within; the rest is easy; the rent is continued on both sides down other sutures, until a door is opened, whose inner walls suffer no harm to the delicate creature struggling to escape.

Next, the butterfly leaves its "encasement," finds "a friendly twig," and "sitting in the sunshine dries its moist quivering wings, gently fanning them up and down, until, full of new life and courage, it ventures forth—a thing of beauty and a joy forever."[9]

Life stages inspired an entire generation of naturalists to breed and study butterflies and moths, among them Emily Morton, Caroline Soule and Ida Eliot (who worked together), Thomas Bean, William Wright, and Samuel Eliot (who bred more butterflies than even William Henry Edwards). Will Doherty, a talented young naturalist from Cincinnati who would become the country's greatest collector of tropical Lepidoptera, modeled his investigations on "eggs and larvae" after those of Scudder and Edwards, intending to write a "useful supplement" to their labors.[10] Scudder drew on his work in volume 2 of *The Butterflies of the Eastern U.S. and Canada*.[11]

Evolutionary thinking transformed the work on stages, above all for Edwards, captain of American breeders, who was fascinated with how the multiple forms in nature emerged, evolved, diverged, and fought for precedence. By the late 1870s, with a little help from the brilliant German zoologist August Weismann, he had charted the extent of polymorphism in butterflies, narrating the life story of the zebra swallowtail, once thought by Europeans to be *three* distinct species instead of *one* species in three *forms,* which Edwards had discovered it to be. In a series of cold experiments, modeled after Weismann's, Edwards traced how temperature shifts from spring to summer changed the zebra's morphology, essentially reproducing evolutionary history, the "primary," and smaller, spring form having arisen in the glacial period, and the "secondary," larger summer form in a period of climatic warming. He made the summer form revert back to spring's by placing pupae, for a time, in a cold environment (such as a cellar or icebox). Edwards wrote glowingly to his favorite audience, Henry Edwards, that his "experiment" produced "results much more satisfactorily, than did Dr. Weismann's. If all comes out, it will be triumphant," and it was, with Edwards broadcasting his triumph in the late 1870s, through the many journal articles he published, from Berlin to Boston.[12] John Lubbock, one of England's most respected naturalists, in his 1881 book, *Fifty Years of Science,* ranked Edwards as a great pioneer of biological science, on a short list with only four other men: Weismann, Henry Bates, Alfred Russel Wallace, and Charles Darwin.[13]

Edwards's volume 2 of *The Butterflies of North America* gave the same treatment to the tiger swallowtail, a butterfly strategic to his analysis, and the first American butterfly ever pictured: in a painting by the Englishman John White and rendered as a woodcut by his countryman Thomas Moffat, in 1634, in his *Insectorum sive minimorum animalium theatrum.*[14] The tiger flew throughout North America, from central Alaska to the summit of Mount Washington, its first brood in the Atlantic states in May, during a promise of warmer days, and its second as a summertime visitor, darting and swooping in arcs of energy among the trees or sailing gracefully down like falling leaves from roosting sites into the gardens or meadows below. Wild cherry, birch, ash, tulip poplar, and basswood were among the many food plants for its larvae, and the adult was remarkable for "its peculiar dimorphism," Edwards said, "which, so far as is yet known, is without parallel among butterflies." At its most northern point, all the females and males were yellow with black vertical stripes on the wings (the females with much blue in the hindwings, the males usually black). At its most southern point, the females were all black, or melanic, without any stripes, the males yellow and striped. As with the zebra swallowtail, naturalists had for many years seen these forms as separate species, not as the same butterfly. Edwards exposed their mistakes, building on the work of earlier Americans. He relied on Weismann to determine where the black form came from and why it was black. Drawing on Darwin, Weismann had asserted, in 1872, that "the yellow is the ancient and original form, the black a much younger or more recent form able to perpetuate its type through its descendants till it has become common, sometimes, almost to the exclusion of the yellow and original form." Edwards carried this argument further by outlining an ecological "belt of dimorphism" or realm of "intergrades" between the northern and southern extremes along the northeast corridor; in this area of swirling incoherence, several butterflies (pearl crescents, blues and purples, and wood nymphs) struggled over millions of years to achieve species identity. Here, too, the melanic and yellow female swallowtails competed, according to Edwards, with the blacker form being more tenacious.[15]

But why did the black form have the upper hand (or wing)? Weismann thought sexual selection was the answer—that yellow males, for some reason, preferred black females. Edwards, however, had observed many yellow males impregnating both yellow and black females "on the wing"; moreover, he noted, "the males may be seen coquetting with the yellows as freely as with the blacks." He proposed, instead, that the black female was succeeding (or would succeed over time) because it had an "astonishing energy" and because its blackness mimicked the blackness of other swallowtails, especially of the pipevine swallowtail, a mostly southern species with a bad taste and smell that repelled birds. The yellows, on the other hand, "are captured by birds and other enemies during the day" because of "their gay colors," which "render them an easy prey."[16] Edwards illustrated these arguments with three marvelous plates by Mary Peart and Lydia Bowen, more than he ever devoted to any other American butterfly. The first plate showed the tiger's life history (caterpillar, egg, pupa), plus that of the females, black and yellow. The second plate pictured the black female and the "two sexes of the yellow form together," the first such representation published, and the final plate depicted a yellow male in the company of three intermediate females, one half black, half yellow, the other two with suffused mixtures of yellow and melanism, each, perhaps, a creation of the belt of dimorphism.

Edwards always looked to the early stages to establish specific or generic identity, the preoccupation with wing color seeming to give precedence to the imago. "I can't help thinking," he explained to Scudder, "that the first three stages are the ones in which [generic] affinities are to be sought, rather than in the fourth." But he decried other naturalists' reliance for identification of species or genera on butterfly wing venation, on arrangement of veins on the wings, or on, God forbid, the genitals. "I have very little respect for wing classification," he wrote Scudder, who, like Grote, considered wings and genitals worthy of reflection; in 1888, he declared, "One caterpillar is worth fifty genitalia for the purposes of distinguishing obscure species."[17]

Like Darwin and Weismann (and before them Buffon and Humboldt), Americans practiced a broad ecology in which "climate and environment," as Edwards put it, must be taken into account to understand "the descendants of a common parent-form." Grote called it "the phylogeny of species," writing, "The crucial test of our modern idea of species lies in the demonstration of the fact that, in the whole life history, the cycle of reproduction is *now* distinct. To the establishing of this fact repeated observations are often necessary. The whole condition under which the form is produced must be understood." We must deal with "geological conditions" that "have played a part in the evolution of species," he noted. "For students of butterflies and moths the criterion of species must lie in knowledge of the whole life of the insect."[18] Butterflies did not come out of nowhere fully dressed for business, Grote believed, but "adapted by change of habit to altered circumstances," in the face of countless indignities, and in response to volatile weather and temperatures, earthquakes, and shifting land masses.[19] Those insects that met these challenges endured, Scudder wrote, their "advantageous variations perpetuated and intensified by the survival of the fittest, through the law of inheritance."[20]

Edwards saw vast geologic variability in temperatures, cold to hot, writ small in the lives of polymorphic and dimorphic butterflies, the butterflies telescoping evolution, bearing ancient history in their bodies, revealing how modern species emerged. He believed that butterflies were, in Scudder's words, extremely "sensitive to the least influence from the outside." A sudden drop in the temperature might strike down entire populations with devastating effect; a simple sustained frost felled many billions. "Winter, indeed," observed Scudder, "is the prime cause of variety in nature."[21] A shift from dry to wet seasons could kill butterflies, William Doherty observed in India, unless the butterflies could make adjustments.[22] Grote noted the impact of what he called "all the phenomena of climate and temperature" on the geographic distribution of moths and butterflies, and both he and Scudder wrote at length of the great glaciers that dropped down from the north to remake the butterfly universe. But even more deadly dangers lurked to demolish butterflies and moths at every stage of their existence—dangers that were especially harsh for butterflies, which did no harm to any other living thing.

"The life of a butterfly," wrote Scudder in "The Enemies of Butterflies," "is one of imminent danger from birth, nay before birth, to death." He was thinking here of such predatory species as the "creeping crawlers"—mites, ants, and spiders—in pursuit of quick meals, or the "pirates of the air," as Edwards observed, the dragonflies, bees, and wasps that "pounced upon" butterflies "as do hawks on small birds, bearing away their prey to be devoured at leisure."[23] Butterflies had even graver foes, notably the remorseless critters Scudder named "the wandering buccaneers" of nature's realm. Today, scientists call these buccaneers "parasitoids" and believe that close to two million species of them may exist, equal to 20 to 25 percent of all insects, or, put even more impressively, to about one-tenth of all animal species.[24]

Parasitoids are *not* parasites, though that's what nineteenth-century naturalists, including Scudder, called them. Parasites seem almost sweet-tempered by comparison to parasitoids, for parasites do not kill their hosts (the term is Greek for "unwelcome guest").[25] Parasitoids, on the others hand, *are* killers, some adapted to devour only eggs, others to eat the larvae or pupae. Tiny flies lay their eggs on the surfaces of larvae; the hatchlings will later eat the larvae from the outside. Others feed secretly from within, killing the host only when the parasite itself starts pupating. Many have an aptitude for *targeting* prey, like one female parasitic fly described by Archibald Weeks, a Brooklyn moth specialist, in an 1887 issue of *Entomologica Americana,* an influential Brooklyn

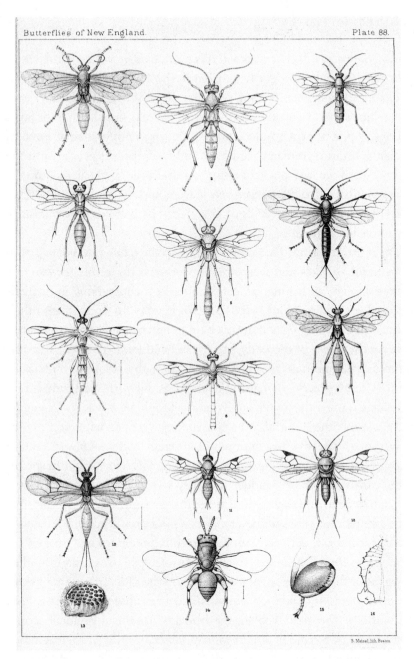

Shown here is an unusual grouping of parasitoids from Scudder's
Butterflies of the Eastern United States and Canada, *vol. 3, plate 88.*
Scudder called them "evil-looking" and believed they were the most
dangerous of all butterfly enemies. Note the ichneumon wasps 1 and 2, as
well as the parasitized pupa of a butterfly (13), full of holes left behind
by recently emerged wasps or flies whose eggs had earlier been laid inside
the pupa, later to hatch and, then, to eat the pupa from the inside out.

journal on insects. Weeks watched this particular fly with a hand lens as it dropped a single egg delicately between the "eyes" of its prey—in this case, a moth caterpillar feeding on a leaf. Soon the fly added other eggs to the same spot; "so gently was this done," Weeks noted, "that the larva did not at first appear to be disturbed, but presently the adhesion of a foreign substance seemed to annoy it and it scraped its eyes against the bitter edge of the leaf in a vain effort to rub off the barnacle-like ova."[26] George Hulst, a Brooklyn moth specialist, compared these insects to the Grecian horse within the walls of Troy. Some parasitoids even parasitize the butterflies' parasites.

The most notorious parasitoids belong to the Ichneumonidae, a populous family of flies and wasps that encompass the genus *Ichneumon.* Slender insects with long, graceful antennae, ichneumons inject their equally long and graceful ovipositors, or egg-laying stingers, into their victims, discharging eggs that, over time, hatch inside and proceed to "live on the juices of the body," to quote Scudder. Ever since Aristotle, naturalists had been aware of parasitoids, though only in the mid-1700s did they learn precisely what these creatures did with their time.[27] Linnaeus had named the whole family Ichneumonidae; it was revised again and again by other naturalists as new species were encountered.[28] By the early 1800s, extensive descriptions of the genus *Ichneumon* were regular features in entomological writings, appearing like some frightful specter at the carnival of life.[29] Scudder said they were "evil-looking." Darwin was unnerved to find that nothing other than natural selection—certainly not "a beneficent and omnipotent God"—had produced such a being.[30]

The term itself, derived from the Greek *ikhneumon,* or "tracker," in turn, from *ikhnos,* meaning "track" or "footsteps," was usually applied to wasps or flies. Alpheus Packard wrote, in the 1880s, that about three hundred species of such ichneumons had been discovered throughout the world.[31] Thorstein Veblen, the brilliant American political economist and caustic critic of American capitalism, writing at the beginning of the twentieth century, seemed drawn to these insects as emblematic of a heartless universe and gleefully told tales about them, especially to impressionable undergraduates at Stanford University, where he taught briefly. According to one such student, R. L. Duffus, later a widely read journalist, Veblen explained that a caterpillar, once stung, "might know that something had happened, but not what. It would go about its business, troubled by sensations which it might attribute to an acid stomach or overwork, until the eggs hatched. Then the little ichneumon grubs would start eating the caterpillar, which, being alive, had kept fresh without refrigeration. The caterpillar would then realize what had happened, but it would be too late." Duffus, sounding like Darwin, wrote that he himself "was inclined to think well of Nature," but "what was one to think of the beauty of the world if such things were going on it. . . . If there was a God, He was letting the flies get away with murder."[32]

William Henry Edwards reared parasitoids from butterflies caught in the field, such as the red-spotted purple, the spring azure, and the hack-berry butterfly.[33] Of a little parasitoid he saw under a magnifying glass inserting its stinger into the eggs of a giant swallowtail, he wrote Joseph Lintner, "One would not think a fly that size would have the strength to push its ovipositor through the shell of such an egg." He described another encounter: "I had an odd thing happen yesterday in connection with *Atalanta*" (the red admiral). "The small larvae which came at the tail of their generation seem frequently parasitized, and the fly comes out of the larva in the second larval stage. . . . Well, in the glass in which I had some cocoons, I observed a swarm of minute flies . . . 3/100 in. long. On examination of the cocoons, I found one which had a minute round hole near the top, and no doubt out of these little flies came. The larger cocoons have the whole top left off. This means that the para-site itself was parasitized." Edwards put together a large collection of both "parasites" and willingly shared them with Leland O. Howard and Charles Valentine Riley. To Howard, the insects Edwards had seen were "hyper-parasites."[34]

Scudder considered parasitoids as so crucial to understanding the chain of natural life that he allotted two separate plates to them in vol-ume 3, the only such treatment he gave to other insects besides the but-terflies. He also invited Howard and Riley, both of the Department of Agriculture, and Samuel Williston, an entomologist from Yale, to sub-mit essays on them for volume 3 of his *Butterflies of the Eastern United States and Canada*. Williston picked small flies as his subject of "The Dipterous Parasites," and Howard four wasp families, including the

Ichneumonidae, with special scrutiny of five or six kinds of hyperparasites he called "secondary parasites." Charles Valentine Riley, one of the most original Darwinian biologists of his day, analyzed many tiny brown and black wasps of the family Braconidae, but he doubted the classification; with so much variation in the species, he felt, one could make no "well-marked divisions." Study confirmed for Riley "the idea of the nonexistence of species as such in nature." Years before, in the late 1870s, he had even tried to persuade Scudder—at a time when Scudder could not be persuaded—that nature could not be split into clear-cut species and genera: "I feel convinced that the more we understand our species, the more we are acquainted with their biology, the more we shall find they run into one another. In a *natural system* we never shall be able to draw arbitrary lines."[35] Still, Riley named a few, among them a tiny parasite of the red admiral Edwards had sent for him to determine: he named the species *Apanteles edwardsii* in honor of Edwards.[36] Riley, along with Williston and Howard, did the most authoritative work on parasites and parasitoids by Americans up to this time.[37]

Scudder reported, in "The Enemies of Butterflies," that "certainly ninety-nine one-hundredths of every brood" of a given butterfly "perish before maturity" from parasitism, "and even after maturity is reached a very considerable proportion of the remainder must come to an untimely end within a day or two of birth." The waste of life was incredible, but the most egregious damage came from the insect parasitoids. At the egg stage, Scudder observed in an essay, "The Perils of the Egg," "the struggle for existence" becomes acute; "excessively minute" wasps and flies beset the butterfly, and out of thousands of eggs laid "by say thirty females" (and here Scudder quoted from a letter received from William Henry Edwards), "hardly twenty butterflies result." The larvae, too, had its enemies, notably several parasitoids that had evolved to kill only larvae and nothing else. "Nearly one-third of all the butterflies described in the body of this work," Scudder wrote in "The Enemies of Butterflies," "are known to be attacked . . . by these horrid fiends." They treated "a caterpillar" like a "peripatetic banquet hall."[38] Milbert's tortoiseshell butterfly, a lovely small, orange-banded northern species, had many suitors, among them a fly that consumed nearly all the larvae produced at the end of the summer. "In one instance," Scudder observed of his experiment on this "parasite," *Apanteles atalantae,*

THAROS, dim. form MARCIA.

Var.	A. 1-2.	a, a^2	Egg	magnified
	B. 3-5.	b	Larva (young)	,,	,,
	C. 6-9.	c	,, after 1st moult	,,
	C.D. 10-11.	d	,, ,, 2nd ,,	,,
	D. 12-14.	e	,, ,, 3rd ,,	,,
		f	,, mature.	
		g	Chrysalis.	

17. Drawn by Mary Peart and colored by Lydia Bowen, this plate of the pearl crescent butterfly was the first life history of this common American species ever published, depicting the butterfly, caterpillar, chrysalis, and egg in magnified form.

CHARITONIA, 1 2 ♂, 3 ♀; 4. 5. 6, vars.

a E. qq. a² micropyle, both magnified.
b ‥ d Larva, young to 2ⁿᵈ moult ,,
e ,, 3ʳᵈ ,,
f ,, mature.
g — g³ Chrysalis.

Drawn by Mary Peart.

L. Bowen. Col.

18. This portrait of the zebra long-wing (or, in Edwards's preferred Latin, *Heliconia charitonia*), a semitropical species in Florida belonging to a mostly tropical butterfly genus, was created by Edwards based entirely on specimens sent to him by Annie Wittfield, the young invalid daughter of William Wittfield, a Florida physician.

19. Portrait of butterfly artist Mary Peart by Peart's niece, Caroline Peart. Note the butterfly at the top of the image. Peart visited cat shows in Philadelphia with William Henry Edwards. *Courtesy of Franklin & Marshall College and the Permanent Collections of The Phillips Museum of Art, Lancaster, Pennsylvania.*

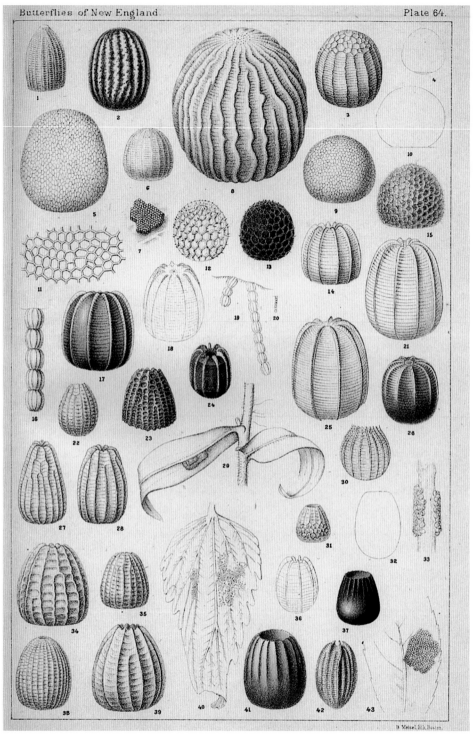

B Meisel lith Boston.

20. One of six plates devoted entirely to the eggs of butterflies in volume 3 of Scudder's *Butterflies of the Eastern United States and Canada*, this plate is unique in its whimsically arranged and precisely rendered surface details. But it is special in another way: the biggest egg in the ensemble, number 8 in the top row, comes from one of the smallest butterflies, the White Mountain butterfly. How are we to understand its hulking size, with a magnification of 40:1, except as an expression of the pride Scudder felt in his discovery of the butterfly's caterpillar on the top of Mount Washington?

21. Several of these images of caterpillars in Scudder's volume 3 were based on paintings by Mary Peart and by Mary Edgeworth Blatchford, Scudder's sister-in-law, who tended his household after the death of his wife, Ethelinda. Among the caterpillars are the zebra swallowtail (14), the giant swallowtail (16), the black swallowtail (17), and the pipe vine swallowtail (20).

PAPILIO.
III.

3.

2.

1.

e.

f.

b.

a.

f².

d.

d².

c.

c².

g.

Drawn by Mary Peart.

L. Bowen, Col.

TURNUS, 1. 2 ♀. 3. Var. GLAUCUS ♀.
a. Egg magnified.
b. — f Larva.
g Chrysalis.

22. A true-to-life, magnificent portrait of the eastern tiger swallowtail at all stages of development, with caterpillars of several sizes feeding on tulip tree leaves. Nothing like it had ever before been published.

MELITAEA.

I.

PHAETON, 1.2 ♂. 3.4 ♀.

Drawn by Mary Peart   L. Bowen. Col.

a.a².	Cluster of Eggs.	e²	Larva at 3ʳᵈ moult magnified.
b	Egg .... magnified	f	„ „ 4ᵗʰ „
c	Young Larva „	g	„ „ 5ᵗʰ „
d	Larva at 1ˢᵗ moult „	h	Chrysalis
e	„ „ 3ʳᵈ „	i	Completed web.

23. Another splendid portrait by Mary Peart and Lydia Bowen, with every feature of the life history of the baltimore checkerspot on display, distinguished by what Edwards called the "webs of phaeton," a community of caterpillars in a common nest, living on behalf of a common aim. Edwards thought the webs were unique in the butterfly world, but Scudder later showed that they marked the life histories of other species as well.

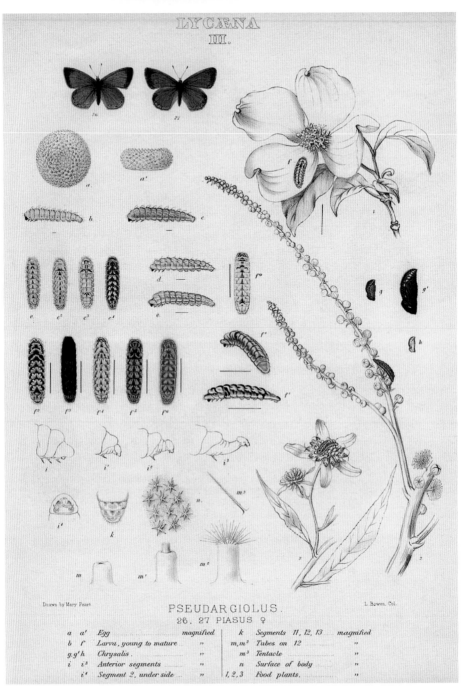

LYCÆNA
III.

PSEUDARGIOLUS.
26, 27 PIASUS ♀

a	a'	Egg	magnified	k	Segments 11, 12, 13	magnified
b	f'	Larva, young to mature	"	m, m²	Tubes on 12	"
g, g' h		Chrysalis	"	m³	Tentacle	"
i	i³	Anterior segments	"	n	Surface of body	"
	i⁴	Segment 2, under side	"	1, 2, 3	Food plants	"

24. This plate of the spring azure, in volume 2 of *Butterflies of North America* by William Henry Edwards, reveals a comprehensive mapping of the life history of a favorite butterfly of his; as the earliest butterfly on wing, it always announced the coming of spring, emerging "on the first sunny day of March." Note especially k or the magnified segments, 11–13, of the blue azure's caterpillar, which friendly ants visit in the summer, tapping on them, as Edwards observed, "like the thrumming of a piano," in an attempt to draw forth from m and m2 (the magnified tubes on 12) an irresistible sweet liquid. To ensure that they get the liquid, the ants seem to stand "guard" over and *protect* the larvae from the dangers of parasitic wasps, an example of a remarkable phenomenon in nature—the cooperative or symbiotic relationship of two very different species.

of the twenty-five larvae which I placed in a breeding cage, only five became chrysalids. From the body of each of the others, when fully grown, a number of worms emerged and spun themselves up into small, white cocoons placed with perfect regularity side by side, forming a compact bundle, usually round in form, made up of from twenty to sixty cocoons, the whole enveloped in a cottony substance. The cocoons were in every instance spun underneath the larva; as the mass increased, the body of the exhausted larva above it was raised up from the leaf or stem on which it rested, and embraced the bundle in its curve. The larva presented us with an instance of great tenacity of life; even when every portion of its body had been honeycombed by the escape of the large number of parasites which it nourished—sufficient, one would suppose, to produce speedy death.[39]

Another of the Milbert's enemies, as lethal as this fly, "calmly suck the caterpillars in one of their retreats." The common red admiral, wherever it roamed, was likely to come face-to-face with wasps eager for its company, one wasp attacking the larva before it "completed its very first nest," and another overtaking it so completely that it "is often difficult," Scudder wrote, "to procure a single butterfly from a large number of larvae taken in the open air; they are crammed full of parasitic enemies, which suddenly emerge through the skin of their victims when full grown, and spin their pure white cocoons beside the now collapsed prey." Still another "crowded" the butterfly's larvae "full of worms," producing "coal-black chrysalids," that "stood erect in their hinder ends around the corpses they have destroyed, like black tomb-stones in a cemetery."[40]

Life for butterflies was a horror show, and Scudder, along with Edwards, Grote, and others, marveled at the butterflies' capacity to endure it. Edwards pondered: How did the rarest of American hairstreaks, a diminutive thing with turquoise underwings banded irregularly with red and white spots, manage, eons back in time, to chart a path across America, from Maine to Arizona, where it "seems to have found its true home"? "It is wonderful when we think of it," he noted, "that a delicate butterfly, expanding scarcely more than one inch, should have found its way through a whole series of States, even into Canada."[41] Or how did caterpillars, riddled by parasitoids or nibbled to near death from within, find the energy to go on living? How did female butterflies know to deposit their eggs on the right larval food plants? Edwards wrote eloquently about this habit of the female, characteristic of all butterflies and moths:

> Nothing seems more wonderful than that the egg should invariably be laid on the food plant proper to its caterpillar; for very few caterpillars are omnivorous, but nearly all will feed on two or three, and often on one species only of plant, and if they do not find the right plant they die of hunger. It would seem as if the butterfly has a remembrance of her former caterpillar state. Now she is as different as possible, a creature of the sun and air, eating no solid food, for she has no mouth, but lives on liquids drawn up through a tube; then she was a crawling worm, and voraciously fed on leaves, cutting them with powerful jaws. And between these stages there has intervened another that would have seemed to have divided them completely, certainly to have extinguished all recollections in the butterfly. And yet she seeks the particular plant her caterpillar must feed on, and finds it.[42]

Grote thought it even more "wonderful" that *moths* found the right food plants, since they did all their searching "in the dark."[43]

Many butterflies managed to cope—even conquer—the nightmares nature dealt them. In every case, Scudder, Edwards, and Grote maintained that how these insects related or failed to relate to their complex environment made all the difference. *Adaptive* ability, in other words, sharpened and honed by chance (not by choice) over many, many years, sealed their fate. Scudder said sweepingly what the others also believed. "Observe," he wrote in his excursus 64, "Postures at Rest and Asleep," "how wonderfully alike are the actions of butterflies of the same group, i.e., descendants of the same stock; their habits have become ingrained by repetition through the ages; habits which it were almost certain destruction not to obey, since *in nearly every one protective meaning may be found*."[44] Every facet of a successful butterfly's identity, from egg to metamorphosis and adulthood, was an outgrowth of the butterfly's need to fight off assault.

Some butterflies survived environmental dangers because they had acquired hibernating abilities. Others became multibrooded. Still others produced eggs and pupae sturdy enough to withstand even the most deadly winter sieges. Sometimes, as both Scudder and Grote showed, geological shifts carried butterflies off to new homes far from their origins, where they lodged themselves, such as the White Mountain butterfly, in New Hampshire. Other lepidoptera acquired another skill: they migrated when things got tough. Scudder wrote of sudden butterfly swarmings and sudden disappearances, and of interlopers descending from the north and invaders from the south, such as the giant swallowtail, an ordinarily southern species, "now established in the extreme southwestern corner" of New England, and the buckeye butterfly, with an apparent "tolerably secure foothold in southern Connecticut."[45] He recounted the life history of an "invasive" colonizer par excellence: the cabbage butterfly (*Pieris rapae*), a small white European pierid that, let loose on a grand scale, became a major agricultural pest, one of the few such in the butterfly world. Said to have been introduced into the United States from Quebec in 1860, it had spread southward to the gardens of New York City by the decade's end. A keen-eyed Theodore Mead reported, "I myself found it swarming about parks in the heart of the city in June 1870." By 1880, the European pierid had taken up residence nearly everywhere in North America, an achievement Scudder mapped from beginning to end in one of the first visual records ever conceived of such activity, using data from his many correspondents.[46]

If *Pieris rapae* was a colonizer, not a migrator, the monarch butterfly was the *real* thing, migrating south in the fall and back again in the spring. Scudder named it the monarch (rejecting "storm butterfly" and "milkweed butterfly," among other names) and helped lead the first investigation of its movements, variations of which have continued to this day, with literature more vast than that on any other butterfly.[47] Monarchs must have been flying back and forth for many years, but not until the 1860s and '70s, after Americans had embraced Western natural history that created a *context* (context is *all*, after all) in which *looking* at such things made sense, did Americans suddenly begin to observe the monarch's habits. Everywhere men and women reported monarchs flying by the millions in September in gigantic, undulating waves extending for miles; like the passenger pigeon migrations of the age, these swarms sometimes obscured the sun, blurring day into night.[48] Monarchs still migrate this way, but nowhere near as phenomenally as in the nineteenth century.

Unlike Edwards, Grote, Scudder, and Strecker, Charles Valentine Riley was an insect generalist throughout his career. But he did original work on butterflies, including being the first to describe the monarch's life history, in 1868, in the *American Entomologist,* which he edited with Benjamin Walsh. He was also the first to see, if with limited proof, that the butterfly traveled south and returned in the spring. Scudder stated his own views in his life history of the insect in *The Butterflies of the Eastern U.S. and Canada* and in his essay "The Swarming and Migration of Butterflies." "I believe," he wrote, "there is over the entire extent of the country inhabited by it, at least east of the Rocky Mountains and north of the Gulf States, periodic movement of the butterfly, in the south in the season which corresponds to the end of September in New England, and to the north in the time of the first season of egg laying."[49] He based his conclusion on the ample evidence he received in response to the hundreds of letters he sent out to "many intelligent observers stationed" around the country. "I find that the milkweed butterfly is beginning to swarm this autumn," he wrote his friend Charles Fernald in Maine. "Could you send a few statistics on the point, such as an estimate of the numbers seen within a given period; the general direction taken by the butterflies; how much variation there appeared to be from that direction; whether there was any reversal of direction," and so forth. "It only needs that you should have your weather eye out." At the same time, Scudder called for "systematic and concerted study over a wide extent of territory before a satisfactory solution can be expected" (not achieved, however, until the 1970s, nearly one hundred years later). Scudder believed that the monarch lived longer than any other butterfly (up to sixteen months) and that, in New England (and everywhere else as well), it was mainly single-brooded. But that, too, begged for more "systematic and concerted study."[50]

Not everyone agreed that the monarch migrated south, like the birds. Edwards, for all his observational acumen, thought the idea ludicrous, ridiculing Scudder for even considering it and for generalizing, as he so often did, Edwards charged, on the basis of "one or two facts." Rather, the monarch hibernated where it was born and bred (so to speak), going nowhere; it had a short life, said Edwards, like every other butterfly, three or four weeks, with the exception of insects that hibernated as larvae, eggs, or pupae. Moreover, the monarch was not single-brooded, as Scudder contended, but in places like Edwards's home, the Kanawha Valley, of West Virginia, triple-brooded, and elsewhere possibly "more-brooded." One shouldn't generalize, he insisted. In the end, Edwards proved Scudder wrong about only one thing: the number of monarch broods, a mistake Scudder conceded.[51] On the other questions, Scudder proved victorious (although Edwards may have never acknowledged it). Many adult butterflies *did* live longer than three or four weeks, and, of course, the monarch migrated. While other butterflies died from the impact of icy cold, this butterfly, for some unknown reason, had learned to adapt by going south.

Of all the trickery evolved by butterflies to survive, none aroused naturalists' curiosity more than the way they modified their bodies and wings to frighten or mislead their predators. Many hid from view under rocks or brush, in holes or nests. Alarmed by something, they might drop to the ground and "feign death," as Grote suggested, although he preferred the phrase "keep still," since lepidoptera, like other animals, had no idea what death was and could hardly "feign" it.[52] ("Whether insects can have any knowledge of death, as such," Grote observed, "may be a matter of opinion, but I would as soon credit them with a knowledge of history.")[53]

Butterflies also sometimes "swarmed" or "assembled" in ways to bewilder or confuse, as Grote put it, or, being "shy," appeared in public infrequently, ready to flee into the shadows at the least hint of danger, as Will Doherty speculated.[54] Many species used "Timidity as a Source of Protection," Doherty observed in a letter to his mother, wondering why Darwin and Wallace had "left" this "subject untouched." "You see in cold countries," he explained, "the struggle for existence is chiefly carried on against adverse climate; the cold weather kills ten butterflies where their enemies kill one. But in tropical climates animals chiefly have each other to contend against. They are always chasing or chased."

Several butterflies, almost "dying out," had survived by "being shy," he noted. "They are peculiar to these hills [the Kanan Devan Hills in southwest India] and are exceedingly timid and hard to catch."[55]

But butterflies found even better ways to deceive: they acquired the ability to mimic the forms, shapes, and colors in the natural world around them, so far as to change, over extremely long periods of time, the very character of their physical identity. Mimicry is most commonly found in insects rather than in other animals, and has been most often studied in butterflies, the term itself first appearing in the 1815 *Introduction to Entomology*, by William Kirby and William Spence.[56] No naturalist took the subject seriously, however, until Henry Bates, the English explorer and naturalist, propounded, in 1866, the first and most enduring theory of it, fruit of his fourteen years on the Amazon River and his study of numerous tropical butterflies and profoundly influenced by his understanding of Darwin's theory of natural selection. Bates observed that many *palatable* species, attractive to predators, gain protection from their enemies by successfully imitating or mimicking *unpalatable* butterfly species. But how did such a remarkable thing occur in nature? And how, moreover, did some species succeed in their resemblances, while others similar to them and living in the same locality did not? The answer: "Natural selection, the selecting agents being insectivorous animals"—and these could be anything from birds and lizards to other insects—"which gradually destroy those varieties that are not sufficiently like unpalatable" forms to elude predation.

"When the persecution of a variable form is long continued, the indeterminate variations naturally become extinct; nothing remains in that locality but the one exact counterfeit." The insectivorous animals, in other words, actually do the work of God by "destroying variations unsuitable to the locality" or for survival, thereby allowing the fit or successful butterfly counterfeiters—or those immune to predation—to emerge, thrive, and reproduce.[57]

Bates's breakthrough essay, with its unprecedented empirical proof of a theory that Darwin himself had not effectively demonstrated, released a wave of thinking on mimicry throughout the West and the far reaches of the Western empire—in South Africa, with the work of Roland Trimen; in British India, with the studies of Lionel de Nicéville, Major G. F. L. Marshall, and James Wood-Mason; and, in Nicaragua, where the British gold mine engineer Thomas Belt, reporting on mine practices, analyzed the mimicry of the heliconian butterflies.[58] In England itself, Alfred Russel Wallace observed, among other things, that females won most by imitation ("probably due to their slower flight, when laden with eggs") and that the bright colors of many insects served as warning signals to predators to back off or suffer the consequences. Another Englishman, Edward Poulton, endowed Wallace's warning concept with a name: aposematic (derived from Greek *apo,* or "away," and *sematic,* or "sign"), a term widely used today; he proved, with a series of light experiments, how readily chrysalids mimicked the colors of their background. Scudder wrote that Poulton obtained "almost at will chrysalids of different colors, according to the tints with which he has surrounded them, and so has opened a new field of experimental inquiry."[59] An ardent Darwinian, Poulton believed that "the principle of Natural Selection explains the origin of all appearances except those which are due to the subordinate principle of Sexual Selection."[60]

Americans, too, were influenced by Bates's mimicry theory, which goes far to explain why they were so interested in predators, singling them out for their selecting abilities. Like Darwin and Bates, American butterfly people saw nature as a chaotic whole, generating over millions of years a staggering community of organisms, some strong enough as species to withstand the fiercest buffeting. The American interest in mimicry began in the late 1860s, with Charles Valentine Riley, who wrote about it in his own *American Entomologist,* and continuing through the 1890s with Abbott Thayer, who invented a "beautiful new law" in nature: "countershading" or "obliterative shading," whereby many animals produced a misleading blend of colors, shapes, and lines. A collector of tropical insects in his youth and a Darwinian like Riley, the Yankee Thayer wrote adeptly of mimicry in larvae and adult imagoes, at the same time criticizing the concept as too limited to explain how nature "protected" animals from predators. Influenced by Poulton's thesis that "natural selection dictated that all animal coloration prove useful," he proposed, instead, a law of "obliterative coloration," which asserted that *all* butterflies (not merely the inconspicuous ones) elude their enemies because their "surfaces" reflect the patterns and colors of their backgrounds, "causing them to pass for empty spaces through which the background is seen."[61]

In the time between Riley and Thayer, other Americans were studying how butterflies and moths at every stage of the life cycle imitated the world around them to deceive what preyed on them. The underwings of many American butterflies were notorious for mimicking leaves, as American naturalists pointed out, but the *entire* wing surface often mimicked the background. Grote, drawing on the work of the American D. S. Kellicott, described the moth *Rhodophora florida*'s imitating "the withering blossoms of the Common Evening Primrose," hiding during the day by copying, exactly, the colors (yellow and pink) of the flower. "Sometimes," Grote reported, "the pink of the wings is not wholly covered but the tone of the continuous colors is such that the harmony is complete."[62] Caterpillars, such as those of the spanner moths, the hawk moths, and the bagworms, misled predators by looking like leaves, as Joseph Lintner and Grote observed. "When discovered, a little colony were hanging head downwards," wrote Grote of the spanner species, and "were remarkable for mimicry of withered leaves."[63] Pupae, Scudder noted, such as those of the giant swallowtail and the falcate orangetip, reflected the surrounding world, the former with the appearance of "a broken bit of rough bark," the latter disguised as a "doubly sharpened stick."[64]

Butterflies protected themselves, too, by mimicking other organisms, from insects to birds. Herbert Smith, a young American explorer from Manlius, New York, who would nearly equal Will Doherty in his skill as a tropical butterfly collector, first encountered this copying in the Brazilian forests; Smith had a collection of such mimics packed away in closets in his Brooklyn apartment and, in the late 1880s, offered the collection to the Museum of Comparative Zoology. Beetles and moths that "mimicked wasps," "ants that mimic bees," and "butterflies and moths that mimic Heliconians" were included. "I have myself made special study of these forms," he told Samuel Henshaw of the Museum.[65] So had Doherty in India, observing butterfly ocelli, or eyespots, in the Kumaon region; they formed on the wings during the wet season and replicated, apparently, the eyes of owls and other predators. Ocellation protected butterflies "during the rains when insectivorous birds" were "especially numerous."[66] Near Rangoon, Doherty caught a species of *Thaumantis* that "mimicked when flying a large cicada. It swarms also in Borneo, where I have often mistaken [it] for the cicada, and *vice versa*."[67]

Scudder probably wrote more about mimicry and protective resemblance in the 1880s than anyone in America—another indication that he had turned away from Agassiz on matters of evolution. In his many life histories and essays, he observed how concealment and deception functioned in the butterfly life cycle. Thus, female butterflies sometimes laid only a few eggs to escape predators, or the eggs themselves duped ichneumons and other parasitoids by growing "flexible filaments" or hairs "so that the poor bewildered madame must struggle through a weary chaparral before she can attain the barren grounds at the summit and find a spot to readily insert her sting." At the caterpillar stage, the butterfly was driven by the need to "conceal," resorting to feeding at night and hiding by day, dropping to the ground and curling up when frightened, and moving "rapidly in motion forward and backward" to frighten interlopers; at different molts and in the pupal stage, the insect might look like "a lump of bird dung," discouraging birds from picking it off.[68]

Most interestingly, a few North American butterflies resembled other butterfly species in appearance, notably the viceroy butterfly, an orange-and-black insect that copied the monarch, a noxious insect avoided by birds and other similar predators (see color plate 4). Scudder coincidentally coined both vernacular names, monarch and viceroy, the second species, according to him, belonging to the genus *Basilarchia* (that term failed to stick; today the accepted generic grouping is *Limenitis*). This genus contained two other well-known eastern species besides the viceroy: the red-spotted purple, flying in the southern part of the range, and the white admiral, in the north, especially in New England (at least, Scudder thought they were separate species; today, experts think otherwise).[69] The red-spotted, in particular, was one of the most sought-after butterflies by novice collectors, as well as hard to catch. "Well do I remember a chase for this butterfly," recalled Frederick Clarkson of New York City in 1885, "the first that I had ever seen on the wing. It was a royal game of tag, with hide-and-seek variations. We see-sawed up and down a ravine for nearly an hour. By the time I had worked my way down over the rocks and through the briers, it was spreading its wings on the bank I had just left, and when I returned it was away again to its favorite leaf on the other side. Tired and heated, I gave up the chase, when the *arthemis*, in a most provoking way, lit upon a shrub beneath my very nose."[70]

To Scudder the purples were "the very queens of butterfly society," especially the white admiral, which he had seen in his youth crowding roadsides and wooded paths by the thousands, and streaming into the kitchens and parlors of country homes. Though visually different, all three "species" had a great deal in common: the same kind of eggs, the same larvae and pupae, and the same wing structure; they shared so much that they belonged, indisputably, to the same genus; only color and pattern differed, although the two purples also showed a good deal of orange, in the form of spots in rows along the hind wings. These orange spots were critical, explaining how and why the viceroy butterfly split from its purple chains. Both purples, Scudder claimed, hybridized at the margins of their ranges, where, millions of years ago, their progeny began to display a growing number of red spots and to fly "in the same region and at the same time" as the monarch butterfly. Slowly, the new variant capitalized on its new color until it began to resemble the monarchs and so could "venture more often into the open country than its allies, and thus gain wider pasturage and surer assistance." "If one has the slightest advantage in the fight of life," Scudder wrote in "Mimicry and Protective Resemblance; or, Butterflies in Disguise," in volume 1 of *Butterflies,* "how small soever this difference may be, it must, by the very laws of natural selection, be cherished, perpetuated, increased by slow but sure steps. Knowing what we know of the laws of life, mimicry of favored races might even have been predicted." The viceroy was a remarkable example of butterflies mimicking butterflies. It demonstrated how resourceful "natural forces" remade organisms into "protected species."[71]

Scudder (along with Edwards and Grote) often commented on how butterflies and moths concealed themselves and took different forms throughout their lives, virtually the whole of their life histories marked by shifting adaptations. At every molt, a butterfly larva took a new form; so, too, many butterflies were sexually and seasonally dimorphic or polymorphic. Pupae of the same butterfly might vary in color and form. In a sense, the real identities of lepidoptera were hidden, invisible strands passing through a series of disguises. "All the metamorphoses and especially such complete changes as are undergone by a butterfly during its varied life from egg onward," Scudder wrote in his remarkable excursus "Hypermetamorphosis in Butterflies," "are acquired characteristics, gradually gained in the struggle for existence by adaptive devices."[72] To put it another way, the identity of butterfly species was an outgrowth or consequence of the need to defend in order to survive. That butterflies (and other similar animals) were not what they appeared to be was, in fact, what they were.

Butterflies bonded with one another or with other species to protect themselves against environmental dangers, proof of which came for Edwards when he studied diverse butterfly nests, called hibernacula, woven by many larvae working together in behalf of a community of caterpillars. For months on end, in Coalburgh, he watched the nest building of a lovely black, white, and red nymphalid, "the Baltimore checkerspot" (*Euphydryas phaeton*); he may have been the first naturalist to piece together its life history, for many years considered unknowable in its entirety.[73] Edwards began studying it in 1868 after a Fraser boy in Coalburgh brought him several pupae from near the Fraser house, which, henceforth, Edwards called "Fraser's swamp." The boy had found the butterfly's favored food plant, *Chelone*, with hundreds of black larvae clustered about the stems, eating them down to the water level. From that point on and for the next fifteen years, Edwards paid regular calls on the swamp to see the caterpillars erect their nests, some rather large and filled with many occupants, divided into groups or squads, each caterpillar executing different tasks. At night, after closing up the exits and entrances, "to keep the spiders out," the caterpillars "stayed within the nest." On the third molt, they stopped spinning in preparation for hibernation, which would last from August to the end of winter. Edwards was convinced that the caterpillars spun their webs because "they anticipated storm[s] and were providing against [them]," and that their web-building habits could be found nowhere else in the butterfly world.

Much debate over Edwards's claims in entomological circles preceded the publication of the baltimore checkerspot's portrait in his *Butterflies of North America*. Both Alpheus Packard and Scudder argued in print against his idea that the baltimore was unique in its nest-building habits, the tone of their criticism so sharp as to offend thin-skinned Edwards. "The inference was plain," Edwards wrote to Scudder, "that I was an ignoramus in these things, and did not begin to know anything."[74] Scudder remained firm, publishing in volume 3 of *The Butterflies of the Eastern U.S. and Canada* a plate illustrating the nests of several species. On the other hand, his own life history of the baltimore checkerspot cited Edwards freely and uncritically, recognizing the obvious excellence of the portrait. Edwards wrote with wit and eloquence about this insect. "How do these creatures communicate with one another?" he asks at the very end of his account. Does one "master" caterpillar oversee and coordinate the work of the web, or do *all* the caterpillars have "something akin to the knowledge and judgment of far superior beings which leads each one to see what is needed, and to do it without compulsion and without conflict or interference from others? I wonder if all is really harmony; if some do not shirk their duties; if there be not bickerings and fightings and larvicides! Let us hope not. They seem to dwell in peace, and we will assume that they do, and go to them for a lesson as to Solomon's ants or Sir John Lubbock's wasps."

Edwards observed another gregarious pattern in the butterfly world: the partnerships or symbiotic relationships a few butterflies formed with other insects, protecting them from parasitoids, those predatory "fiends" that fed off of their hosts until they died. One such bond existed between the spring azure—or *Lycaena pseudargiolus*, to use Edwards's Latin name for one of his favorite butterflies—and ants.[75] After Theodore Mead found the larva in 1876 and Edwards himself the food plant (rattleweed, or *Crotolaria retusa*, an aromatic flower), Edwards wrote up a full life history for the *Canadian Entomologist*, a trial run for his 1884 portrait in volume 2.[76] In both versions, he related how, on a walk along the edge of a wood near his home, he had seen ants scurrying over the tops of the rattleweed where several blue butterfly larvae were feeding. The ants ran up and down, "caressing" the larvae—behavior the larvae "no way resented, not even withdrawing their heads from the buds they were excavating"—and stopping to "linger about the last segment" of the caterpillar, tapping it gently like the "thrumming of a piano." Edwards thought that this segment might have an organ of some kind that emitted a sweet liquid the ants found irresistible. He was not sure. Again and again, for five or six years on warm summer days, Edwards looked for rattleweed or dogwood (also a blue butterfly food plant), taking a "hand glass" with him, until, on a bright June late afternoon, he saw on the hill behind his home, radiated in light, ants on rattleweed feeding on two tubes on the twelfth segment of the caterpillar. "The ants fastened greedily on the tubes. I saw sometimes two put their heads onto them and drink to the last morsel," he reported.[77] Edwards sent some larvae to Joseph Lintner, in Albany, and Hermann Hagen, in Cambridge, for their analysis; both confirmed what he had seen. "The greatest discovery I have made this year," he told Henry Edwards in 1877, "is that of the ants attending larvae of Pseudoargiolous."[78]

But what sort of adaptation, wondered Edwards, was this bonding? Surely the blue butterfly was not doling out sweets just to be kind to ants. There had to be an adaptive purpose. A letter from August Weismann broached a simple suggestion: "Observe what enemies the larvae have. It is conceivable that there are such enemies as are afraid of ants." Sure enough, when Edwards looked around with his magnifier, he spotted at least "three species of parasites about these larvae," then, a fourth, an ichneumon fly, which, he learned from Ezra T. Cresson, at Philadelphia's Academy of Natural Sciences, was adapted to attack the larvae only "at the last two stages." Ordinarily, these parasitoids took no prisoners, but in this case, the ants stopped them in their tracks. On one June day, "in the woods," Edwards watched as an ant defended a caterpillar:

> I saw a mature larva on Rattle-weed, and on its back, facing the tail, stood a large ant. At less than two inches behind, on the stem, was one these [ichneumon] flies, watching its chance to thrust its ovipositor into the larva. The fly crawled a little nearer and rested, and again nearer, the ant standing motionless, but plainly alert and knowing of the danger. After several advances, the fly turned its abdomen under and forward, thrust out the ovipositor, and strained itself to the utmost to reach its prey. The sting was just about to strike the extreme end of the larva, when the ant made a dash at the fly, which flew away, and so long as I stood there, at least five minutes, did not return.

So here was the secret to the blue butterfly's immunity from the evil one: a valiant centurion was prepared and determined to beat off all barbarians.

Will Doherty discussed this symbiosis in his 1886 work "A List of Butterflies Taken in Kumaon" (the Kumaon is a mountainous region in northern India near Tibet), two years after volume 2 of *The Butterflies of North America* appeared, writing that he knew of only one other naturalist who had observed the relationship, Frederic Moore in his 1880 *Lepidoptera of Ceylon*. Doherty was unaware that Edwards had done anything on the subject, although Edwards knew of Doherty's work, having heard about it from Lionel de Nicéville, an expert on Indian lepidoptera and the resident curator of the Natural History Museum in Calcutta (later the Indian Museum). Doherty had much to say on the subject, including discussion of the eleventh segment of the lycaenid larva, with a "tubercle exuding a viscid juice." "It exists in all the Lycaenidae known to me," he declared. "It is peculiarly attractive to ants, which at all hours surround the poor caterpillar and, by stroking and tickling it with their antennae, induce it to yield up this sweet liquid." He added another fascinating detail: that the ants pick up the caterpillar and "deposit" it, he wrote, "in an open space just within the mouth of their nest, whereupon the latter immediately attaches itself to the bark and commences its transformation. I have counted as many as thirteen chrysalids so attached in one nest at the foot of a kind of babul tree. All were uninjured and produced perfect butterflies. The instinct which induces the ants to preserve these caterpillars in their nests, thus sacrificing a large present supply of food to the possibility of a future supply of sweet juice they are so fond of, strikes me as one of the most remarkable things in nature."[79]

Edwards may not have been the first to note this romance between ants and blue butterflies (indeed, two Germans had observed it one hundred years earlier than either Doherty or Edwards), but he was the first to have it fully depicted, in a magnificent color plate by Peart and Bowen showing not only the tubes on segment twelve but every other feature of the butterfly's life history—food plant, egg, pupa, and larvae in their various molts. And he may have been the first to understand the character of the symbiosis: "The ant saved the larva, and it is certain that Ichneumons would in no case get an opportunity to sting so long as such a vigilant guard was about. It seems to me that the advantage is mutual between the larvae and ants, and that the former know their protectors, and take satisfaction in rewarding them."[80] This life history beautifully revealed the butterfly people's conviction that an understanding of butterflies required knowledge of how they *related* to other animals, or, to quote Scudder, of how they made "friends and associates" out of other species.[81]

How eye-opening it must have been for ordinary, educated men and women to have found out that several butterflies, once understood as species, separate and distinct, were really the *same* species at different seasonal times or in different geographies. Or that you could not know the identity of any one butterfly simply by inspecting the adult insect; you would need to examine all the forms it took throughout its life. Or that evolution could be glimpsed—in Scudder's and Edwards's portraits of the red-spotted purple, viceroy, tiger swallowtail, and related species—through the wings of a butterfly. Nature was not only a violent and fierce place full of devouring flies and wasps but also a vast reserve of strategies and maneuvers to aid butterflies and other animals in their battle for survival. Nature was a destroyer and a protector in equal measure. At the same time, the achievement of the butterfly people invites further speculation. Why, for instance, did they—both Americans and Europeans—see so much protective resemblance and mimicry in the natural world when, before the 1850s, the subject had never come up except in relation to human behavior? For thousands of years, no one had scrutinized the rest of nature and seen what Bates, Wallace, Müller, or Scudder now saw. Certainly, the spread of interest in natural history throughout the middle classes, to say nothing of the scientific revolutions themselves, with their new ways of seeing and observing, had a pivotal impact, as did Darwin's theory of evolution, an influence noted by Scudder in 1889 and, ever since, by naturalists. Another likely cause may have been the increase of travel and exploration by Westerners throughout the tropics, from the eighteenth century onward, step by step exposing people to a staggering variety of beings never known before, and, in particular, to mimickers whose numbers *far* exceeded anything found in colder climates.

The invention of photography, in the 1830s in Europe, may also have had an influence; by the 1850s it had spawned thousands of studios, where people went to sit for their portraits, promoting a new fascination with the human image. If people could mimic themselves—make copies by machine, as it were, that might outlast them and assume an independent life—could nature not "photograph" species so that one might look like another, with a similar aim in mind? In a brief 1888 essay, "The Origin of Ornamentation in the Lepidoptera," Grote used the word "photography" to express the effect produced when "wing patterns" were "reproduced in some species exactly, and in some whole families in the style of a rougher copy." All arose, he said, from a single "primitive band produced by an *outside* process, the effect of light and shade on the wing itself." "Under the murky skies of the Carboniferous the colors of the insect remained dull. Upon this plain wing, the first shade or marking may have arisen by a process comparable to photography, the action being produced by the same chemically acting ray of light."[82]

Yet another condition, paralleling and shaping all the other conditions, may help to explain why so many were ready to find protective trickery everywhere. By the 1830s, industrial capitalism had changed England, and after the 1840s—especially with the expansion of the railroads—much of the United States, into a market economy. Dependent on expanded educational access for all peoples, marked by great innovation and invention, but nearly unregulated and unchallenged, capitalism altered the character of human destiny.[83] Before such a revolution, most people belonged intimately to local places, usually in agrarian settings, and did not readily circulate; to quote the historian Karen Halttunen, "All reacted on each other in a hundred ways" and "knew each other by innumerable means." But as capitalism converted both land and labor into fungible commodities, mobility replaced rootedness as a hallmark of human relationships, and millions of human beings migrated to growing cities for employment, to encounter people they would otherwise never know and often had reason to fear. The only clues to the identity of others beyond the immediate neighborhood would be visual, and many began to worry that city dwellers could not be "trusted," that they hid behind masks or dissembled to get what they wanted. An entire literature grew up around this concern, climaxing in the United States in Herman Melville's great novel of 1857, *The Confidence-Man: His Masquerade*.[84]

It would be a mistake to overemphasize economic culture as causal in matters of scientific change. Human mimicry is not of the same order or the same kind of disguise found in nonhuman nature. People who "mimic" others might act intentionally and, often, unethically; animals behave unintentionally and never unethically, as Scudder carefully pointed out in his essay "Mimicry and Protective Resemblance." There is "no intention in the case so far as mocker and mocked are concerned." But as Scudder also implied, naturalists may have begun to employ such words as "mimicry" and "trickery" because the words—as well as the reality—had become such an intrinsic part of the everyday life of human beings. He warned of the dangers of employing this discourse, because it tended to obscure what was actually going on in nature. " 'Imitation' and 'mimicry' both imply intention," Scudder wrote, "but the limits of our language compel us to use figurative speech; we have no word to express unconscious mimicry." At the same time, reliance on this language may have actually helped open the eyes of naturalists to mimicry in nature; after all, it *did* exist, if not in the same degree or in the same way.[85]

It is a curious thing that whenever Scudder and others put together the "lists of enemies" that butterflies had to deal with in order to survive, they failed to mention human beings. This is especially notable for Scudder, whose books were big, and whose lists covered birds, wasps, spiders, numerous other animals, the weather, glaciers, and floods but omitted people. How far had these naturalists come, if they had come at all, from the earlier Humboldtian outlook that affirmed a wholeness of nature strong enough never to break apart or fragment? Did they believe that humans had no harmful impact on other species?

The proof of human harm was growing, and if it hadn't worked its way into the lists, all the butterfly people suspected it was happening. They believed, for instance, that the Xerces blue, a pretty little lycaenid butterfly found only in and around San Francisco, had vanished as a result of habitat destruction caused by development. Although the theory was later proved incorrect (the insect was seen still flying in the 1940s), many butterfly people, drawing on their own experience, judged that the Xerces blue had disappeared, including Strecker, who was told about it by Hans Hermann Behr, a resident of San Francisco.[86] Strecker feared the erasure of what he called the "paradisiacal spots" that had given rise to much of butterfly life, yet year after year, as large-scale farming advanced around Reading, Pennsylvania, such spots were "becoming more rare," he lamented. "It has cut me to the soul many a time to see just such places burnt over, strewed with lime and ploughed up to raise wheat to make bread, to keep the worthless souls in the worthless bodies of worthless human beings which live and die without leaving the slightest vestige of a footstep 'on the sands of time.' "[87] Strecker told amateurs how to kill butterflies, but, just like Grote, he knew the value of the totality of life out of which the butterflies came. Save one and you save the other.

In *Genesis I–II: An Essay on the Bible Narrative of Creation,* written in 1880, Grote bemoaned the "American conceit that he who has the most money is a great man. . . . We value Science chiefly for what it will bring in money and comfort. The results are that our industrial enterprises take the form of monopolies, our lands are falling into the hands of fewer owners, and we are wasting our natural resources." Ten years later, he concluded that "everything degenerates at the hand of man," a paraphrase perhaps of Rousseau's "everything degenerates in man's hands," although Grote did not acknowledge its source in *Émile.*[88] Sounding a bit like Strecker, Scudder mourned, in an excursus titled "Local Butterflies," that "our cultivations have made much havoc with our butterflies, for as one spot after another, is brought under drainage, the plants become for that locality extinct and with them butterflies depending on them for food." And in another excursus, "The Spread of the Butterfly in a New Region," he described the "influence of man, and particularly of civilized man," who "is everywhere upsetting the arrangements of nature, directly or indirectly exterminating all forms which cannot endure his presence or withstand the baleful influence which follows in his train." In yet another reflection, "The Best Localities for Collectors," he noted "the rapid and wholesale changes wrought upon the face of the land by our irreverent civilization," which have despoiled "the sanctity of certain special spots passed down by successive generations of butterfly hunters."[89]

Edwards, too, had experienced the adverse impact on nature of human beings, after years of perhaps never thinking twice about it. Sometime in the early 1880s, the U.S. Corps of Engineers completed construction of a system of movable dams and locks along the Kanawha River near Coalburgh that, when opened, released a flood of water, lifting and floating vessels on an "artificial tide." Part of one of the most impressive public works projects in U.S. history, and backed by coal mine owners to make river transport competitive with rail, the new system replaced the old, inefficient towboat introduced by William Henry Edwards around 1870.[90] A rapid influx of water into Fraser's swamp, along the banks of the river near Edwards's home, destroyed the habitat of one of his most beloved butterflies, the baltimore checkerspot. Over time, the old water level in the swamp returned, but the butterflies were gone, so, at six a.m. one May morning in 1884, Edwards restocked the swamp with adult checkerspots he had bred himself, setting free six females; the next day, he released fifteen males and an additional female. The experiment was a "great success," he reported to Scudder, and Scudder concluded that the butterfly was "more enduring than most species as is proved by its requiring more violent means to extinguish life."[91]

And what of the beauty of butterflies? Were all the aesthetically pleasing features of nature—the symmetries of line and patterns, the forms and colors human beings had come to associate with beauty itself—were all these merely the fruit of the need of butterflies and other organisms to defend themselves and adapt? Was living beauty reducible only to function or utility? In 1870s and '80s, the glories of the butterfly wing were the subject of engrossing interest, and many Western naturalists—William Henry Edwards, Augustus Grote, Alfred Russel Wallace, and Samuel Scudder, among them—claimed that, yes, natural selection *did* cause the beauty. Scudder, in particular, seemed to be emphatic: The "more we contemplate natural selection, the more we comprehend how powerful an element it has been in the development of the varied world of beauty around us." Like Darwin and Wallace, who much influenced him, Scudder believed that the morphology of butterflies and moths was a response to their menacing environment. "A very large proportion of the colors and patterns upon the wings of butterflies, far larger, I believe, than is generally conceded," he wrote in his excursus "Color Preferences of Butterflies: The Origin of Color in Butterflies," "must be looked upon as protective and to have originated in the simplest possible manner through natural selection. . . . It seems that we shall have to concede to the same laws of development which have moulded the structure and form of all organized beings, the power to develop that wonderful display of color and pattern on the wings of butterflies which appeals so powerfully to the aesthetic sense of every human being."[92] Agassiz was, it seemed, dead. Or was he?

Nearly all his life Scudder had viewed the visual complexity of the lepidoptera as "a synonym for all that is delicate and exquisite," beyond anything in the insect world or, perhaps, in *all* nature; he considered the wings especially exceptional, because they contained features not explicable "as purely for the purposes of the ephemeral creature itself."[93] Part of this beauty was *invisible,* most notably the tiny scales on the scent glands of many male butterflies, called androconia. (That term, a Greek compound word invented by Scudder in the 1870s meaning "male cone-shaped figures," is still in use today.) Scudder, perhaps more than any of his contemporaries, championed the microscopic study of butterfly morphology and of the "inside" life of natural forms, examining the eggs, parts of the caterpillar, and the antenna and proboscis of the imago. The androconia occurred as a black slash on the forewings to attract females, and differed "marvelously from ordinary scales in the variety of their form and exquisite structure," as Scudder put it in one of his last essays, "Sexual Diversity in the Form of the Scales."[94] Scudder, who wrote four excursuses having explicitly to do with the color of the wings, was moved and perplexed by this spectacle, observable only through the microscope. Why were the scales so ornate and detailed when neither the eyes of predators nor the eyes of butterflies themselves (whose color vision, in any case, was weak, able to apprehend only masses of color, not specific images) could perceive them? "Who is to see and benefit by them?" Scudder asked. "Assuredly not the insects themselves; they may profit, indeed, by their function, and no doubt natural selection has perfected that to the uttermost, even beyond our ken." But these "objects" are "invisible to them." "Is there not here a beauty of form and structure which is an end in itself, subserving no material end, of no possible profit to the possessor?" Yes, he answered, and it has nothing to do with natural selection but, instead, arises from some "infinite and eternal divine force, guiding all forces, an infinite, uplifting power, which we may trust," and which human beings have "not yet comprehended."[95]

All this applied equally to the *visible*, or outside, surfaces of butterflies, for even though Scudder had recognized that, from the point of view of the insects themselves, natural selection had "borne its part in the work," he believed it insufficient to cover all the ground: "There has

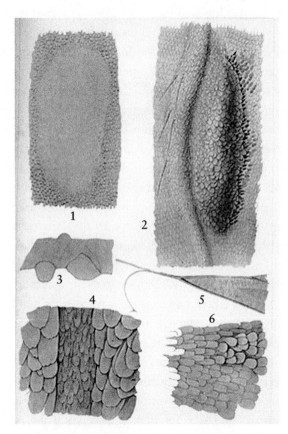

This plate, with images magnified at 150/1, reveals the scent glands of a few male butterflies, with their scales shingled one upon the other in undulating folds. Scudder called them androconia, from Greek for "male cones or scales." Note number 2, showing the monarch vein on the forewing, with accompanying pouch, concealing the density of scales of the butterfly; see, too, the androconia of the regal fritillary (number 4), which are "mingled with the scales covering the vein" on the hindwing. For Scudder, these scales had a "beauty" that seemed to serve no adaptive purpose but derived its power from some "mysterious," unknown cause, or existed, perhaps, merely for the sheer delight of human beings. See Butterflies of Eastern United States and Canada, *vol. 3, plate 44.*

not yet been brought forward one single line of evidence to show that natural selection can produce that harmony of tint and design which each of the whole tribe of butterflies displays on its individual surface; a harmony so infinitely extended when comparisons are begun that the eternities would not suffice to exhaust them." Here, too, Scudder cited a "preordaining plan" or some "higher law, which has other ends for beauty." And what were these "other ends"? *To please human beings,* Scudder asserted, in a down-to-earth, unscientific sort of way. But rather than say this in his own words, he quoted from an exponent of German aesthetic entomology. "I cannot do better," he wrote, "than translate from Adolf Werneburg, *Der Schmetterling und sein Leben* (1874): 'When we consider,' Werneburg says, 'the remarkable splendor of color which is not only peculiar to Lepidoptera in a far higher degree than to any other group of insects, but which is also displayed before the eyes of the observer in a remarkable way; and when we further remember that in many cases the color is not of the slightest use to the creatures themselves, but rather of disadvantage by its luster and brilliancy, we cannot forbear to enquire into the meaning and purpose of such a phenomenon. And here I at least am unable to find any other reply than this: that the beauty of butterflies serves to enliven and embellish, and thereby, like all other beauties of nature, to do its part in the cultivation of the human mind and heart. With this agrees the fact that it is precisely those that fly by day, when man is most in the open air, and beauty can be more readily observed, that are the most beautiful.' "[96]

In a context that was, by the 1880s, very pro-Darwin and pro-functionalism about natural forms, it was brave of Scudder to take this position. Nevertheless, earlier in this very decade he had also equivocated, attacking early on the position he would later defend. Scudder criticized Edwards, in 1877, for not "appreciating the real beauty of Nature."[97] Yet shortly thereafter, he argued in his 1881 book, *Butterflies,* in a chapter called "Ancestry and Classification," that those naturalists with a passion for the color and designs of lepidoptera actually jeopardized the cause of entomological science. "The progress in the classification of butterflies," he observed, "or the appreciation of their true interrelationships has been grievously checked by the very charm which so often attracts men to their study. There is such a rage for their collection by amateurs, enchanted only by their exquisite beauty, that their scientific study has been largely abandoned by those who are best fitted for this work by special scientific training." Eight years later, Scudder republished "Ancestry and Classification" in *The Butterflies of the Eastern U.S. and Canada,* as excursus 7, but under the title "Ancestry and Butterflies." Much of the text was *exactly* the same as had appeared in the earlier volume, although the passage above had been omitted. Why had Scudder changed his mind? Was it because he had come to see the injustice of his claim, that, among other things, it pretty much encompassed nearly all his contemporaries, that it blamed the fanatical lovers of beauty (read: Strecker) for the failure of other

This portrait of Scudder was done in 1884 by Sarah Whitman, an accomplished Boston artist inspired by Art Nouveau, a movement just emerging on the art scene. Whitman transformed Scudder into a poetic dreamer (and perhaps, at times, he was).

Americans to embrace natural science, and that it disparaged his own love of beauty as the genesis of his vocation? Possibly all the above, and also possibly the fact that, for whatever reason, he may have undergone a conversion or, at the very least, a resurgence of the confidence in things he may have learned earlier in his life. Perhaps, too, Scudder's attitude toward beauty had changed because he had fallen under the influence of the modern arts and crafts movement that affirmed beauty as a central part of everyday life. Nearby his home in Cambridge was the studio of Sarah Wyman Whitman, a follower of that movement who specialized in iridescent colors and created stained-glass windows in the manner of John LaFarge. Whitman painted Scudder's portrait in 1884, giving him the look of a colorist himself, poet of dreams and reveries.[98]

Scudder belonged to a minority of naturalists at the time who took the "charm" of butterflies and of all nature seriously in its own right; rather than shutting one door (the aesthetic) in order to open another (the scientific), he seemed prepared, in the late 1880s, to open both doors, ready to accept beauty both out of context, as it were, standing alone, in the service of the advancement of human beings, and *in context,* or in relation to everything else, from the parasitoids to the androconia, nature in all its "multiplicity and mystery," to quote Humboldt once again.

Still, Scudder was not alone. Alfred Russel Wallace, the cofounder with Darwin of evolutionary theory and exponent of natural selection, contended that natural selection did much to explain the appearance of moths and butterflies. He also believed that the sensuous surfaces of the butterfly wing and of other organisms existed simply to delight people like himself. "What could be the use of the butterfly's gaily-painted wings," he wrote, "except to add the final touches to a world-picture, calculated at once to please and refine mankind?"[99] Herman Strecker, a Humboldtian agnostic, had no need to wrestle with this question; he was merely grateful for all the glory that flooded in upon him from everywhere on the planet and, like his British mentor, William Hewitson, let no law of nature interfere with his gratification. William Henry Edwards, a Darwinian committed completely to the scientific method, "reverenced the facts and meant to abide by them," as he put it to Scudder, yet he rejoiced over the plates of Mary Peart and Lydia Bowen that enhanced and gave meaning to his volumes, and, presumably, to the nature they reproduced.[100]

Augustus Grote also shared this view. A secular agnostic and an advocate, like Edwards, of natural selection, he nevertheless hoped to salvage the emotional side of religion and often proclaimed the artistic rewards of studying moths. He would have heatedly disputed Adolf Werneburg's belief that the "most beautiful" lepidoptera are "those that fly by day." The ornate wing scales, especially on moths, Grote observed, possessed more diverse "soft and delicate colors and patterns" than even the "gaudy day flies," more than any artist could possibly desire.[101] "Entomology," he observed, "has its aesthetic side, although we are aware that the beauty is not in the object itself, but in the effect which we perceive in it. This is one of the enchanting deceptions of a world we none of us can really understand, though most of us believe that we have understood it, and many even that they have succeeded in it."[102] The beauty, Grote seemed to be saying, is *in us*, and because of that, we see it in the butterflies. Grote came close to a position held by his contemporary George Santayana (and by Santayana's teacher William James), who claimed that the *presence*, or reality, of nature's "ornaments" related to something within human beings, something at the core of the subconscious self that expressed itself as something "out there," not only as a protection from enemies or as serving any utilitarian aim, nor as traceable to Scudder's "eternal divine force," but simply as beauty, the thing itself. All we need do, Santayana suggested, is open to the subjective, emotional side of our own being, and it will happen— we will see the beauty of the world.

The butterfly science of the 1880s was a fascinating mixture of scientific confidence and human longing. Scudder, Grote, Strecker, and Edwards all saw aesthetics as bettering civilization, as beneficial to science, as something meant to satisfy or meet human needs, not necessarily only as a consequence of natural selection. An ally of science, not its competitor or usurper, beauty drew Americans to lepidoptera, setting them on paths to lifetime fulfillment.

The natural history of Scudder, Edwards, Grote, and Strecker formed a high point to a remarkable moment when, through the lens of the butterfly, many Americans became aware of all the shapes, colors, patterns, and surfaces around them that nature was capable of creating, the existing species and genera, real or mimicked, solitary or communal, the variations, dimorphisms and polymorphisms, the emergent new organisms struggling for identity and stability, in a kaleidoscope of ever-changing relationships revealed to Americans, in part, by the Darwinian revolution and, in part, by their own eyes. This experience would grow over the years, as a result of the impact of two other major changes. First, Americans would come to know not only their own butterflies but those of the rest of the world as well, a change due to the commercial expansion of the country and by its entrance onto the world stage as an imperialist power. A new breed of collector would appear, first in England and Germany, then in the United States, influenced by Otto Staudinger of Dresden, who, after 1885, expanded his collecting empire far beyond anything that had existed earlier when he began selling lepidoptera. The leading American collector would be the tragic Will Doherty, who brought the butterflies home, as it were, feeding the enormous collections of tropical and exotic species in the United States. At the same time, Strecker would come into his own as a butterfly man, embracing completely the joys of tropical butterflies and moths and serving as a portal through which such insects would pass into the hands of numerous Americans, poor and rich alike, thereby helping to transform their approach to the natural world and to life. William Holland, a wealthy Pittsburgh minister, would also be changed (and charged) by this new climate, shifting his collecting zeal almost wholly to nonnative butterflies and inspiring others to do so. At the end of the century, he had amassed a huge collection and become the director of the Carnegie Museum of Natural History, the most advanced such American museum of the age, a model for all the others, founded and funded by Andrew Carnegie, America's greatest steel magnate, builder of steel rails for railroads.

The second major change of the times was the industrialization of the country, which brought with it the production of an artifactual diversity and an aesthetic power nearly equal to that of the natural world. Industrialization helped change or redefine the way Americans viewed the whole of the natural world. This was an extraordinary moment: beauty of a natural kind appearing in tandem with beauty of an artificial kind, each competing for dominion over the American soul.

Part Two

ENCOUNTERS WITH THE
BUTTERFLIES OF THE WORLD

SIX

In the Wake of Empire

Before 1875, the market for butterflies in the West was essentially a domestic affair. To be sure, many affluent collectors bought tropical species and composed influential books about them, among them Dru Drury and William Hewitson in England, and Jean Baptiste Boisduval in France. Poorer collectors, too, such as the actor Henry Edwards and the stonecutter Herman Strecker, in America, turned to whatever means they could—some perhaps unscrupulous—to procure exotics. Yet despite this trend, most people collected only native insects. In Germany, according to a dealer from Ebensfeld, the "love of exotic butterflies was very insignificant, with most collectors looking with contempt upon every insect not born within the bounds of our continent."[1] For the British, even the European butterfly was beyond the pale. "No one in England would care for a general collection of European Lepidoptera! Were the specimens all truly British, it would sell well!" wrote one English expert in 1872 to the German butterfly authority Otto Staudinger.[2] Americans, too, had these tendencies. "The world is full of beautiful butterflies, but those that fly at home are the best," said Augustus Grote. Theodore Mead told Henry Edwards simply, "I give preference, of course, to our native insects."[3] Even those who could have collected abroad confronted a world closed to strangers, with the Asian subcontinent beyond the reach of most naturalists, China cut off except for the port city of Canton, and sub-Saharan Africa "just a series of coasts," the interior untouched by Westerners, its "vastness never entomologically explored."[4]

But in 1877, the acclaimed English evolutionist Alfred Russel Wallace opened his luminous essay "The Colours of Animals and Plants," published simultaneously in England and the United States, with a litany of all "the natural objects" that had come to human beings "from the hidden recesses of the earth" by the "progress of discovery." "We have presented to us," he wrote, "an infinite variety adorned with the most beautiful and most varied hues. Flowers, insects, and birds, are the organisms most generally ornamented in this way; and their symmetry of form, their variety of structure, and the lavish abundance with which they clothe and enliven the earth, cause them to be objects of universal admiration. The relation of this wealth of color to our mental and moral nature is indisputable."[5] Twenty years later, Robert Rippon, a Canadian butterfly man, visited Tring, the personal museum of Walter Rothschild, on a vast landed estate thirty miles northwest of London. Rothschild, a young naturalist of the famed European banking family, possessed an unrivaled collection of tropical lepidoptera. Rippon arrived to study Rothschild's birdwing butterflies for his great catalog, *Icones Ornithop-terorum.* His account covered all the known species of birdwings, distinguished by Rippon's own original colorful plates, including one of *Ornithoptera dohertyi,* almost totally black, the blackest of birdwings, which Will Doherty, America's greatest tropical collector, had caught in what may have been one of his blackest moods, on Talaud in Malaysia.[6] It was endemic only there, and so rare that a skilled German collector sold it to the Royal Museum of Natural History in Berlin for more money than any other butterfly. In 1903 Rippon renamed it *Troides dohertyi,* and that name still stands.[7]

Butterflies, Rippon wrote in his catalog, in words more ecstatic than Wallace's, were "one long vision of beauty in form, variety of pattern, and delicacy and splendor of colour, illustrated by tens of thousands of species ranging from a few millimetres in expanse to ten or eleven or even more inches; their patterns simple in the extreme, or so intricate and complex as to bewilder the eye." The "splendor of color, varied and wonderful" included "every imaginable tint, from black or white to dazzling crimson, scarlet, blue, green, pearl, silver and gold; with markings sometimes resplendent with apparently pure gold, silver, copper, aluminum, and the colour and flashing of all precious stones— prismatic, silky, velvety, diaphanous, quite transparent, intensely white, or intensely black, or ivory-like; with colour reflections in the most unexpected places; with changes of colour hardly dreamed of by the artist, yet so beautifully harmonized as to create astonishment in any sensitive mind." "Some of these glorious things" could be "found in all climes and at all altitudes, from far within the Arctic circle to the Equator, and from sea-level to 18,000 feet of mountain height! A truly royal

Divine gift to the earth is this one order of animals alone! But the glory of it all is that we only begin to dream of the wealth of creative wonders and beauties as we contemplate these."[8]

From the late 1870s, ordinary Germans had been able to buy a single *Morpho menelaus* for two Prussian talers, or $1.50.[9] "Twenty years ago," observed Otto Staudinger in his own enormous 1889 catalog, *Exotische Tagfalter* (Exotic Butterflies), "blue Morphos and magnificent birdwings were a rarity in a private collection. Today tropical treasures are accessible to naturalists who have at their disposal only modest means." Any amateur can now "command an overview of the whole butterfly fauna of the tropics."[10] In the Linnaean age of the mid-1700s, naturalists knew of little more than 270 species of butterflies, and 460 moths, but these numbers were very quickly overturned by the course of events, as explorers and collectors discovered more species. "The number of these beautiful animals is very great, though Linnaeus reckoned" only several hundred, wrote the Comte de Buffon in 1793. "The catalogue is still very incomplete. Every collector of butterflies can show undescribed species; and such as are fond of minute discovery, can here produce animals that have been examined only by himself."[11] By 1880, according to Samuel Scudder, the number of recognized or described butterflies rose to ten thousand.[12] The availability of such "creative wonders" to more and more Americans, as well as to people throughout the West, represented something new under the sun.

The market for exotics dawned unexpectedly and rapidly in the last decades of the nineteenth century, as Western powers swallowed up tropical societies in Asia and then in Africa.[13] England, the archimperialist, annexed outright, among other places, North Borneo, Brunei, the Malay States, Sarawak, Egypt, and India's last vestiges of autonomous lands.[14] Such a power grab relied on the extended telegraph lines and on new canals (the Suez opened in 1869), on a revolution in small arms weaponry, on the invention of large steamships, and, above all, on the railroad, "the main engine of imperialism," as historian Ronald E. Robinson has called it. By the 1880s, nearly every significant city in India had its own rail depot.[15] The railroad allowed for the rapid and safe shipment of military personnel, managers, goods, money—and butterflies. And with the railroads came the spread of actual roads, built especially by British engineers, who drove them through the densest forests to obtain the mineral wealth of India and Burma. In 1886, Randolph Churchill, secretary of state for India, gave Upper Burma to Victoria as a birthday present; three years later, a British mining company began a sixty-four-mile road from Irrawaddy, Burma, to the Burmese ruby mines, once run by Burmese emperors but taken as booty by the British.[16]

Many naturalists condemned (and many condemn it even more today) the ecological havoc inflicted by these agencies of improvement, but they readily took advantage of them, and complained when roads were clogged or nonexistent. The American Will Doherty was "quite crazy" about "the grand virgin forest" in northern India near the Himalayas, "its beauty beyond description." He deplored Malaysia's "scarcity of true forest" and trees "cut down" in Celebes, "except in the most un-get-able of places." Yet he told his mother about abandoned coal mines in Pengaron, Borneo, with "numerous roads open to them—an inexpressible advantage to the collector." And he happily traveled the road from Irrawaddy to the ruby mines seeking butterflies.[17]

In England, a turning point for butterflies may have come on the day in 1872 when Stevens's Auction Rooms in Covent Garden put Alfred Russel Wallace's renowned collection of East Indian butterflies up for sale. Stevens's, founded in 1776 to sell mostly rare books, had developed into the center for the auctioning of everything from live animals to the skulls of criminals. In the early 1800s, natural history sales took over, insects attracting the affluent naturalists of London, especially the sale of rare British butterflies, which commanded high prices for the rest of the century and beyond. The auction of Wallace's exotics that day, unusual given the English obsession with native lepidoptera, set off a buzz throughout naturalist circles.[18] A few years later, London collectors were fighting "to obtain species from all parts of the world, especially from Africa"—so William F. Kirby, a respected author of the first catalog on world butterflies, wrote to Strecker.[19] New merchants prospered, such as Watkins & Doncaster, soon rivaling Janson's, at 44 Great Russell Street, set up in 1852 to sell natural history materials. London, more than ever, served as the world's magnet for the exchange and sale of butterflies.[20] When Bates and Wallace collected for museums in the 1850s, they never veered from their primary purpose: to learn about the natural world. The new generation of tropical collectors, however, had more interest in selling than learning. William Hewitson may have pioneered this trend by dispatching his own men into Bolivia and Peru, as he and other Londoners fought over rare acquisitions like the female of the Colombian emerald butterfly, a vernacular name then popular for a species costing ten guineas a specimen on the market.[21]

British naturalists financed their own collecting trips into tropical jungles. Among the most imposing of the amateurs was Henry Elwes: six foot eight, red-bearded, and preposterous, the epitome of the worst of British imperialism, dismissive of the "inferior" peoples and hateful of Jews. (After knowing Berthold Neumoegen for a time, Elwes "wanted nothing more to do with him," he wrote Strecker. "He will find plenty of his own kind in Germany, but not here I hope.")[22] In 1879 Elwes invited the Oxford-educated millionaire Frederick Godman, a match in money and lepidoptera, to journey with him to his tea plantation in Darjeeling, in northeastern India on the border of Sikkim, where hundreds of Europeans lived and visited—many, like Elwes, in quest of rhododendrons and azaleas and the insects of the Himalayas.[23] Together the two men probed for rarities along the outskirts of the town, with an entourage of two servants, several ponies, a bird collector, a plant collector, and twelve coolies, the cheap manual labor of the day.[24]

The British market carried living as well as dead insects. One of the leading dealers in pupae and eggs, a self-described expert on the "living insect," Alfred Wailly of Clapham Road in London, bred moths in his backyard in an "immense number of cages."[25] He "distributed pupae and ova to my many correspondents in Europe and America." His transatlantic business, specializing in a family of large silk-producing moths, exchanged his progeny for insects bred by foreigners who sent pupae for him to sell. Once, from an American (probably Strecker), he received seven thousand cocoons of spectacular American moths—so many that he had to sell them cheap or destroy them.[26] In the summer of 1880, the green luna moth, an American exotic by continental standards, was in such demand that many European dealers tried to breed thousands, threatening to depress the insect's market value. Wailly reared his own supply, obtained from Americans, and placed the larvae in nut trees near his house to feed, though the bad climate endangered them and the sparrows ate them up.[27] He relied on nearly fifty individuals to send him cocoons from Ceylon, Madras, Bombay, Calcutta, and the Cape of Good Hope. He received thousands of one generic group that boasted the world's largest moth, the *Attacus* species from the Himalayan forests, where the thick foliage stayed fresh for so long that many caterpillars grew to five to six inches. After an earlier shipment pupated on their way to England and died, Wailly put his "imports" in icehouses aboard ship or packed them in double cases with ice inserted between the sides.[28]

The Europeans—above all, the Germans—helped build the basis for the market in exotic and tropical lepidoptera.[29] The French interest in exotics apparently never went far beyond Paris, and even by the early 1890s, Will Doherty wrote his father, "French traveling collectors scarcely existed." And to his mother, in 1888: "the French naturalists are not so cosmopolitan as the Germans, that is, they care for the French things only."[30] The Germans, on the other hand, already had a flourishing trade in the late eighteenth century, serving the aristocracy and the early museums; by the mid-1870s the activity had widened, due to international commerce, an influx of missionaries into foreign lands, and the launching of many German scientific expeditions.[31] In the port city of Hamburg, the wellspring of the business, firms specialized in exotics. Hans Godeffroy, sold, among other things, Indian and Chinese butterflies, and from Leipzig to Munich, *Insectenhändlung*, or people who sold insects, cultivated the foreign trade.[32] *Händler* peddled from house to house, hotel to hotel, displaying showy specimens to their potential customers.[33]

By the 1880s, Otto Staudinger was one of the most respected butterfly men in Europe with a worldwide reputation as a dealer of lepidoptera, "all others guiding themselves more or less by him," as Strecker said, admiringly.[34] He ran a transnational business far removed from his customers and more far-reaching than Godeffroy's in Hamburg, Deyrolle in Paris, or even Watkins & Doncaster and Janson's in London. Staudinger published yearly price lists and attempted to standardize prices across boundaries, often visiting London to compare his valuations to those set by the traders in rooms encircling the Natural History Museum at South Kensington. Competitors sometimes undercut him, but all measured the "worth" of their own insects by Staudinger's determinations. He sold by installments and dispensed generous credit to reliable customers, letting them "pay next year, if it is not now convenient to you," with discounts up to 50 percent to "all Americans"![35] But he would hound anyone who failed to pay, even beyond the grave. After one American customer died, he wrote to the man's wife that if "she did not explain things about the settlement of her husband's Estate he would take

Otto Staudinger.

other measures." When, in 1893, Berthold Neumoegen died without having paid up on a sale, Staudinger pursued his wife for the money.

Staudinger had cultivated special relations with poor but eager clients like Strecker, who would do anything, short of killing, to get a butterfly. Nearing sixty in 1894, Strecker asked Staudinger, "Why did God implant in us unquenchable desires, and then deny the means of gratifying them? Can there be a worse punishment?"[36] Strecker tried to resist all offers of credit from Staudinger on the grounds that "I owe not any man anything now and if I can, I want it to remain so." "I must laugh a little at myself sometimes when I think that in years past when I had nothing in the world except my collection, how I used to run up bills of 100 or so and now, when I am finally fixed with good assets in real estate and other excellent necessities, I watch every dollar I spend."[37] Then, almost in a blink of an eye, Strecker ordered several new birdwings and swallowtails, at more than $100. Staudinger, who had waited, knew his man.

A decade earlier, Staudinger had moved into grand quarters, which he'd named Villa Sphinx, built in a spacious park in the Blasewitz section of Dresden. The family occupied the top floor, while two enormous lower floors were reserved for handling, packaging, display, and storage of his unparalleled collection of insects.[38] As his quarters grew, so did his reputation, consolidated, by the early 1890s, by his publications on butterflies and by his close and strategic relationship with

Kaiser Wilhelm II. Staudinger sold large quantities of exotic material to the Royal Museum of Natural History in Berlin, among the largest and oldest in the world, founded in 1810.[39] The bond with Wilhelm II, more than lucrative, also gave Staudinger access to the Royal Museum's curators, who advanced their own careers by identifying and describing what he sent. In his last will and testament, he donated his massive collection of Palearctic—chiefly European—butterflies to the Royal Museum, with the stipulation that it remain undivided "*bis zum Jahre* 2001." The museum consented; even today, the collection is listed as "the Staudinger collection."[40]

Staudinger had, in 1871, introduced exotic butterflies to his catalog, though they consumed only a tiny part of a five-page listing.[41] In 1878, a small African section appeared; by 1880, the catalog was nearly one-third exotic, though he still advertised in English journals that he sold "principally Palaearctic" species.[42] Reflecting the trend set partly by Alfred Wailly in 1882, he added living pupae and eggs. By 1888, exotics nearly matched the European insects in a now twenty-three-page catalog, and by 1891, they had achieved parity in a catalog thirty pages long. He even issued a price list on insect larvae, many of foreign species.[43] Staudinger came close to monopolizing the market in exotics by sending a small brigade of collectors abroad, beginning, in the early 1870s, with Heinrich Ribbe, who went to Panama and Peru, catching for his boss ten thousand butterflies and moths, among Staudinger's first acquisitions of this magnitude.[44] The shift into the non-European and tropical seemed inexorable and would have astonished old-time clients, but it illustrated how many hundreds of new species had been "discovered" and described since 1856, the year of Staudinger's first catalog.[45]

Henry Stainton, an admired English naturalist, warned Staudinger that enlarging his business to incorporate the tropical would be too costly and too draining, and Staudinger agreed, in 1880 hiring Andreas Bang-Haas, a young Dane, to buy and sell the butterflies, so he might have time for his systematic work. "I hate very much the Entomological trade," he told Stainton.[46] He was now freed to do "the higher thing"—to locate and describe new species, all tropical or exotic, and to publish scientific catalogs such as his 1889 *Exotische Tagfalter.* Staudinger still had personal contacts with very select customers (such as the Americans Strecker and Neumoegen, as well as Henry Skinner of Philadelphia), extending them credit and tempting them with new "discoveries" at "50 per cent off on $100 lots."[47] But all the while Bang-Haas readied the business catalogs for distribution, carried on his own extensive trade, and proved an excellent salesman in more ways than one; soon he would marry the boss's eldest daughter, Carmen (named for her birthplace near the Alhambra in Spain).[48]

Germans pursued and studied nature throughout the world, many molded by a grand liberal cosmopolitan tradition that marked German culture from the late 1700s on and that acted as counterpoise to a late-blooming, strident nationalism.[49] A remarkable number of Germans lived throughout Asia and Malaysia, working as civil servants for the Dutch, once a serious colonial power but, after the 1880s, second fiddle to the British, who chose strategically to let them rule the Malay Archipelago.[50] The English, too, employed German naturalists in India and elsewhere, out of respect for their scientific education.[51] Altogether, they formed a geographical base for European and American collectors and dealers, one sustained in the Asian subcontinent by a string of German hotels and boardinghouses stretching from India into Malaysia, erected near railways and in port cities.[52] In Penang, in 1886, Doherty stayed at the Nederland Hotel, run by Germans, staffed by Chinese, and catering to "Europeans," as "all the white people" were called, by Doherty's account. At four in the afternoon he had tea and toast; at half past seven, the hotel put him "next to Fraulein Grunberg," and the dinner guests conversed "in six or seven languages" to the accompaniment of a music box playing "La Donna è mobile" and "Robert, toi que j'aime."[53] Doherty flirted with the Fräulein, as he did with other young Western women who, for whatever reason, trailed along in the wake of imperial rule.

Among the other Germans who collected in the tropical or exotic regions was Hans Fruhstorfer, whom Doherty would befriend and who was self-taught in five languages. Fruhstorfer bought and sold butterflies to support his systematic studies. By the early twentieth century, he would become an admired authority on tropical butterflies.[54] Fruhstorfer commanded the arc of tropical trade, from Brazil to Java, as few ever did. As a youth he had looked at pictures of morphos and read about them in travel books, but he wanted to see for himself those living "shimmering light-blue mirrors," as Vladimir Nabokov would later call them, the apex for Fruhstorfer of all that was good in nature, flying most abundantly in southern Brazil "in the height of the tropical summer."[55] As a youth of twenty, he collected in 1886 along the Capivari River near Rio, where, in a "lonely" valley, he saw hundreds of morphos "floating along as if lost in a dream," searching for the "tangled shores of a crystal waterfall," where they briefly settled down among other "butterflies of all sorts." "Just as wild animals tread down regular tracks in order to reach water," Fruhstorfer wrote, "so the Morphids assembled here daily. They did not come in crowds but singly, floating along quietly. And how patiently one waited, until after some minutes of silent expectation a second iridescent form appeared, to be captured with the almost unfailing certainty of long practice as soon as it ventured within reach of the net." Sometimes, in the course of a single day, he took easily one hundred morphos.[56] He made a big profit selling them through his father's business in Berlin, enough to finance the next stage in his career, in Ceylon and Java, by which time he was doing a profitable business directly with Berlin's Royal Museum of Natural History.[57] "I collect only to make money from the butterflies," he wrote Strecker, but "I am not a dealer of the size of Staudinger."[58]

The tropical butterfly business also attracted Americans, who benefited immensely from the British and German achievements. After 1885, the appeal of exotics grew, expressed best by Henry Skinner, the editor of the popular *Entomological News,* whose enthusiasm for butterflies beyond American shores was intense. In 1889, the year the *News* first appeared, he wrote Strecker that the journal would be "devoted to the species of the world," far beyond anything known before in the United States.[59] And when the American assault on the Philippines began in 1898, he glimpsed exciting prospects for collectors. "Now that the U.S. will add to its territory, the question arises what will entomologists do in the matter?" Oppose the war? No. Stick to American fauna? No. "The proper plan for students to adopt would be to ignore political and geographical lines and take up for study some genus, family, or order of the world."[60]

Long before this time, a few brave Americans had sought butterflies below the equator, notably, Titian Peale from Philadelphia and also Thomas Horsfield of Bethlehem, Pennsylvania, who, in the 1820s, had traveled to Java under the aegis of England's East India Company to make the first major collection by a Westerner of Javanese butterflies.[61] But neither he nor Peale cared about making money from insects. Not until the 1880s did such a quest for gold begin, when some Americans, risking body and mind, set out for South America, where "trained collectors" from England and Europe had "made thousands of the dollars from collecting," as Willis Weaver from Salem, Ohio, told Strecker in 1878.[62]

Weaver, untrained and naive, could not bear "the difficulties of climate and climbs." "Heat, fatigue, hunger, and the desire of finding a night's lodgings swallow a man up so that he is no longer a responsible being," he wrote.[63] Fred Knab, a young German-American from Chicopee, Massachusetts, funded by subscribers, took a steamer to Brazil, but he suffered from "clouds of mosquitoes" that tortured him at every turn, and stinging ants that hid in his shoes and clothing.[64] Lacking a compass, he lost his way in a pathless woods on a cloudy day, getting out only by following a streambed. But in two years he captured two-thousand specimens, many in the towns, where butterflies clustered "everywhere about the houses," "alighting under the verandas and often entering rooms," in a rainbow of colors.[65] In the town square of Santarém, where some Americans ran a distillery that catered to a large German population, he once saw hundreds of morphos—"*die grossen Blauen*"—pooled on a spot on the road so saturated in schnapps, a strong ginlike drink flavored with fruit, that he could pick the lovely insects up with his fingers.[66] Another, even more successful, German-American collector was Oscar T. Baron, who worked episodically as a railroad field engineer in California, selling insects as a sometime lucrative backup. He knew a lot about the food plants, habits, butterfly life histories, and "would raise all [he] could merely for the love of the matter."[67] In the Andes Mountains, crossing a "summit range of about 13,000 feet," Baron netted numerous butterflies, which, combined with his haul from Mexico and elsewhere, reaped $4,000.[68]

But Baron, Knab, and Weaver could not hold a candle to three other collectors: a married couple, Herbert and Daisy Smith, and Will Doherty. Herbert Smith, an all-around naturalist, explored the Amazon in the early 1870s. In 1879, in his mid-twenties, he married Daisy (her full name was Amelia Woolworth), a lively inquisitive Yankee born in

Brooklyn who shared his yen for nature. An expert on tropical birds and a good taxidermist, Daisy, like Theodore Mead, felt no pangs about slicing up or boiling birds. For two years just after their marriage, they traveled the Amazon region, supported by subscriptions, and with Daisy shielding her husband, who was deaf, from the possible depredations of jungle prowlers, they accumulated a great mass of faunal material, including 15,000 butterflies and moths of 1,800 species.[69] Two years later they were back again, this time serving two wealthy Englishmen, Osbert Salvin and Frederick Godman, for a massive biogeography of Mexico and Central America, with subcategories (insects, mammals, fishes, and so on). *Biologia Centrali-Americana* would reach sixty-three volumes to become one of the great studies of faunal life; its data on many groups have not been superceded to this day.[70] Near the Veracruz mountains in Mexico, the Smiths took 250 new species of Hesperiidae (skippers), the most abundant of all butterflies.[71] Henry Elwes called them "the two best collectors in the world."[72]

Will Doherty thought of himself as "first and foremost a lepidopterist" who *specialized* in exotics, on which no one could beat him.[73] As complex a man as any of America's complex butterfly people, Doherty was fluent in many languages and had an impassioned interest in classical music and in the great literature of the West, nurtured by a mother who raised all her children to be readers. Doherty, though a generation younger than America's foremost butterfly people, was really one of them, as much a Romantic as Strecker or Grote, and as talented a naturalist as Edwards or Scudder. Had he stayed home, he would have much enlarged the American achievement, but he lived and published mostly abroad, and in 1886, in a fateful decision to become a commercial dealer, he virtually ended his career as a naturalist. This decision would one day cost him his life.

Doherty's father, James Monroe Doherty, the director of a railroad company and a naturalist, had helped introduce Cincinnati to both the first successful electric transport and the Linnaean Society, the city's first natural history institution. Will was a sickly child, and stayed at home under his mother's care until 1868, when he was eleven; she tutored him and watched over him more than she did her other five children. He attended the University of Cincinnati for three years, and then, at twenty-one, about to matriculate to Yale, he managed to get invited

These two photos were taken in 1900, at the end of Doherty's life, near or in an encampment along the Uganda Railway, the main means of transport to his collecting grounds. Left, he overlooks a ravine, armed against lions that threatened him and his men; right, he sits in his tent, holding a cigarette in his right hand, with his left hand gently on the neck of a (probably blind) young woman, likely one of the many hangers-on who trailed along behind explorers or collectors like Doherty, appealing or waiting for a handout. Courtesy of the Natural History Museum, London, England.

to the 1878 Paris Exposition as an attaché to an entomological exhibit organized by the U.S. Department of Agriculture. Soon thereafter—around 1879 or '80—he went off alone on a fateful tour, carrying letters of introduction and his father's promise of financial support. In three months, on foot or horseback, he traversed Holland, Belgium, and Germany, crossed the Alps, much of southern Europe, and ended up for a time in Persia (present-day Iran). There he met Wilhelm Petersen, a young cosmopolitan German naturalist, who would one day write a respected catalog on the butterflies of Estonia. Petersen taught Doherty how to identify, collect, organize, and preserve butterflies.[74]

Collecting draws people into the world, and even the most sedentary or homebound by instinct are exposed to new life, new experiences. This was profoundly true of Doherty, who carried his net throughout southeast Asia, into Malaysia, and down, finally, to Africa. Around 1882 he first entered India and found a "fairyland." "I leave a sort of train of butterfly catchers behind me wherever I go, like a comet," he wrote his mother.[75] Once he ran barefoot after a rare satyrid in a streambed flowing through bamboo thickets, "the confounded insect disappearing as if by magic, among the impenetrable inter-lacings of the bamboo stems." In a dense forest, just as he prepared to pounce with his net "on one of the most beautiful of all butterflies, a *Thaumantis*" with wings of purple and Prussian blue, an "awful dog" dashed out from nowhere "with such a scattering of leaves that I never saw where the *Thaumantis* went to." Near Animudi, the tallest peak in southern India, he found what he thought to be "new species every day or two." "It is impossible to describe to you how glorious it is when one finds an absolutely new organism," he wrote his mother, in that extended correspondence of a lifetime.[76] He boasted he could "distinguish every genus of butterflies in South India by some peculiarity of flight alone." In 1883, he settled in Calcutta; visited the Calcutta Museum of Natural History, built by the British in 1814, the oldest museum in Asia; and became acquainted with the gifted Lionel de Nicéville, the museum's naturalist and the coauthor of *The Butterflies of India, Burma, and Ceylon,* the first catalog on subcontinental butterflies, one indebted to Agassiz, Scudder, and Edwards.[77] De Nicéville taught Doherty about the life histories of butterflies, about geographical conditions and distribution, and how to use a microscope.

"I feel wonderfully privileged to have all this before me, and go about in a mild ecstasy," Doherty rejoiced. He got up every day at five-thirty and worked eight hours straight through, "studying and dissecting in the insect room with de Nicéville," with two hours for exercise. He learned Darwinian theory and began "original work" on the "eggs of butterflies," with "a very strong bearing on the theory of Descent."[78] He felt, he wrote, "like a Naturalist"; "I have passed my apprenticeship." He aimed to "enter life as a professor of Natural History" and write books and articles, including for Scudder's Boston Museum of Natural History. Money didn't matter. "My habits are simple and I don't mind that," he told his father. The key thing was "not to throw away my gift."[79]

Over the next two years, from 1883 to 1885, Doherty explored northern India, though repeatedly ill, once gulping down bottles of quinine. Yet, "you see," he wrote his mother, "so long as I am at my entomology I never know an idle or unhappy moment."[80] Setting up camp in the Chittagong Hill Tracts in what is now Bangladesh, he studied eggs under a microscope; in Kumaon, he sought out butterflies at all elevations, including *Papilio machaon,* the swallowtail with nearly global distribution that William Henry Edwards had wondered about a year or two before in the journal *Papilio.* It stunned Doherty, as it had Edwards, to encounter *P. machaon* at many "zoological zones," from 2,000 to 14,000 feet. He was even more surprised to find a "painted lady" butterfly of the genus *Vanessa* sipping nectar at 2,500 feet in the Kali Valley *and* on the summit of the Lepu Lek, far above the snow line at 18,000 feet, thus contradicting "all laws of distribution."[81] In a meadow near Almora, he plucked the caterpillar of a *mineus* (probably *Mycalesis mineus*) off some grasses after observing it "resting for hours with its chin strongly retracted, and its short, ear-like horns bending forwards looking much like a cat." He took it back to his Almora bungalow and for five days watched it slowly metamorphose into "a clear, transparent green [chrysalis], unmarked except by the black dots of the spiracles."[82] "It's quite surprising how the most elementary points of structure of butterflies have been neglected, owing to men studying only the dried cabinet specimens instead of those just caught," he wrote. "I am achieving great things with the microscope."[83]

Back in Calcutta, Doherty wrote a long paper called, rather dryly and misleadingly, "A List of Butterflies Taken in Kumaon," soon published in the *Journal of the Asiatic Society of Bengal,* the most admired science periodical in India, founded by the British, a showcase for the best work of both British and Indian naturalists.[84] Doherty's list dealt with themes that preoccupied Edwards and Scudder in the 1880s: seasonal polymorphism and sexual dimorphism, geographical distribution, ecological zones, symbiosis and mimicry.[85] An original blend of natural history and biology, the paper made Doherty famous, if mainly in Germany and England.

Doherty compared himself to American businessmen who made "millions and employed thousands." What he did "seemed trivial," of course, but he favored people who "bent heart and soul in purposes still vaguer and more impractical than mine." "I recognize them as brothers."[86] But in 1886, when he was twenty-nine, a crisis tested the limits of Doherty's philosophy. After years of apprenticeship, spending hours peering through lenses in museums or in the field alone in his tent, he chose, out of pressing necessity, a path other than becoming a naturalist. His father had lost his job and could no longer support his family in Cincinnati or send his son money, as he had been doing since 1878, even borrowing to do so. Doherty had hated the paternal dole, and he now scraped by, writing articles for American magazines to meet his own "simple needs." Yet it distressed him that his family at home might be impoverished. Then, suddenly, he had a flash of insight. In Penang, a central port city on the Malay Peninsula, he had heard about a man named Kunstler, a German butterfly merchant in nearby Perak Province who made thousands of dollars selling butterflies to well-off collectors and scientific societies, and to other persons as "parlor ornaments" (butterflies in glass cases). He learned, too, that Eduard Honrath, a wealthy Berlin art dealer who had built a villa in the city explicitly to house his collection of tropical butterflies, financed many of Kunstler's ventures.[87] Doherty sailed off to Perak to introduce himself and found out immediately that all he had heard was true. Kunstler was an affluent man who "kept" his wife and children "in good style," Doherty wrote his mother.

He had "found his vocation," as he called it.[88] Doherty knew plenty about butterflies, rarely tired, and, unlike Kunstler, who refused to collect beyond Perak, feared going nowhere. Back in Penang, he walked about for hours in a high state of elation. Six months later he was writing to Arthur Butler, the curator of insects at London's Natural History Museum and a potential buyer, that he planned to "sacrifice the privilege of writing myself naturalist and gentleman and become a mere dealer." The decision was extraordinary considering that, only a year or two earlier, he had insisted that he would never "throw away my gift," declaring that "selling my butterflies would be like selling my own toes." Privately, he still had misgivings about selling his toes (only "duplicates," he now insisted) and seemed certain he would go on writing scholarly essays on butterflies. But the die had been cast: to save his family he would sacrifice the world of value for the world of price. "If Kunstler could get money," he wrote, "imagine what I could do."[89] Butterflies contained a new elixir, "dollar bills waving their green wings."[90]

Doherty began his new life within weeks of writing Butler. It was a propitious moment. The demand for exotic and tropical lepidoptera was at a high point in the West. At first, he did his own collecting, learning quickly how to mount and pack insects and how to protect them from pests and molds. He experimented with net sizes and invented traps and baits as clever as anything Theodore Mead came up with in America;

one of his best, "a dreadful thing," "the Quintessence of Putrescence," attracted many unusual moth species.[91] He hunted in gorges, on plains thick with elephant grass that punctured his skin, in marshes glutted with leeches, and, in Macassar, "in the beds of the mountain torrents": "I acquired the quite goat-like facility of jumping like a goat from boulder to boulder."[92] Once, he pitched into a deep, cisternlike, tiger pit set by natives; it was covered with logs, except for a single treacherous opening.[93] While his competitors usually resided in one place for a year or two at a time, he rarely stayed anywhere for more than two or three months. He was the first Westerner to collect in Engano, Sumba, Sumbawa, and Biak Island, the first in Humboldt Bay, or northern New Guinea, and among the first in northern Burma.[94] He visited the eastern Himalayas, "richer in butterflies than any other spot in the Old World." By early 1888, when he was still only thirty-one, tens of thousands of butterflies had fallen to his net, with more than two hundred species and in "considerable duplicate" for sale to collectors.[95]

But Doherty had even more daring plans, and to execute them he needed to draw on a range of resources to overtake his chief competitors, the Germans. He beat them all, partly because he sold to anyone with a buck, with no reservation about person or place.[96] "I sell in the best market and do not mix politics and business," he told his father.[97] He excelled, too, because he used better than anyone else the imperialist grid of hotels, boardinghouses, roads, and railroads, and he depended on the most skilled native collectors on the subcontinent, the Lepchas of Sikkim, people with authoritative knowledge of the natural world. Henry Elwes had recommended the Lepchas, who gave names to more living things "than nearly any other" people, creating a sophisticated non-Linnaean taxonomy steeped in religious and totemic meaning, which an English colonel transcribed and published in the *Dictionary of the Lepcha-Language*. Butterfly names filled more space than any other word, save one: "demon" (for every demon, perhaps a butterfly existed to counteract the demon's evil power).[98] Naming had a spiritual purpose, but with the side effect of helping Lepchas discriminate among various butterflies, Doherty reaped an enormous benefit. He hired three Lepchas and a fourth native, not a Lepcha, who served him for years as a majordomo and cook. "Everything depends on these Lepchas of mine," he wrote his mother.[99] When he decided to collect in Malaysia and found that two Lepchas—Pambu and Kanchu—were unwilling to go far away from home, he gave them big advances, higher wages, and pensions to cover burials and funeral feasts should they die on the job. He also honored their customs and religion. Together the men collected five thousand specimens in Celebes, and twenty-five hundred on the remote island of Wetter, many of rare or new species, and all preserved in perfect condition. They surveyed the entire extent of the Malay Archipelago, from Penang to New Guinea. Pambu, whom Doherty much admired, would stay with him for another seven years.

Before he began collecting for money, Doherty had lived in India, mostly in European quarters, and was "always interested in humanity," studying people as well as butterflies. He read a good deal and usually had "two novels going at the same time," until he pushed further into Malaysia, "forgot about Anglo-Indian life, and dropped the novels." He spoke Lepcha, never English, with his "collecting men," except Malay with his cook, Chedi, the only non-Lepcha. He got tired of talking with Chedi about nothing except "curry and rice."[100]

In 1888, on a recuperative trip to the Perak Museum, in Taiping, Doherty befriended Ernst Hartert, a charming, red-bearded German army officer and ornithologist who had just arrived from collecting in Perak, which was, as he put it, also an *"ungemein reiche Schmetterlingswelt,"* or an "extraordinarily rich butterfly world." (The museum had the first library in Malaysia, a magnet for Western collectors).[101] Earlier, at a moment of high imperialist theater, Hartert had joined a German naturalist expedition up the Niger River to Sudan, the sixth such venture in Africa by Germans since 1849; with him had been a young Paul Staudinger, on a mission to collect butterflies and beetles for his father, Otto.[102] The expedition had shifted gears unexpectedly, abandoning its naturalist agenda in favor of a "geographical exploration of Central Africa," reflecting the decision of German chancellor Otto von Bismarck to match the British in the scramble for the African interior, and thrilling young Staudinger with the chance to serve "our immortal hero, Kaiser Wilhelm." Appointed by the ship's captain, Paul assumed the job of seducing tribal leaders to the German side, enlisting "helpers" in the imperialist mode: forty-three porters, two horse-tending men, a personal servant for himself, and a "Yoruba 'boy' " for Hartert as steward.[103] Years later, in an essentially nonimperialist account, Hartert

Ernst Hartert at the foot of the Gunong Ijau river, Perak, Malaysia, shortly before meeting Will Doherty in the Perak Museum. From Ernst Hartert, Aus den Wanderjahren eines Naturforschers *(Berlin: R. Friedlander, 1901–2).*

bemoaned the "outrageous horror of slavery" throughout the region, which "christian Europeans" had imposed on Africans.[104]

On the first day of their meeting Hartert borrowed and soiled one of Doherty's best suits and consumed all his stationery. "He is the most exigent fellow I ever saw."[105] Yet, famished for friendship, Doherty proposed, almost on the spot, that they collect together in northern India near Tibet. Hartert agreed, aware that German naturalists admired Doherty's work on butterfly classification, and at the end of July, the two took a steamer through the Bay of Bengal to Calcutta, with, in their tow, thirty-six suitcases, a cook, and a servant.[106] They boarded an English rail train, fitted with a luxury dining room car, that crossed the rushing immensity of the Ganges River and veered north toward the Himalayas.[107] Once in camp, in the same tent—Doherty never shared a tent with any other man—they talked in German and English about science, philosophy, and politics, disagreeing seldom except about Kaiser Wilhelm, to Doherty a "ridiculous figure."[108] Hartert considered Doherty "no ordinary man," "thoroughly educated . . . to an astonishing degree," and "an instructive talker on almost any subject."[109] Near the Dikrang River just north of Sadiya, Hartert collected specimens of "a new *Apatura* species" and a small chrysalis that pupated into "a beautiful *Poritia*" butterfly, later described and named by Doherty as *Poritia hartertii*. Near Margarita, the two men caught many male skippers, the scents of which Doherty could identify, "half vanilla, half heliotrope, as they flew over our heads." They studied mimicry in the jungles and also on the wall of their dining tent, where Doherty spotted a piece of bird dung ("*Voegelschmutz*") that turned out to be a noctuid moth.[110]

In November 1888, Hartert's military leave ended, and he returned to Germany to marry. Doherty was devastated. "Since he left me I am intolerably lonely," he wrote his mother. "How I managed to exist before I met him I don't know at all, nor how long I shall manage to exist without him. I don't want *much* society—he was quite enough, a solitude à deux is quite lively enough for me, but I really can't stand being by myself much longer. There is no hope of our meeting again." (He was wrong here; they would meet again under quite different circumstances.) Hartert's companionship reminded Doherty how crucial it was "to want somebody for my very own as it were, who shall regard all other people about us as outsiders." "Hartert and I got as far as that," he wrote. "We were a great comfort to each other and both were quite aware of the fact."[111]

Over the next eight years, Doherty worked his way doggedly through Malaysia, collecting for well-heeled Westerners. He returned to America twice to recover his health and to find "some lovely girl" to marry. He wanted one who had "conceived a romantic affection for me upon reading my articles on the Lycaenidae and the Hesperiidae," a kindred spirit like Daisy Smith, who "enjoyed amazingly" roaming with her husband, Herbert, "through the wildest parts of Central America."[112] Doherty always thought he was unattractive, but many women disagreed. To his cousin May he admitted, "Yes, I really must marry. We shall be very happy, my wife and I. We will migrate like the swans and swallows, and live sometimes in the Amazon Valley or wherever butterflies are to be found, and sometimes in America where people pay for them."[113] Just before sailing back to London in 1895, to resume his work, Doherty went to Cambridge, Massachusetts, and knocked on Samuel Scudder's door. "I tried to see you twice," he wrote to Scudder, after failing to find him, "because ever since our little correspondence, and my first sight of your great work, I have been anxious to meet you, and now I am going back to the Far East in six weeks or so."[114]

Doherty's much shorter second trip home, in 1897, was devoted almost exclusively to recovery from the years of great stress. He had endured deafness in one ear, influenza, and typhoid fever, painful infections of the lip and tongue, insomnia, near paralysis of his legs and arms, and boils on his body and face that prevented him from lying "down in any position without great pain."[115] His mind, too, had been assaulted, by the sight of dense primal forests in northeastern India (Talliar) laid waste by venal British planters, leaving "acres of charred chunks, sometimes piled high on each other," a reminder of "the burnt forest in the North of Pennsylvania, all black trunks still standing. I don't know which is the dreariest."[116] He encountered imperialism's miserable wreckage: a dentist from Philadelphia addicted to morphine; a slim blond German living with his mother on the island of Wetter in "great state"; a "dipsomaniac young woman" who sang lachrymose German songs in her hotel room until three in the morning; Christian missionaries in New Guinea who owned slaves.[117] After Hartert left, his loneliness had worsened, and many of his letters home from that period reveal a growing hardness and paranoia. Influenced by that old anti-Semite Henry Elwes, he now spotted Jews everywhere, writing his father in 1890 that "Staudinger's partner Bang-Haas is said to be a perfect Jew" and believing, incorrectly, that both Strecker and his neighbor and fellow butterfly merchant Levi Mengel were Jewish. Bang-Haas may have been Jewish, but his son, Otto, became a card-carrying Nazi in the 1930s, ending his letters with "sieg heil," the only evidence, one way or another, about his ethnic identity.[118] Doherty canceled his subscription to *Harper's Weekly* because the magazine's editor welcomed Jewish immigration.[119] Now alienated from America, he scorned the "stock-gamblers and silver kings" who "imported half a million paupers and outlaws from Europe in order to reduce wages and hasten the general starvation."[120]

In the spring of 1896, he journeyed to Darjeeling to rehire Pambu and Chedi, then sailed to the Dutch East Indies for four happy days with Hans Fruhstorfer, a "fiery, hardworking little man," who told him "they are wild about you in Germany. If you want to know what *Schwärmerei* (romantic rapture) is, just go there."[121] Doherty caught some rare butterflies in Batchian, specimens of *Ornithoptera croesus,* a splendid golden birdwing, which he mailed in tins to Charles Oberthür and Edward Janson, of Janson's in London.[122] "My position is now very strong," he wrote his father.[123] But the gloom returned a year later, when he began collecting near New Guinea and enraged natives killed Pambu, "the best of my men," he wrote to his mother, "equally good with gun and butterfly net, and altogether more reliable than the others, cheerfuller, too. My position is greatly weakened. I was really attached to him."[124]

But what depressed Doherty most in this bleak time was his own wretched performance as a butterfly collector. He had success with birds, but almost none with his chosen interest, and he could ill afford that, since his reputation and self-respect depended on his skill with butterflies. "I am first and foremost a lepidopterist," he told Hartert, who had tried again and again to convert him to birds. "If I fail in insects, I fail in everything."[125] Repeatedly his collecting suffered when he misjudged the seasons or moths refused his baits. He seemed powerless to control the course of his own life.

Doherty had come to see "luck" as one of the "governing factors of the world": "the stars in their courses seemed to be fighting against us, and none of us doubted for a moment that we were under a curse."[126] In a letter to his mother about Pambu's death, he wrote that "all is for the worst in the worst possible of worlds, but there is no need of saying so."[127] He advised his brother Harlan never to become a butterfly dealer—it "leads to nothing and unfits one for everything else." From the island of Biak, he confessed, "My mind seems always going round in the same circle. My ideas are getting ridiculously fixed and dogmatic on subjects of which I know very little—simply for want of anybody to set me right. I have entirely forgotten how to talk." "Perhaps it is simply terror of Civilization, which is a very awful thing when you come to think about it. Or perhaps it is because I am tired and worn-out."[128] He took refuge in books to combat a "gibbering idiocy," sending home appeals for cheap editions, reading and rereading Henry James, George Eliot, Jane Austen, Rudyard Kipling, and, above all, Robert Louis Stevenson, whom he thought the finest writer in English. "Reading is about half of life, isn't it?" he asked. "I sometimes wonder which are the real people, the ones I have known for a week or a month here and there, or the ones in books, that I have known for years and can meet again whenever I please."[129] As the outside world grew uglier, he turned to an inside one, a magnificent substitute and an antidote to the aloneness that threatened to engulf him.

Yet his moments of commanding success as a collector compensated, at times, for his misery and, perhaps, for his sacrifice of his vocation as a naturalist. His "big line of business" saved his family from poverty, and he gained fame and glory. "I am a personage," he wrote his father in 1896. "Collectors are fighting over me."[130] "The market for me is limitless."[131] As Americans pursued him, so did the great collecting houses in London, Janson's and Watkins & Doncaster, as well as rich Englishmen like Frederick Godman and Tory Party member Edward Grey (otherwise known as Lord Walsingham), who specialized in microlepidoptera, the tiniest of moths, and Charles Oberthür, with one of the biggest fortunes in France, a man Doherty greatly respected. In 1889, Arthur Doncaster of Watkins & Doncaster offered to buy *all* his insects; the sale would have freed him from dependence on private collectors. Otto Staudinger tried to entice him into his army. The wily, bigoted Henry Elwes guaranteed him £500 a year because he had "done more than any other single man" in "collecting" and "studying the materials he obtains," adding, "You can have your own house." Doherty rejected that offer (as he did the others), intent not "to be kept or bossed by anybody."[132]

Ironically, Doherty's warmest patron was Jewish: the young Walter Rothschild, who showered him with money, praise, and data in thrilling doses. On a visit to London in the spring of 1893, Doherty rode by train to Tring, the Rothschild family estate outside London, for his first meeting with his best "correspondent"—he preferred that term over "client" or "customer." "He seems burning up with zeal," he wrote his mother, "and I think will do great things for science before he dies."[133] Three years later, Rothschild—a giant of a man with a beard as red as Ernst Hartert's—escorted Doherty about his greenhouses and butterfly cabinets, hovering over him, citing aloud, word for word, passages from Doherty's own articles, and "staggering" him with his "wonderful fluency." "I never saw such a man," Doherty informed his father. He will be a "really great naturalist, as he ought to be with his powerful mind."[134] During both visits, Doherty saw Hartert, now curator for

Rothschild's bird collections. Hartert, Doherty related, "was very glad to see me." He introduced his wife, Claudia, and Doherty sat down with him and Rothschild and together they charted out a new phase in his collecting career that would include collecting birds as well as butterflies.[135] Doherty would see Hartert frequently over the next years.

Doherty savored his butterflies in Rothschild's fourteen hundred cabinet drawers, many of which he had sold to Rothschild for sometimes extravagant sums; he had risked everything to obtain them. Rothschild had already attached *dohertyi* to the names of countless butterflies and moths, including *Attacus dohertyi* (the Atlas moth), which Doherty had caught in 1895 in Timor; he saw it in Tring a year later.[136] In London, Henley Gross-Smith, a wealthy lawyer, and William Kirby, a curator at the British Museum, were "working up" Doherty's butterflies from Timor and Sambawa, "publishing them monthly in pamphlets with magnificent plates," to his great enjoyment. "They are enormously dear."[137] In London, Doherty met Arthur Doncaster, his leading butterfly agent and the co-owner of Watkins & Doncaster. "He is deaf and dumb," Doherty reported to his mother, "but we wrote down the longest possible conversation with one another, and he introduced me to various people—on paper." Then Doncaster took Doherty on a tour of Doherty's own captures through the aisles of Watkins & Doncaster, all "beautifully arranged, set and pinned, a great pleasure to see, and also whole rooms full of others of mine, the cases piled up to the ceiling."[138] The gloom was gone.

Back in America, Herman Strecker and others were busy making their own collections of exotic lepidoptera. As usual, after dinner each evening, Strecker fled to his "semi-monastic den," as John Morris referred to his refuge, where his *Sehnsucht* banished all but butterflies. If it was still light out, he could see, through a window facing east, Neversink Mountain, where he had reveled as a boy in a wild array of butterflies.[139] Late in the evenings, he would sit for a long time, supplicant before his specimens, reflecting on them: What an emerald green color this one has, lavishly filling the outside edge of the forewings, and the interior of hind wings. How it glows against the velvety black of the long, lustrous forewings. And this one with all the blue *inside* the wings—and such a blue, such a diaphanous blue—with black holding hostage the margins and the tails. The universe seemed packed into it, a reminder to Strecker, perhaps, of Humboldt's "vault of heaven, studded with nebulae and stars."[140] What forces of nature converged to make it seem a gateway to some distant galaxy? And when he was through looking, he read his letters, like the one from a New York dairy farmer who envied his evening pleasure, writing, "What a glorious thing it must be for you, tired with work and trouble to sit down alone with your cases, and glorious flies. I think that should be a splendid way to die, right in front of them with your arms spread out to take the whole glorious collection to your breast."[141] Yes, glorious, splendid, glorious, indeed. "Generally with me everything winds up there 'like the circle that ever returneth into the selfsame spot.' "[142]

By 1890 Strecker owned nearly a hundred thousand specimens, three times his 1870 collection, and by 1900, two hundred thousand lepidoptera, including a large portion of North American species, but mostly exotic and tropical specimens.[143] In his own "vault of heaven" was a rare African butterfly colored orange-red and black, with a nine-inch wingspan, the male form of *Papilio antimachus*—the rarest, then, of all Papilionidae. (By the early 1890s, Strecker had obtained several male specimens, through Watkins & Doncaster.)[144] The female, even more rare, was hidden within the forests, and would not be identified until the century's end. Another rarity in Strecker's cabinet was the male *Bhutanitis lidderdalii*, a beautiful black swallowtail striped vertically in white, with blue, red, and orange together in the tails, named for the English doctor in an Indian regiment who'd discovered it. Its picture, first drawn and colored by William Hewitson, drove numerous collectors into northern India, armed with nets, guns, money, and local coolies. Will Doherty first caught one in the Naga Hills, and in 1890, Elwes mailed Strecker his very own specimen, also probably caught by Doherty.[145] *Ornithoptera paradisea* was snared by one of Staudinger's men in New Guinea around 1891, green and black in the long forewings, with much gold in the hind wings, the ends of the tails curled delicately, as if molded by a jeweler's hand. Staudinger himself had described and named it. "I saw the picture of this heavenly thing for the first time in the *Iris* [Staudinger's journal]," Strecker wrote, "*gewiss hat es den rechten Name es sieht aus wie ein Schmetterling aus dem Himmel*." (Surely it has the right name, as it resembles a butterfly from heaven.) He pleaded for an affordable pair, male and female.

There were also his "stock of monstrosities," "my especial mania," as he called it—insects with three wings; a female moth without wings or legs, unlike its mate, which had both; a silver butterfly from Chile, its wings, body, head, and antennae coated entirely in "burnished silver."[146] And what of his female tiger swallowtail with "the left wings and half the body *black,* and the right wings and half the body, *yellow; it is all* female though, not a hermaphrodite. I acquired another female that is all mixed up, neither black nor yellow but both. There is some rule for these things, but we have not yet got to understanding them."[147] "If you want to see lepidoptera especially that mimic others and other things," he informed Samuel Henshaw of the Museum of Comparative Zoology in Cambridge, "you ought to see my collection."[148]

Other Americans, too, after 1885, were making collections almost wholly from the tropical or exotic trade, key among them, William Holland, the Pittsburgh minister whose destiny as a collector would become intertwined with Doherty's. Wealthy, ambitious, and eager for the approval of those with power and status, Holland saw butterfly collecting—as he did everything else—as a route to high social rank, collecting primarily because that was what men did who had the means to command some of the most unique creatures on the planet. But he was wealthy through no account of his own.

He'd been born in 1848 on the island of Jamaica, at a Moravian missionary station run by his parents, devout Moravians who wanted to "bring Christianity to the needy" and to "undertake any task assigned to them by their church," to quote their son.[149] His father collected butterflies and other insects, and a cousin of his mother's, Thomas Horsfield, had studied butterflies on Java in the early 1800s. After the family migrated to a Moravian settlement in Salem, North Carolina, Holland caught his first butterflies and, like Edwards and others, learned to skin and stuff birds.[150] Following the Civil War, he attended Amherst College, where he bragged to his parents that his cabinet of specimens "beat" Amherst's cabinet "all hollow."[151] Then, out of loyalty to his parents, he enrolled in a Moravian seminary near Bethlehem, Pennsylvania, hating it all the while, because he believed it would trap him in poverty and self-sacrifice. Unbeknownst to his parents, he applied to and got accepted at the Princeton Seminary in New Jersey, run by Presbyterians and the most prestigious theological institution in the States.[152] The leader of an affluent suburban church near Pittsburgh interviewed him for a job. "Half-a-dozen millionaires" lived within only a "rifle-shot of the church," he wrote his parents; the church attracted "the very best and most substantial kind of people."[153] It was an irresistible overture.

Holland decided to become a Presbyterian, as entrepreneurial in religion as he would become in butterflies.[154] In an 1873 letter home, he explained, "I am given

William Holland. Courtesy of the Heinz Library and Archives, Pittsburgh, Pennsylvania.

to calculating my chances very closely and have done so heretofore"; against the "prominent influence" of his new pulpit, a Moravian ministry would be an "insignificance." Besides, "my friends among the rising young men of the country are mainly in the communion of the Congregationalist and Presbyterian churches"; "to leave them" would "go hard with me." Whereupon Holland's accommodating father apologized for having "protested against you receiving the call" and insisted that his son "follow the light that God gives you, and He will bless you!"[155] Holland despised people he considered losers, including his younger brother, Daniel, and perhaps even his own father and mother. "The trouble with Dan," he wrote his parents, "is that he isn't anything. I mean he has nothing that he can really do." He must learn that "brains are always marketable. Culture, if of the right sort, always commands money."[156] William had learned that lesson himself and, at twenty-six, became the minister of a large Presbyterian church at a comfortable starting salary of $2,500, with "a prospect of increase" and a parsonage. A few years after that, he married Carrie Moorhead, the daughter of John Moorhead, the owner of the Pittsburgh Iron and Nail Works and the director of the Exchange National Bank. A little later, after the deaths of both

John Moorhead and John's wife, Holland became the executor of the estates of both, with "practical oversight and control," as he put it, of more than $1 million.[157] It was his wife's money that allowed him to go on a butterfly shopping binge, first for American insects (all of Theodore Mead's, all of William Henry Edwards's) and then for tropical fauna, a quest that took him a little longer.

He began by advertising his desires—as Strecker had done—in the relevant magazines (including *Papilio*) and by seeking out Strecker himself for advice and guidance.[158] He inquired of Strecker how to set prices and how to "trade" butterflies with Europeans. At Strecker's invitation, he spent a day and night at the older man's home in Reading in May 1883, mesmerized by his "truly wonderful collection," and both humbled and exasperated by comparison. "I have now nearly 1500 species," he told his parents. "He has over 70,000 fine examples of species from all parts of the world." Holland was secretly determined to close the gap between them, and he obsessed so much about it that he contracted a "terrible facial" tic that drove him "almost insane"; he tried to treat it himself with morphia, cocaine, and chloroform, becoming so numbed he could scarcely move. William Henry Edwards sympathized wryly: "I was sorry to hear of your illness but I should think with all the insects of the world in your brain, you would take to your bed permanently. It would kill me."[159]

In the same year Holland saw Strecker's insects, he started to buy from Staudinger. Five years later he jubilantly added a marginal note to the latest list he had received from Staudinger: "all insects marked off with red ink are in my collection." The red was smeared over virtually every species, and there was nothing left for him to buy.[160] (In 1892, Strecker, who could hardly match Holland's buying power, attached a double-underlined comment in the margin of a letter from a friend: "*Holland is a shark.*")[161] As for Will Doherty, Holland had heard of him in 1886 from an Anglican missionary in Penang, and soon thereafter he became Doherty's first big customer, "offering" the "most liberal terms," along with a $100 check for openers. But, in fact, Holland paid his bills late, sometimes aggravating Doherty, who planned his movements in relation to payments from his customers. "Holland writes at great length about trifles," Doherty told his father in Mount Auburn, Ohio, in 1890, "and then when I particularly want to hear from him there is a silence of six months or so. He sends me money now and then, without telling me what it is for and how much more he owes me."[162] As Holland became aware of Doherty's upper-class clients, he behaved toward Doherty as if he were Doherty's *only* patron and exhorted him, on patriotic grounds, to get *all* the butterflies for *him*. Doherty refused, writing home that Holland "is tremendously provoked to see my best things [butterflies] go to Elwes, Oberthür, and regards himself as my especial benefactor, though up to his last payment, he systematically underpaid me."[163] Doherty wrote his parents that "Oberthür is worth fifty men like Holland."[164]

Besides Holland, there were many other Americans in debt to Doherty. Berthold Neumoegen, flush from his new success on Wall Street, in the late 1880s began "vying with the greatest collections in Europe," as he put it. He, too, devoured Staudinger's lists and, on a trip to a sanatorium in Silesia for the treatment of consumption (tuberculosis), which would one day kill him, spent two weeks in Dresden at that merchant's villa. There he bought a copy of his *Exotische Tagfalter,* bound in two volumes.[165] In gratitude, Staudinger got him elected to the Entomological Society of Berlin.

Staudinger and Neumoegen arranged with the Berlin art dealer Eduard Honrath to send a collector to Ceylon and Java for three years, and Neumoegen employed collectors throughout the world.[166] He tried, like Holland and other rich men, to "take" Doherty "over" as his personal insect provider, claiming, in Doherty's words, that "there will be no end of the money I can make out of him, and from this time forth I may live a prosperous gentleman."[167] Doherty had liked Neumoegen, but he grew tired of his bravado and his chronic delays in payment, driving Doherty, in the end, to a rare fury because he desperately needed the money.[168] All the same, Doherty sent Neumoegen many fine specimens, the best of all a black-and-metallic-green swallowtail endemic to Sambawa, one of many green lepidoptera he obtained from the greenest jungles in the Malay Peninsula and Borneo. "The brightest metallic green is, I think, the latest developed color among butterflies, and decidedly the most conspicuous," Doherty observed in the article "Green Butterflies," published in Scudder's *Psyche* in 1891. "No one who has not seen it can imagine the brilliancy," he said. "The brightest of the metallic blue butterflies look dim beside it."[169] After Neumoegen received *all* the Sambawan specimens in one mailing from Doherty, he picked the swallowtail to send to Honrath in Berlin—failing to mention Doherty's name—for confirmation of its uniqueness. Honrath, excited beyond words, begged to describe and name it, and did so, calling it *Papilio neumoegeni,* for the wrong person; the name still stands. When asked its value, Neumoegen responded, "Who can say? It is the only one in the world. Supposed you offer me fifty pounds, which I would certainly refuse, for money can not buy it."[170]

Neumoegen reported to Strecker on his progress, arousing the stone carver's envy but never coming close to the size of his collection, as did no other American in the nineteenth century, not even Holland. Strecker, until his death, was still cutting and decorating gravestones, some on a grand scale, such as an 1887 Civil War memorial for the Union dead in Charles Evans Cemetery in Reading; it overlooks Strecker's own undecorated gravestone. Clients complained occasionally about his work: "The cross is not exactly perpendicular to the base of the stone. It *leans a little forward,*" wrote one. "There are ugly marks—eyesores—all over the stone, and the lettering at the head of the stone is very indistinct. As we have had an experience with one imperfect stone, you will well understand our feelings."[171] Strecker was proud of his handicraft, but his heart and soul were in his lepidoptera, and by the 1890s, he not only had a gigantic collection but had become engrossed in the business side, described by other dealers as a "large handler of exotics."[172]

By 1895, many merchants in the United States were doing a full-time business in insects, while others operated on an informal or part-time basis; it was impossible to establish exactly the total figure, since the line between formal and informal dealing was always unclear, despite published directories of naturalists and dealers.[173] Strecker himself crossed the line, selling throughout the country and abroad, notwithstanding his constant demurrals about business. "I hate business people, their ways, their slang, their tricks, their crookedness," and he was not a businessman in any conventional sense, dealing, as he had always done, to get money only to buy more glorious butterflies for his own collection.[174] Like Staudinger, he printed a price list, practiced a liberal credit policy, and advertised; although unlike Staudinger, he never had an Andreas Bang-Haas to take over the business.[175]

By 1890, Strecker was, perhaps, the most prominent dealer in America, at the center of a democratic circuitry of distribution, doing more than any other to bring the tropical and exotic world to American collectors, poor and well-off alike. By the mid-1890s, commercial dealers had downgraded what had once been Strecker's "*haupt* desiderata" into ordinary duplicates, underlining the degree to which market prices had come to determine the "value" of lepidoptera, a trend Strecker earlier tried to steer clear of but now had to negotiate, struggling to make a profit before the market turned even lower. He stashed the duplicates in his fourth-floor "selling" room, twenty thousand in 1885 ready for circulation and probably three times that number ten years later.[176] Suites of previously rare birdwings and papilios stocked his cabinets, as did the blue *Morpho menelaus,* the blue *Morpho sulkowski,* and the blue *Morpho cypris.* He tantalized young Charles Dury of Cincinnati by mail with a "brilliant Blue Morpho," an insect Dury had "*never* before seen." Out of curiosity, Dury visited Strecker's home; like so many others, he came away dazed. "I have *thought and dreamt* of *huge butterflies,* since I saw your collection. When I wake up at night I fancy I see them flitting before me and spreading their wonderful wings."[177] In 1880, Adolph Eisen, a young leather cutter in Coldwater, Michigan, received several exotics from Strecker in a package, crowned by a *Morpho cypris* and a *Morpho sulkowski.* "I can not find proper adjectives to express my admiration of these Insekts." But Strecker's insects "aroused the whole neighborhood, everybody wanted to see them." "Some people even asked me if I had made these Insekts myself and I could not convince them that they were natural specimens."[178]

After unwrapping a package from Strecker, a manufacturer of rope buckles and anti-friction pulley locks in Newark, New Jersey, thanked him for the "great joy" such "really beautiful things aroused" in him.[179] "They were a beautiful lot," said a sheep farmer from Dutchess County, New York, of an 1891 Strecker mailing; "I never tire of looking at them. Those yellow flies from Florida and the Orange one from Mexico are the brightest things I have ever seen."[180] Strecker helped celebrate the birthday of a lawyer in Volga, Dakota (then a territory), whose wife wanted to "make [her] husband a present, and I know of nothing which he would enjoy more than butterflies." Strecker dispatched nineteen swallowtail species from South America on Christmas Day.[181] He mailed morpho butterflies to Emily Morton, the young naturalist in Newburgh, New York, who used them at the center of a collection she was making for a friend. He sent some also to Adrian Latimer, a young drugstore employee in "straitened circumstances" in the rural village of Lumpkin, Georgia, working a sixteen-hour day in unhealthy "confinement wrapping packages, filling prescriptions, and manufacturing nostrums" for $12 a week. "It is perfectly magnificent," the young man wrote back. "I did not dream of there being such splendid things in the world. Since getting [the box of *Morpho cypris*] I have been scarcely fit for work, being almost unwilling to take my eyes off it." They represent "something to live for" besides "money-making."[182] Strecker delivered morphos, too, to Frank Snow, a leading Midwest naturalist who pioneered the science curriculum at the University of Kansas at Lawrence; he'd told Strecker that he wanted no tropical insects, but Strecker sent some anyway, and Snow reversed himself, saying, "I have broken my rule." During exam time at Princeton University, an undergraduate, Elison Smythe, later a respected entomologist, received a radiant blue wafer. "I believe that the Morpho has cost me about ten points on my last exam, for while I was studying I had to stop every now and then to take a look at it."[183]

"When i opened the box [of *Ornithoptera*]," a young man who worked in a Pittsburgh glass factory since his thirteenth year wrote Strecker in 1886, "my hart leaped with joy to think i am a possessor of such wounders of nature." A month later, after examining another box of "beautiful flying gems," he told Strecker he "never had the least idea that I should be the fourtunate possessor of such wounderful things as i have been getting from you."[184] He couldn't spell, but he was intelligent and witty and had poetic talent, writing "of nature's sweeter cup" of flowers "as they peep up through the sward" and "show the beauty of our god." He practiced what he called "Butterflyology," "not for ambition's sake, but because i love the beautiful in nature." "If I had a million dollars, I would use it for butterflies."[185] The man's name was George Ehrman, and despite repeated layoffs and bouts of poverty, he would amass one of the most remarkable collections of tropical butterflies in America and acquire his own string of correspondents. Strecker opened doors for him, sending him gorgeous insects, sometimes in exchange for native rare moths Ehrman caught at the fiery gas wells in and around Pittsburgh, features of the industrialization sweeping through that city, which drove many of the industrialists themselves into the suburbs. Ecologically devastating, the wells produced "a smoky atmosphere"—as one butterfly man put it—"which penetrates almost everything unless hermetically sealed" and "prove[s] very injurious to these tender insects."[186] The wells lured hundreds of thousands of moths from the nearby forests, many of which Ehrman captured before they got swallowed up in the flames. Once he walked thirty miles to the wells to get Strecker an assortment of "Sphingidae, Bombycidae, Noctuidae, and Geometridae in aboundance." A grateful Strecker delivered more insects, even *Papilio antimachus*—in Ehrman's words, "that wounderful insect."[187]

The commercial collecting of Will Doherty and Herman Strecker, one in foreign lands, the other at home, had a double effect. On the one hand, it degraded the character of the collecting experience, obeisant more than ever to Doherty's "dollar bills waving their green wings," rather than to nature itself, with all its sensual potential. Both Doherty and Strecker were aware of this fact and hated it, Doherty perhaps more tragically than Strecker, who seemed to have few regrets about persuading his customers to give up collecting native for foreign species, which almost always meant putting a price on an insect, in disregard of its natural place and intrinsic value. The commerce of these men, along with that of the Germans and the English, sometimes served science, but as time went on, it also acted as an ever-growing countervailing power. Imperialism fostered commerce; the entire market in living things (or in once-living things) may have emerged when it did because of imperialism. Yet the transfer into America and elsewhere of ever-cheaper specimens, from nearly every corner of the earth, did more than exploit nature commercially or demean collecting; limited by its emphasis on what the market would bear (thereby excluding, at one time or another, many lepidoptera), it nevertheless brought the fecundity of exotic nature, superimposed over native nature, to many Americans, introducing them to a "new wealth of beautiful forms," in the words of George Santayana.[188] Exotic butterflies—the "ornamented organisms" of Alfred Russel Wallace, the golden birdwings of Robert Rippon's tribute in *Icones Ornithopterorum*—came to symbolize the transnational character of the American experience. Such an exposure was the fruit of the ongoing migration of Westerners into a paradise of animal color, pattern, and form.

SEVEN

Butterflies at the Fair

Millions of moths, once hidden in darkness, swirled through the electric lights of the 1893 world's fair, more blinding in their brightness at the time than any artificial light on earth. Butterfly people who attended the Chicago Columbian Exposition that summer, along the shoreline of Lake Michigan, must have wondered over them. Samuel Scudder, who, in *The Butterflies of the Eastern United States and Canada,* had noted the impact of artificial light on insects, was there. So was Theodore Mead, who brought with him other members of the Edwards family but not William Henry, who stayed home, captive to his caterpillars.[1] Official displays of lepidoptera were scattered throughout the grounds; they had been orchestrated by one of William Henry Edwards's friends, Selim Peabody, who years earlier had exhibited his own collection of butterflies at an "exposition" in the Chicago Academy of Sciences. Now he was chief of the fair's Department of Liberal Arts, with more exhibits, according to him, than any other department of the fair, housed in the "largest building in the world under one roof ever erected."[2] Illinois State was displaying a big array of butterflies native to the region, arranged by George French, the author of the country's first butterfly manual; David Bruce, an alpine butterfly expert, had mounted the Colorado display in that state's "historical department"; and New York State had invited Herbert Smith, the tropical collector, to present his selection "of showy insects."[3]

Most of the butterfly collections at the fair were of native species, although the Pennsylvania managers had tried hard to persuade Herman Strecker and William Holland to put their thousands of specimens, tropical and domestic, on public view. Holland had been the state managers' first choice ("they almost entreated me," he told Scudder), but he'd refused, too busy and fearful of damage to his insects.[4] Strecker had also refused, lest he lose his job at the Reading marble yards, and despite a nice money offer. The Pennsylvania fair managers had pleaded with him not to "miss the greatest opportunity in [his] life," and so he'd reversed himself. Soon thereafter, however, the managers had reversed themselves, probably because Holland had insinuated that Strecker might embarrass the state.[5] "I understand that, failing to get my collection," Holland wrote Scudder, "the Commissioners will request Strecker to exhibit himself and his collection at the Fair. One of the Commissioners told me that he regarded Strecker as the larger curiosity of the two, a remark which you and I probably would fully appreciate."[6] Pennsylvania's big space reserved for butterflies went unfilled.

Historians have seldom written of the Chicago world's fair as nature-friendly, but it was. The butterflies were there, not in splashy numbers, though enough to be noticed. There had been exhibits of them earlier, at local and state industrial expositions, but the 1893 fair may have been the last fair to show lepidoptera, and much else in the natural world. Tropical fish swarmed in the vast aquarium in the Fisheries Building, "never before seen in any exposition," according to the *Official Guide*. Mounted reptiles, birds, and animals, some "rapidly becoming exterminated" (according to the guide), filled the U.S. Government Building. Live exotic animals were caged in the Midway, presented by Hagenbeck and Co., Germany's inventors of the modern zoo. There were tree specimens and tree parts in Forestry and eight greenhouses in Horticulture, with rhododendrons, roses, carnations, and orchids (18,000 orchids!), plus geometrical beds of tulips and velvety pansies outside, belonging to "a revolution in the pansy world." A stream of flowers, ferns, palms, and shrubs, wild, domestic, and tropical, snaked through to Wooded Island, the ten-acre retreat landscaped by Frederick Law Olmsted for weary fairgoers.[7] Nature had been harvested from everywhere on earth in this imperialist time, and even the smells of family farms from the hybrid rural landscape still *inside* the city of Chicago itself reached the fair. The older natural history tradition had grown more vigorous, enlisting more and more advocates over the last one hundred years.

Far more visible than nature's diversity, however, were the exhibits of man-made artifactual diversity, at one end of the scale paintings from Constable to Cassatt in the Arts Building, many lent by Americans who'd bought them from abroad. "Never has there been so brilliant a showing of modern works of art as here assembled," claimed

*Night illumination at the 1893 World's Fair in Chicago, showing
a searchlight sweeping the darkness. From John P. Barrett,*
Electricity at the Columbian Exposition *(Chicago: Donnelly, 1894).*

the American guide.[8] The fair organizers talked up cultural enrichment
("unspoiled by commercialism," insisted Peabody's daughter in her
biography of her father). But there were other signs of human creativity
in exhibits of technology, arrayed alongside nature's specimens in one
gigantic revelation of the prowess of Western nations. Henry Adams
observed of the fair of 1893, in his *Education,* "The majority at least
declared itself, once for all, in favor of the capitalistic system with all its
necessary machinery."[9]

Fair managers did nothing to hide the imperialist purpose of this
achievement. The *Official Guide* called it "the first Exposition ever
held" in which the "foreign nations and their colonies are represented in
all the great department structures of the World's Fair." Selim Peabody's
Liberal Arts Building allotted prime floor space to Germany, England,
and France, shoving other nations to the periphery. At the urging of the
"Kaiser himself," German inventiveness could be found in the "grandest
display of industries and arts ever made at any foreign exhibition," perhaps most forcefully by the Krupp armament company, which displayed

the world's biggest military cannon, with enough firepower to unleash twenty-three-hundred-pound shells at a range of sixteen miles.[10] American engineering included, ironically, a model exhibit of Dam 6 on the Kanawha River in West Virginia, one of ten in the system of movable dams that had contributed to the flooding of William Henry Edwards's population of baltimore checkerspots in Fraser's swamp. It was the only exhibit chosen to illustrate the "involvement" of the U.S. Corps of Engineers "in large public works."[11]

The railroad overshadowed everything else, in all its multiple designs, from the great feeder line, the Illinois Central, which discharged visitors by the millions into the fair through a massive depot at the densest edge of the city, to the latest rail inventions in the British and German exhibition halls and the parade in Louis Sullivan's U.S. Transportation Building of the entire "evolution of the World's Railway," assembled by the Baltimore and Ohio Railroad.[12] It was at the White City that Americans first became aware of the fetish potential of electrical light, which turned the fair into one huge flaming orb at the edge of the lake. The 1876 Centennial Exposition in Philadelphia, by comparison, had shut down at night. Architects wired the 1893 fair for three times more electric lights than Chicago itself, the most electrified city in the country.[13] Tens of thousands of lights outlined every major building, beamed forth from the incandescent Tower of Light of the Electrical Building, radiated through the big Columbian Fountain near a prime fair entry, and flashed at the top of the fair's tallest building. Searchlights, modeled after German prototypes, penetrated points deep in Chicago and scooped out of the night a flying horde of lepidoptera.[14]

On view, too, at the French and German Pavilions and at the U.S. Photographic Section, were the latest advances in color photography.[15] The Germans, in particular, dazzled with an exhibit on "a new world of color," so said the *Guide Through the Exhibition of the German Chemical Industry.* Four prominent German companies had brought their products to the fair, expanding the chemical rainbow by adding synthetic indigo, deltapurpurine, azoblue, chloramine yellow, heliotrope, and so on, plus a whole range of new reds, new oranges, new violets, and new grays, many flourishing overt signs of imperialism, such as "Congo-red," "Congo-orange," "Congo-brown," "Nzanza-black," "Tabora-black," and "Guinea-green."[16] Americans came to gaze at these technologies, as well as at all the goods produced by them, from the Brownie cameras of Eastman Kodak to the mountains of fruit, vegetables, and grains in the California and Illinois expositions, much of it grown by a new industrial agriculture. Human creativity—high art at one end and railroads and piles of industrially generated fruit at the other end—clashed yet also blurred and merged into a whole wild weave, human commerce, human culture, and nature's abundance tangling at so many points as to seem one and the same, a comprehensive and marvelous intermingling.

The Columbian Exposition telescoped a unique moment for Americans, beauty of a natural kind side by side with beauty of an unnatural kind, giving pleasure in a way perhaps never known before, certainly not for so many people at one time.[17] George Santayana, a philosopher of that moment, in his *Sense of Beauty*, written only three years after the fair, hoped Americans would step away from their driven utilitarianism to acknowledge the full aesthetic life before them. "To learn to domesticate the imagination in the world," he wrote, "so that everywhere beauty can be seen, and a hint seen for artistic creation,—that is the goal of contemplation." "Art and life exist to be enjoyed," Santayana would later say, "not to be estimated."[18]

There was another, darker side to the story. Man power had the means to despoil, even devastate, nature power (even as nature, unintentionally, could harm humans). After 1893, man power would emerge on such a Faustian scale as to suggest to many superiority over nature power (to say nothing of God power), with nature no more than a human source and resource. The great French naturalist Comte de Buffon wrote that "all the inventions of men, whether they be necessities or conveniences, are only grossly executed imitations of that which nature makes with the utmost finesse." At the end of the nineteenth century, the American sociologist Lester Ward observed, "Applied sociology proceeds on the assumption of the superiority of the artificial to the natural. . . . The great fact is that man *has*, from the very dawn of his intelligence, been transforming the entire planet he inhabits."[19] As the United States moved at a breathtaking pace into an industrial order, its experience with nature—with all the magnificent butterflies and moths—reached a precarious stage. Even the most artifactual pillars of the new economy—electrical lights, photographic innovation, railroads, synthetic chemistry, industrial technology, and the exploding commercial market itself—had the wherewithal to complement the human experience of the natural world.[20] On the other hand, at every point, the opposite was true. This tension had a singular meaning for pioneer butterfly people, who reaped many of the rewards of the era. Still, the new context diminished them, too, along with the amateur tradition from which they came, even as it raised up others like William Holland of Pittsburgh, who had amassed an immense collection by outspending everybody in America. Step by step from his early days on, Holland had maneuvered himself into strategic positions, until, by the late 1890s, he had become commanding general of the Carnegie Museum of Natural History in Pittsburgh, itself a new version of an old institution, one lavishly funded by the steel magnate and capitalist incarnate Andrew Carnegie. Twenty-six years younger than Edwards, Holland would change the lives of all the leading naturalists, old and young, and not for the better.

Encounters with butterflies were enhanced, in many ways, by American breakthroughs in technology. In Europe, railroads had cut through already built areas, most with ancient histories, but in America, railroads *preceded* development, advancing through sparsely populated spaces and introducing those who rode them to the edges of wilderness or to a reality they could have seen in no other way.[21] New kinds of landscapes appeared in the wake of the incursions; the historian Robert E. Kohler has called them "twilight zones" or "inner frontiers," separating the developed (or town centers) from the undeveloped (the wild country).[22] In time, Americans would recast these spaces or zones into farms, then into suburbs or new cities, in the process weakening or erasing their nature-rich character. Until then, for many decades after 1880, they were allies of the hybrid rural landscape, acting as avenues into nature, both wild and humanized, and inspiring the creation of a new and ever-expanding popular literature on nature. "For a period of some four or five decades," Kohler observes, "the landscape of North America afforded an unusual intimacy between settled and natural areas. Densely inhabited and wild areas were jumbled together. Areas of relatively undisturbed nature, with much of its original flora and fauna intact (except for the larger game and predators), were accessible to people who lived in towns and cities."[23]

At least from the early 1870s on, railroads abetted the naturalist's systematic zeal, delivering Thomas Bean to Laggan, Canada; Samuel Scudder, David Bruce, William Wright, and Theodore Mead to the biologically diverse regions of Colorado; and Will Doherty over the Ganges to the Himalayas (and, at the end of his life, through Kenya). The great butterfly collections of Neumoegen, Holland, and Strecker owed their existence, in good measure, to the Chesapeake and Ohio, the Pennsylvania, and the Canadian Pacific.[24] Once Strecker realized how fast freight moved across the continent, carrying insects to him in only a few days, he became addicted to the mail, waiting impatiently, in throes of *Sehnsucht* unleashed by the railroads, ecstatic when boxes arrived at the post office. When the mail was delayed, he behaved like a child denied sweets, erupting in tantrums. The railroad was a drug, a poison, and an elixir.

William Henry Edwards's experience with the railroad was perhaps more complex than that of any other butterfly man or woman, a romance with both a capitalist and an entomological dimension. Even though railroads (and speculation in them) may have led to the collapse of his business, he never ceased being impressed by their economic march across the country. "The Chesapeake and Ohio," he wrote Mead in 1888, "are making vast preparations for a big coal trade to the Northwest against the completion of their bridge over the Ohio at Cincinnati, and of their new track from Ashland to Cincinnati along the south side of the Ohio River. This will be done by Christmas, and then the coal trade to the far North West even to Dacota begins." And entomologically his excitement rose even higher, since at the end of every new line, new butterfly treasures, hitherto hidden, might be found. "On Wednesday I went to the end of the RR track with an excursion party, two miles along the Falls," he told Mead in 1872, of the "progress" of the Chesapeake and Ohio in a place quite distant from Coalburgh. "The track ended under a tremendous cliff and everything about was picturesque enough," and "I could not but think with pleasant anticipation of an excursion to that region next summer, with a walk of a few miles along the ground under the cliff and of the possible interesting butterflies one might find there."[25] In 1878 he wrote to Lintner: "I see the Southern Pacific RR from the Southwest is advancing fast and Arizona will be open to us before long."[26]

Edwards's research for his third volume of *The Butterflies of North America* depended more than ever on the railroad. For years he had longed to go west with his crony and helpmeet Henry Edwards, to "bathe my entomological soul" by studying the swallowtails and other butterflies on his own, and to "rob a few eggs," as he put it. His relative poverty prevented that, until, in the year of the fair, his son, Willie, flush from profits in gas in West Virginia, seemed ready to help.[27] "My Gas Co. is in good shape," Willie wrote Theodore Mead, his brother-in-law, and "I now control almost all the region, rights to the city, etc."[28] Father and son tracked the progress of Willie's wells and drilled on what was left of the father's land and on property Willie bought back from his father's creditors.[29]

By the mid-1890s, Willie could "foot the bills" for his father's entire trip west.[30] Healthy and spirited at seventy-two, William Henry Edwards arranged with David Bruce to go by rail to Glenwood Springs, a mecca for naturalists at the headwaters of the Colorado River, "the richest spot for rare butterflies known to us," Edwards told Wright.[31] Edwards retraced a route Mead had taken more than twenty years earlier, and arrived in Denver four days later, thence to travel to Glenwood Springs, where he took a room at the new Hotel Colorado, "large, very costly, and well-equipped." In a month's time, he returned home by train, with all his work on swallowtail polymorphism in a valise and his caterpillars in pasteboard boxes bound together by a shawl, and arrived at Coalburgh, "the larvae bearing the journey well."[32] The Glenwood Springs experience occupied a place in his memory that matched his 1848 Amazon journey; "I set it all down in my notebook," he recounted, "written up every day, and it stirs my blood now to read of it."[33]

Artificial light, like the railroads, revealed much in nature hitherto concealed. Americans and Europeans had long been relying on light at night (other than moonlight) to collect moths. "When you go after moths in the night," the English entomologists Kirby and Spence advised collectors in their 1819 *Introduction to Entomology*, attach a lamp to your stomach, "made with a concave back, and furnished with a reflector." "If you hold your expanded fly-net before this," they explained, "you may then entrap a considerable number." Light traps to ensnare insects were invented in the United States in 1860, probably relying on intense limelight, induced by the focusing of a jet of mixed gases on a lump of calcium chloride (lime). The Texas lepidopterist Gustav Bel-frage collected moths and butterflies with limelight in the 1860s and '70s, and some naturalists collected at gas lamps on the streets. "Most every favorable night I go out to the gas lights and look for moths," wrote Charles Dury of Cincinnati to Strecker in 1873.[34] Among the earliest naturalists' accounts of collecting by arc or electrical lights occur in letters from Richard Stretch to Henry Edwards in 1881 and 1882, when Stretch netted hundreds of moths under arc lights at the hydraulic mines in Cherokee Butte, California, and near the three electric lights in Nevada City. City lights, especially, invited much bug hunting. Since the "introduction of electrical lights into our cities," Samuel Scudder observed in *The Butterflies of the Eastern United States and Canada,* "entomologists have made use of them for the capture of insects, many nocturnal animals being attracted from the surrounding country by the brilliancy of the light."[35]

"The electric light is quite an invention to get flies," George Franck, a traveling Brooklyn hat merchant, informed Strecker in the summer of 1884 from Council Bluffs, Iowa. "I am getting some every night, and some times in great numbers. Last night I caught a Sphinx entirely unknown to me." George L. Hudson, an organist from Plattsburgh, Pennsylvania, hoisted a ladder up to city streetlights to reach the moths that fluttered frenetically around the "globes of light." A journalist, Edward Warren, caught "some beautiful little moths with mugs like silk" at the electric lights in Pittsburgh.[36] In Washington, D.C., a daring young naturalist climbed the capitol dome, recently lighted by electricity, later reporting his trove of insects to the entomology club in Washington; John Morris from Baltimore was there to hear it: the "young fellow showed a lot of insects of nearly every order, which he caught on the dome of the capitol, where they were attracted by the electrical light. The number thus drawn together is simply amazing; of moths there is no end."[37] Clear across the country in southern California, Max Albright, a German immigrant and Civil War veteran who lived at the Old Soldiers' Home, paid the fifty-cent train fare to Los Angeles to catch moths in the lights. In the summer of 1894, Albright walked an hour to Santa Monica, with its fourteen "electric lights scattered from five to seven miles which I run through in a zigzag fashion." By eleven-thirty, he was back home, footsore, "in time to wash and eat my meal." "Every day I make the twelve mile tramp."[38]

Before technological advances in printing press technology and photography, the cost of printing and illustration prevented the mass production of nature books, but after 1885 large numbers of books could be produced quickly, with sophisticated black-and-white images, usually created by an advanced photochemical process called the "halftone," which allowed publishers to reproduce photographs. Between 1890 and 1891, the *Entomological News* offered the first photographs of butterflies by the orthochromatic process, a step beyond the halftone, with some clear colors, but not nearly as evocative as the earlier hand-colored images of butterflies.[39] Doubleday and Page in Manhattan, attempting to erase the color deficit, pioneered the printing of photos in three colors from halftone plates. The company's first such book, was printed in 1897: *Bird Neighbors* by Neltje Blanchan, with "birds as they are in life," the author claimed, "each according to its own habit of existence." The house's director, F. N. Doubleday, boasted that he had produced, by 1900, "practically the only popular books illustrated in color photography in this field," books on birds, trees, wildflowers, fishes, mammals, beetles, and mushrooms. In effect, Doubleday had joined together the two aesthetic imageries of the world's fairs, one natural, the other man-made. In 1898 it published *The Butterfly Book* by William Holland, the beneficiary of this iconographical evolution and the most successful book of its kind in America and probably, at the time, in the world.[40]

Holland was in an ideal position to write this book. Not only did he have his wife's wealth and an enormous butterfly collection; by the 1890s, he had also captured a high degree of institutional power. In 1890 he quit his Presbyterian ministry (after strategically embracing it twenty years earlier) to become the head of Western Pennsylvania University (later the University of Pittsburgh), in another ten years converting it from a "high school" (his phrase) into one of America's substantial modern regional institutions, with "income-bearing assets," as he put it, rising from $200,000 to $600,000. He established an undergraduate natural history society, "went out and shot a few birds" with the students, and taught them the secrets of taxidermy.[41] But a few years

later Holland's appetite reached its pinnacle: as a result of his close association with Andrew Carnegie, he became the director of the Carnegie
Museum of Natural History, soon the nation's most advanced museum,
superior even to New York City's American Museum of Natural History, a post he held along with his university chancellorship, until Carnegie persuaded him to give up the latter for the former.[42] Holland had
first met Carnegie in the late 1880s, at a fashionable summer resort in
the Allegheny Mountains patronized by the "very best people" of Pittsburgh; by 1890, Carnegie had committed $2 million to the building of a
library, a museum, and an art gallery just a square block from Holland's
house.[43] "This suits me admirably," Holland wrote Samuel Scudder.[44]
As one of "the Trustees of this enterprise," he consulted closely "with
my friend Mr. Andrew Carnegie" on how "his magnificent gift might
best be applied."[45]

Holland's relationship with Carnegie was the key to all his later success, and he did everything he could to please the magnate, making the
museum a site, above all, for spectacular display (principally of dino-
saurs, Carnegie's favorite fauna). In return for Holland's services, Carnegie rewarded him by advancing his status as an international butterfly
man.[46] He lavished insects on Holland and financed construction of a
large modern iron-wrought gallery in the museum to hold them, with
space for many hundreds of butterfly boxes, turning the collection into
the best institutional one in the country. He invested Holland with complete authority over the museum, so much so that it seemed "not for the
public" but "for the benefit of the one at its head," as the Pittsburgh naturalist Edward Klages (one of the two brothers who collected butterflies
for Rothschild in Venezuela) complained.[47] Carnegie permitted Holland
to use the Carnegie imprimatur to get whatever insects he desired for
himself and for the museum. If Rothschild had his Tring, Holland had
his Carnegie. "My friend, Lord Rothschild and I," Holland would write
an inquiring naturalist in 1902, "have paid higher prices for butterflies
than any other individuals now living."[48]

Backed by Carnegie, who would leave him $5,000 a year in his
will, Holland was a representative figure of the new industrial age, and
his institutional status attracted an offer from Doubleday to produce
a new book of color photographs.[49] Ready to absorb any fashionable
new machine, never looking backward, Holland was among the first in

Pittsburgh to get a typewriter, telephone, and Dictaphone.[50] "It is absolutely impossible for me," he told his parents, "to accomplish all that I have to do, unless I avail myself of some of these labor saving devices, which are being multiplied in this age."[51] In 1897, seeing a windfall in the new color photography, he responded to a Doubleday overture and began negotiations on a guidebook to American butterflies with abundant color photos. Holland already had the butterflies, possessing by this time *all* of William Henry Edwards's insects, plus thousands other Americans had collected, and he had his own money to invest in the book. "Holland must have sunk thousands of dollars in publishing," Edwards told Scudder. "He told me he would have to sell 8000 copies to get his money back."[52] Doubleday had the technology and was "most energetic in pushing sales," Holland informed Scudder, whose own books barely broke even. They had "already disposed of 15,000 copies of *Bird Neighbors,* to which my book, in a certain sense, will be a companion piece."[53]

Holland's *Butterfly Book,* subtitled *A Popular Guide,* looked little like the small European pocket guides of the past or today's glossy paperback guides. Nearly four hundred pages long, and covering most North American butterflies, along with their anatomy, classification, collecting, and historiography, it loosely recalled Scudder's own great volumes, although Holland, knowing how poorly Scudder's had sold, had tried to avoid their scholarly "heaviness." He'd interspersed undemanding, folksy little poems and essays throughout the text (with only a tiny selection from Romantic poetry, most probably lifted from Scudder; there was nothing in a foreign language, and no Shakespeare). The book's special feature was its forty-eight color plates, five of larvae and pupae, again a borrowing from Scudder (with Scudder's permission), presented in vivid colors and, according to one expert on photography, demonstrating the color process better than any other book or periodical of the time, the "textures as well as the tints rendered with rare fidelity."[54] Especially impressive were William Henry Edwards's blue lycaenids; the golden, brown, and black skippers; and the orange-and-black checkerspots, sometimes in straight rows or in angled portraits, looking as if they had settled on the ground all at once. Other plates showed ensembles of swallowtails—in pairs, in the company of other insects, and on flowers, each perfectly scaled in color and size. On one page, yellow and orange sulphurs, in splendid hues and tints, lined up; a few pages later, purple-and-white admirals seemed to be in flight. But most striking were Edwards's fritillaries, particularly the male Diana, of the warmest orange and brown, and the female, in deepest black and bluest blue, as if just taken fresh from Paint Creek out of Edwards's own net.

The two aesthetics, natural and artificial, that had been on display at the 1893 world's fair converged in Holland's *Butterfly Book,* and like all good colored pictures, these fed the imagination. Hundreds of boys and girls who saw them went out to observe and collect. So, too, the photos, despite the extra cost, radically lowered the volume's price, with large printings feeding demand. Edwards's volumes cost $125, Scudder's $75, and Holland's $3, making it "the cheapest book of its kind ever published."[55] Holland was certain that "the best part of the book is, of course, the pictures, and the making of these involved an enormous amount of labor as well as no small expense, for all of which I expect no adequate return whatever, except the gratitude of my friends," or so he told a Massachusetts professor.[56] In the mid-twentieth century, Vladimir Nabokov still found both price and plates exceptional, though he despised the "stale anecdotes, pseudo-Indian legends, and samples of third-rate poetry." A Samuel Scudder Holland was not, but his book went through several editions, including a 1915 pocket *Butterfly Guide,* one of the most durable ever on the market, a volume that could be flipped through literally in seconds.[57] And despite Holland's claims to Scudder about expecting "no adequate return whatever," he reaped a profit (after 1900, about $1,000 annually), encouraging a wave of similar books.

"Following the publication of *The Butterfly Book,* and certainly because of it," wrote Harry Clench, a respected expert at the Carnegie Museum of Natural History in the late 1940s, "the day of the amateur began in earnest, not only in the large increase in collectors and collections, but in the published record as well."[58] The day of the amateur had actually already arrived, and if numbers rose (as they most certainly did), the causes could be found elsewhere as well. Between 1895 and 1900, state after state passed laws ending the collection of birds and birds' eggs, a response to the Audubon Society, led mostly by women sickened by the slaughter of birds to supply feathers to the hat business.[59] The crusade was a remarkable episode in American history, with notable outcomes for butterfly collecting, since naturalists, male and female, who had collected only birds or birds *and* butterflies now shifted wholly to butterflies (and moths), their numbers growing across the country two or threefold by 1900.[60] Holland's book stoked the fever. It inspired a woman from Portland, Maine, to reclaim the joy of her childhood, "the one thing I cared for" which others thought "*queer in a girl*." Now, she wrote, "I want to take it up" again "as a study," and "your *Butterfly Book* is the best book I have ever seen."[61] "I am 55 years old," wrote a German immigrant in Fort Collins, Colorado, "and since my boyhood have collected butterflies and moths at first in Germany, here in Colorado since 19 years. I have your book about our butterflies and I am very glad to have it; it is just what we wanted."[62] "I began collecting butterflies, when I was about nine years old," explained a young man from Neligh, Nebraska, "but lost interest in them when I was thirteen or fourteen. When I got a copy of the *Butterfly Book* last spring all of my old interest was revived, and it is now my intention to form a large collection. The students of butterflies all over the country should have nothing but the deepest gratitude for what you have done for them."[63]

Electricity, color photography, industrial technologies, and the railroads all increased American contact with the natural world. But they had adverse impacts, too. Moth collectors may have benefited from electrical lights, but the insects themselves fared less well, dying by the many millions as they collided with hot streetlamps or were distracted by glare, suffering the same fate as at the fiery gas wells around industrial cities like Pittsburgh, which could scorch or swallow any living thing that approached them.[64] Yet it seems that no naturalist in the nineteenth century bothered to consider the impact of electric light in any extended way; even into our own time, no one has "measured the effects [of artificial light] on moth populations," as Kenneth Frank recently observed in *Ecological Consequences of Artificial Night Lighting*.[65] What evidence exists confirms a negative relationship between artificial light and the life of moths. Victimized by the "captivation effect," moths often lose their ability to find food and cease to reproduce in a predictable and normal way. They fly about madly, in swooping, chaotic, repetitive, and self-destructive patterns; as one early observer put it, "if a new electric light establishment has been installed near your neighborhood, you will be certain of some rare catches and specimens, but after a few years most of the moths and insects in the vicinity will have become extinct, by reason of beating themselves to death against the lights."[66] Two of today's observers, Kenneth Frank and Gerhard Eisenheis, have conjectured that artificial lighting may undercut or devastate small populations also threatened by habitat fragmentation.[67]

In an irony of the time, William Henry Edwards's third volume came out the same year as Holland's *The Butterfly Book,* each with pictures of the same butterflies but differently produced, Edwards's by the human hand, Holland's by photographic technology. William Holland's photos may have attracted a new legion of amateurs, but in time, and in the manner of all such photos, they would also promote distance from butterflies and from the natural world generally by "suppressing" the handwork or the illustrating skills of individual naturalists.[68] In contrast, by depending on their own artistic craft, both Mary Peart and Herman Strecker engaged butterflies more intimately and closely than if they had photographed them, relying on machinery devised by others. "The correlation of nature-study and drawing is so natural and inevitable," Anna Comstock argued in her popular *Handbook of Nature Study,* "that it needs never be revealed to the pupil. When the child is interested in studying any object, he enjoys illustrating his observations with drawings; the happy absorption of children thus engaged is a delight to witness."[69] By enlisting the child's own ability, sketching or painting led to a level of seeing and immersion greater than any reached by those who lacked the skill or discipline to draw or paint.

So, too, photographic machinery made *collecting* butterflies irrelevant or superfluous. For children, however, actual collecting served best as an introduction to butterflies, to the point of arousing the "beating heart." There were dangers in it, as many indicated, but collecting that served some educational or spiritual purpose could be relied upon to open a door. Comstock made this point, and so did naturalist Herbert Smith, in 1897, in response to those who considered all collecting murder. "The collecting instinct," he argued, "is inherent in almost everyone, and through it, nearly every boy can be drawn into the study of nature. There is something tangible about a collection: the possessor feels the richness of it, and his interest continually grows." Besides, "the quick death of a poison bottle can hardly be regarded as cruelty compared with the lingering death caused by an ichneumon."[70]

Charles Valentine Riley, of the U.S. Department of Agriculture, told Scudder in 1888 that he doubted that color photos of lepidoptera would ever exceed in "naturalness, accuracy, and clearness" the great handcrafted images of the past.[71] Ten years later, fewer doubts existed, for color photography had begun to "capture" so clearly the insects themselves as to foreclose collecting, or so two respected naturalists inferred soon after Holland's book appeared. Frederic Clements, a young university instructor from Lincoln, Nebraska, and later one of America's great ecologists, wrote Holland that he had "become too soft-hearted to be able to 'cyanide' a bee or butterfly, and yet I must know them quite well in my work. Your book helps me out very nicely, as I have no compunctions about catching a butterfly, if I let him loose again."[72] And Gene Stratton-Porter of Indiana, a popular author of books for young people, in 1912 discovered her mission in life: to photograph moths in color immediately after release from the chrysalis or pupa but "*before* circulation was sufficiently established for them to take flight." In an earlier novel, *A Girl of the Limberlost,* she had dealt with collecting but had had nothing critical to say against it; in fact, she'd recommended selling moths as one way for poor girls to help pay for their education. Three years later, however, in *Moths of the Limberlost,* she wrote, "The difference between these perfect living creatures, and the shriveled, pin-pierced subjects of the illustration of any moth book I possess is the difference between abundant life and repulsive death." No longer pro-collecting, Stratton-Porter invoked Holland in that book but only to draw attention to color photographs as substitute butterflies. Passages in Holland's *The Butterfly Book,* she charged, led to collecting and death; while she fostered life by "picturing the beauties and wonders of this creation, for people who could not go afield to see for themselves."[73]

Color photography created a tension between nonhuman nature and human nature, and so did chemical colors. In the same decades that British and German *collectors* searched the tropics for exotic butterflies,

British and German *chemists* invented the first artificial colors, synthesized in the 1850s from aniline coal-tar compounds, and later as chemical alizarin. Synthetic alizarin reproduced an organic compound by the same name derived from the madder root, its invention "arguably a more significant step than the invention of aniline dyes," the science writer Philip Ball has argued, since "it showed that organic chemists were becoming nature's equal and that the natural dyes of the past could be replaced."[74] From the 1850s on, a human-made spectrum of color appeared not only in the exhibits of world's fairs but in the latest fashions worn by visitors to the fairs, as well as in machine-made goods transformed by industrial design, dead things endowed with the appeal of living forms. Whether Americans were able to reconcile these two worlds of color for themselves, or whether they even cared or thought about it, is difficult (if not impossible) to establish. Did they prefer the unexpected and perishable tints and shades of living things to the "permanent" and "fast" palette of machine-produced things, the blue in a butterfly wing to the same blue in a magazine ad? Did they perceive artifactual red as more "beautiful" than organic red? Or did they view them as of equal value, together forming a whole no one had ever before experienced?

In 1899, Herman Strecker received from Europe several "magnificent aberrations" of exotic species, all produced artificially, he told Henry Skinner, "by roasting, starving, freezing, etc." He admitted that "they were beautiful, but beautiful as they are, they never will give me the pleasure I have derived from those I possess that nature alone had a hand in."[75]

The railroad had opened many people to a legion of lepidoptera, but it assaulted the environment, perhaps more than any other technological invention up to that time. Throughout most of the nineteenth century, the fauna of Coney Island, at the farthest tip of Brooklyn, were exceptional, as a cofounder of the Brooklyn Entomological Society, Franz Schaupp, reported in an 1880 essay, "Insect Life on Coney Island." Many "white" animals, for instance—white frogs, white spiders, white grasshoppers, and, especially, many rare white beetles, exceedingly local—lived in small ponds "in immense numbers," a mere hundred feet from "the shore rising and falling with flood and ebb tides," Schaupp observed. "No doubt these little insects are the very aborigines of Coney

Island." Many white butterflies flew there, too, although Schaupp, a beetle man, didn't mention these. Yet as early as 1880, Schaupp wrote, as a result principally of the railroad invasion that carried many people to the island and led to its "improvement," these insects were beginning to "yield to the march of civilization, to the cruel merciless pale faces." "Where but a few years ago" white beetles claimed "dominion" over the landscape, threatened "only by a stray [bug] hunter, now the ground is tramped by thousands and thousands, in long files and broad ranks, and the noble [beetles] present but a remnant of their former greatness." By the 1890s, the beetles had departed, along with many other insect species, doubtless among them butterfly fauna.[76]

For William Henry Edwards, railroads were a blessing, but they also changed for the worse the world he had known as an old and a young man. They sealed the fate of one of his favorite destinations, Glenwood Springs, in Colorado, by 1900 a tourist center on a scale far greater than anything in the past, with thousands of Americans vacationing there—among them, the nature lover President Theodore Roosevelt. By this time, environmental conditions had fostered a decline in the varied butterfly and moth life encountered by Edwards during his 1893 visit, a decline caused by greater tourist crowds and the number of collectors. Ernest Oslar, an Englishman who aided Edwards in his work and collected with David Bruce in Glenwood Springs in the early days, chronicled the change in a 1919 letter to another collector, the rich physician, William Barnes: "Most all collectors tell the same story: Glenwood used to be the Mecca for moths, but is now N.G. [no good]. They all say that bugs get less and less every year. I found it myself. I practically took nothing worth having at Glenwood. In Durango I found the same conditions, other old time places nothing in quantity anywhere." Things looked pretty grim, Oslar said; at the same time, he had just discovered another way to collect rare butterflies and moths, but one that would prove even more deadly to nature than the railroads. He mentioned a new vehicle capable of getting him quickly to good butterfly territory: the "auto or light truck" that had just come on the market. "They have excellent auto roads all over the Rocky Mt. districts, N-S, east and west, which would take one to many localities" with insects "never collected before," he noted, urging Barnes to finance him so he could buy a truck of his own.[77] More than likely, Oslar knew of lepidopterists who hunted by truck or car through the West. Otto Buchholz of Roselle Park, New Jersey, took his first cross-country trip by car in 1907, armed with a butterfly net, and Preston Clark, a millionaire manager of gold mines, and his wife "motored" to the Grand Canyon in 1916; he smashed his arm as he collected by mule along the trails.[78]

But the damage done to Glenwood Springs could not compare to that inflicted on West Virginia, symptomatic of the evil William Henry Edwards's great-grandfather, Jonathan Edwards, had known lurked in the hearts of all men and women. By the end of the nineteenth century, mine wars had erupted along Paint Creek, Edwards's old primal hunting grounds. A distant memory was the coal industry as Edwards had known it, consisting mostly of small mines; they'd been replaced by coal companies run by absentee owners in alliance with the big railroads, dictating conditions throughout the industry and herding workers into company-owned shacks at low wages, with little chance of advancement and no prospects for independence. In the strikes that began in 1897, owners evicted workers from company towns, forcing them to live in tents in the Paint Creek area. Mother Jones, the famed labor organizer, arrived to investigate in 1898, the same year Edwards finished his volume 3. His son, Willie, now head of the William Seymour Edwards Oil Company, had helped build this new economy.[79]

The mess had another dimension, one that reached well beyond West Virginia's Kanawha County. In 1880, virgin forest had covered 90 percent of the state. At the edges Edwards had caught his first female Diana fritillary, an otherwise reclusive insect that spent most of its time hidden in the woods. By 1900, all that virgin forest had vanished in the wake of an unseemly collaboration of lumber companies and railroads. The lumber companies had moved in after pulling down the woods of Wisconsin, using newly invented band saws with "enormous power to transform timber into lumber" and relying on "endless belts of saw-edged steel traveling at a high velocity around a great pulley driven by steam or electricity," to quote historian Ronald L. Lewis, author of a troubling account of forest destruction. But most responsible of all were the railroads, empowered by eastern capital; they were capable of meeting "the challenge of mountainous terrain" and of carrying immense loads of lumber nearly anywhere in the country. By 1900, a web of big and small lines had been dropped on the high forest, preparing the way for "sawmills, pulp mills, tanneries, and lumber camps."[80] The rail and the lumber companies would soon evacuate West Virginia, after laying waste to the landscape's many species of hardwood trees and countless other species from all orders and families of animal life, including, one must assume, the lepidoptera. What had taken Europeans many hundreds of years to achieve, Americans had accomplished here and throughout the country in twenty.[81]

William Henry Edwards, with all his ecological insight, must have understood the meaning of the decimation of the trees. He had read Darwin's books three times over, pored over the work of Ernst Haeckel, who'd first clearly defined "ecology," and may have leafed through George Perkins Marsh's *Man and Nature* (1864), with its testimony to the reduction of the forest in Marsh's native Vermont in his lifetime from three-fourths to one-fourth of its original size.[82] Edwards may also have been aware of the despairing views of Alfred Russel Wallace and August Weismann, naturalists he idolized, both of whom bemoaned the way Americans had "ruthlessly destroyed their native forests," the outcome, as Wallace wrote in 1891, of "the fierce competition of great capitalists, farmers and manufacturers."[83]

Only one meager clue exists (there may have been others) that Edwards shared these concerns—and that was when he tried to revive the baltimore checkerspots in Fraser's swamp, acknowledging, at the very least, that some preventable, man-induced calamity had taken place. But what of the Diana fritillary and the zebra swallowtail, original work on which had virtually established Edwards's transatlantic reputation? What clues do we find there? Surely he knew quite well that if you destroy the moist woodlands, you destroy these insects, which find shelter within the forests, and feed and breed only at the forest margins. In some ways Edwards was hostage to his own past but, even more, to his present and to his family—and, ironically, to his own son, Willie, in particular. A successful entrepreneur by the 1890s, Willie never seemed to care much about butterflies or about anything else in nature for its own sake. He preferred mineral extraction over the study of nature; the only "taxonomy" Willie took seriously was that of coal. In 1892 he wrote *Coals and Cokes in West Virginia,* a "hand-book" detailing the local "species" of coals and cokes and where to find them.[84] By 1900 Willie clearly had the upper hand in the family, and his brother-in-law, Theodore Mead, grew to resent him. In 1935, long after Willie died, Mead wrote that taking his advice "was a great mistake as I should have specialized in biology where my inclination lay."[85] William Henry Edwards may have had reservations about the forests' demise, which happened in his own lifetime. If he did, there is no record of them.

If Edwards was trapped in a contradiction, other butterfly people escaped it by backing reforms to protect nature from further despoiling. In 1900, Congress passed the Lacey Act, a federal capstone on the bird-protection movement banning shipment of "wild animals and birds" across state lines; the law helped prompt President Theodore Roosevelt, a lifelong naturalist, to establish more than fifty wildlife preserves.[86] The federal and state governments, too, set aside many millions of acres of land to form parks and reserves to protect the forests and other natural resources, an extraordinary achievement given the national mantra that "the land is a form of capital and must be used to turn a profit," to quote historian Donald Worster. "Without quite realizing it," the country "put together an entirely new kind of commons—*an American commons*—where individuals may go to find natural resources but which no one can take into his or her exclusive possession." The pro-market journalist Peter Huber also holds a favorable view of the history of the park system but from an angle quite different from Worster's. Claiming the mantle of Roosevelt, whom he professes to admire, Huber argues, in *Hard Green: Saving the Environment from the Environmentalists*, that the parks' appeal is that they provide pleasure in the "visible aesthetics" of nature, which most Americans want to enjoy; on the other hand, by confining nature's "aesthetics" principally to the parks and to the big "visible items" ("the redwoods, cougars, and whales"), people can do what they wish with nature or "property" beyond the parks, guided only by "the market." "Everyone knows where public authority begins and ends," Huber writes, given the "well-demarcated boundaries between private and public space." Both Worster and Huber appreciate the parks, the former because they challenge the goal of "development" and moneymaking, the latter because the parks justify development beyond their borders.[87]

Huber has history on his side, for even though the parks and reserves testified to what seemed a magnificent commitment to saving nature, they were not immune to the impact of America's other identity, its commitment to extraction and profit. After 1916, under the new National Park Service, the parks became tourist sites, reshaped by consumerism and riddled by roads.[88] Nor did they do much between 1890 and 1920 to stop development in vaster areas beyond their boundaries, and they may, in fact, have helped give rise to that development; as one reserve or park after another appeared, the country settled complacently into a tremendous industrial expansion. Later in the twentieth century, the railroads declined as automobiles and trucks displaced them, but they still dictated government policy throughout the period of the park system's major consolidation. In 1915, West Virginia opened the Monongahela National Forest in response to the mayhem of deforestation, but, in the words of historian Ronald L. Lewis, the state "did little to check, much less reverse, the destruction of its own environment."[89]

As the park system came into being, so did industrial agriculture and, with it, the gradual demise of family farming, with unfortunate consequences for the country's entomological diversity and natural science. When, in the 1870s, William Henry Edwards and others studied butterflies and moths, they did so to understand the insects' living character and to do justice to them as autonomous beings in the natural world. Twenty years later, hundreds of Americans pored over the life of insects, including of lepidoptera, not with Edwards's eyes but rather to eradicate them as enemies of the people, in the name of economic entomology, a movement to help farmers clear their lands of all obstacles to profitable production.

Throughout the colonial period, European settlers had cleared virgin forests on an epic scale to make way for family farming in New England and in the South, inflicting a simplified uniformity, wiping out many native insects, and making the land vulnerable to foreign species that settlers had unknowingly brought with them and that had no established enemies in their new habitat.[90] Yet in a pattern of secondary succession, it was family farms themselves, with their various husbandry and crops, spreading everywhere north and west, that reconstituted this denuded geography into a new hybrid landscape, fostering the rise of many kinds of insects and insect predators and thereby re-creating, to some degree the natural balance that heedless deforestation had destroyed.[91] Nevertheless as the wholesale cutting down of timber persisted, Americans began to pursue—side by side with small-scale family farming—a new and radically different, capitalistic, mass-production agriculture, single-cropped, mechanized, and facilitated by railroads that canceled out the strengths of the family regime. By the 1890s, mass production had reached California, with irrigated orchards managed on industrial principles, financed by absentee owners who depended on sweated labor in lieu of machines, growing single crops (lemons, pears, seedless oranges, and so forth) in volume for shipment by refrigerated railcars to Chicago and points east.[92] The new industrial order produced an ecological world even more one-dimensional than the colonial one, and brought on infestations of epidemic proportions.

In the monocultural context, and without checks, a single invading insect army could devour many acres of crops in a single sunny day, a circumstance that Charles Valentine Riley, more than anyone, converted into a mandate for a new science. A brilliant, self-educated naturalist like Edwards, Grote, and Strecker, Riley had a similar passion for natural history; he discovered that viceroys mimicked monarchs. He admired the marvelous colors and forms of all insect orders. Yet as Missouri's state entomologist between 1868 and 1875, he embraced the new applied entomology to rid commercial farms of insect pests; they were only 2 percent of all insects but had enough clout to push the other 98 percent out of his mind.[93]

The practice of economic entomology can be traced in America back at least to the late eighteenth century, as agricultural interests came to penetrate every point on the compass, enlisting advances in natural history in behalf of the administration and control of the natural world. So, too, the Europeans, Linnaeus and Buffon among them, treated natural history from the angle of usefulness, with animals and plants seen as servants to human interests.[94] It did not begin, however, to mature as a discrete independent economic science in its own right until the 1860s, when the British and the French adopted it as a necessity, particularly under monocultural conditions. (In 1868, Boisduval, the namer of so many American butterflies, presided over a Paris exhibition of methods for destroying insect ravagers of vineyards.)[95] But it was in the United States that economic entomology triumphed as nowhere else, thrust onto the national stage principally by Charles Valentine Riley.[96]

After Riley successfully used chemical insecticides in Missouri (especially Paris green, an arsenical compound) to stop a potato blight, Congress appointed him, in 1876, to head the new Entomological Commission. The commission focused on breeding grounds and the study of life histories as the means of stopping insect invasions: search out where eggs are laid or what plants the larvae fed on, Riley argued, and banish pests forever. Riley next became the director of the new Division of Entomology in the U.S. Department of Agriculture, a small agency created in 1862 that had rapidly expanded. In 1884, he created a federal government journal, *Insect Life*, the title misleading since the publication mostly had to do with death. A few years later, the Hatch Act funded agricultural experiment stations in every state of the union, giving teeth to the economic control of insects.

In 1889, James Fletcher, Scudder's dear friend and a champion of the new economic science in Canada, was elected the first non-American president of the entomological section of the American Association for the Advancement of Science, doubtless at Scudder's prompting. Earlier in his life, Fletcher had studied butterflies and moths for their own sakes, and his fondest memory was of the "beating heart" kind. Privately, moreover, in a letter to Scudder he scorned "economic entomology as a stupid subject."[97] But in his presidential address, Fletcher heaped praise on the Hatch Act for giving "great impulse to practical science in all lines," foreshadowing decades dedicated to "the discovery, as soon as possible of practical remedies for the various injurious insects which destroy produce." "Such an opportunity for showing the value of Science has never before occurred."[98] Hundreds of specialists were hired to research life histories, insect control, insecticides, and crop practices, their aim "to defeat the insect menace," as Riley's successor, Leland O. Howard, called it. The men received such good salaries as to form "the first market for economic entomologists in this country."[99] In his heyday in the mid-1880s, Riley himself made $2,500 a year at the Division of Entomology in the Department of Agriculture (a U.S. congressman made $5,000); Joseph Lintner, the state entomologist of New York, in his single stuffy office with no staff, made, of course, much less. In 1880, no schools taught insect study; twenty years later, the United States had more schools and experts than all the countries of the world combined. Entomology was now a comfortable career path, leading to big offices and generous salaries—on average, $6,500 a year (with the figure for incoming congressmen still at $5,000), with entomologists employed by land-grant universities, experiment stations, and the federal government, all, in fact, working collaboratively to battle insect marauders.[100] By 1900, the Division of Entomology had evolved into "the greatest entomological organization in the world."[101]

The era of economic entomology, ushered in by the new industrial agriculture, resembled, in its paradoxical effects, electrical lights, railroads, and photography. On the one hand, in giving well-paid employment for the first time to many naturalists, it bestowed a great boon. On the other hand, it reinforced profoundly the utilitarian thrust of American natural science, weakening the purely scientific side, which could not compete as a job creator, and squeezing out its aesthetic aspects. Scudder, Edwards, Doherty, and individuals like them never worked this new vein. Grote and Strecker did dabble in it, and Grote tried, without success, to get a job in it to support his family, even though he abhorred chemical pesticides and disliked the entire utilitarian bias; his favorite moth family was the Noctuidae because it had no "particular economic importance." "There are plenty of [noctuids]," he wrote, "to reward the labors of the collector, and to puzzle the philosophers who believe that everything has its use, and that man himself is the pivot about which all creation turns."[102]

In 1879, when Strecker consented to speak in defense of insect control before the Pennsylvania Fruit Growers' Association, he started off well enough, lamenting "the butterfly foes my horticultural friends are compelled to do battle with" and advising the use of ichneumon flies to "lessen the numbers of the worms of the cabbage butterfly." But then, unexpectedly, he wandered off on the "glorious beauty of butterflies," "wonderful for their diversity," telling two stories calculated to offend his audience. In the first, he reminisced about an old farmer who, on seeing a young Strecker chasing insects in a field near Reading, shouted out in dismay against him. "The most difficult part to make the old gentleman comprehend," Strecker related, "was what the things could be used for after they were caught." In a second story, Strecker recounted a visit to Philadelphia's Academy of Sciences, where he had spotted a picture of a rare butterfly drawn from a recently captured specimen. "I thought, someone else has had the luck to get that wonderful thing;

I don't suppose I'll ever be so fortunate." Too tired and depressed to return to Reading, he decided to search out an old friend, who, as it turned out, actually *owned* that butterfly. Moved by Strecker's enthusiasm, he gave it to him. "And now that wonderful insect is one of the choicest adornments of my cabinet." "I thought of clairvoyance, of second sight, and of more things than are dreamed of in our philosophy." Some of the farmers must have asked themselves why Strecker was on the dais. Who, for God's sake, had invited him?[103]

In the end, the farmers had the last laugh, for the tide was fast turning against Strecker and other maniacs like him, as economic priorities reinforced entrenched utilitarian approaches to nature. In 1881, Henry Edwards innocently published, as the *very first* article in *Papilio,* a speech by Joseph Lintner in praise of applied entomology. Lintner noted how the subject has "assumed an importance in this country far greater than in any other part of the world."[104] By the 1890s, whole columns, sometimes many pages long, on economic entomology consumed space in *Entomologica Americana* and *Entomological News.* In 1897, Henry Skinner swore, in *Entomological News,* that his journal was "guided entirely by unselfish love for our interesting study and [has] no other motive for existence," yet its "Department of Economic Entomology" occupied four pages, more than was devoted to any other subject.[105]

A new wave of amateur activity counteracted these changes, but from a position of historical weakness and in a way that marginalized the amateur. Samuel Scudder, despite his professional training, usually worked and studied at home, and he wrote his books on butterflies in a manner well beyond the century's tedious specialized taxonomy. His three-volume *Butterflies of the Eastern U.S. and Canada* was full of readable and insightful essays, to say nothing of poetry and pictures. Strecker's early work, too, had occasional powerful eloquence, with its artful hand-colored lithographs and engrossing assaults on the practices of others. Grote pursued the same sort of entomology to the end of his life, veering off as he pleased into philosophical reflection, poetry, and thoughts on beauty.

But by 1900, as economic entomology and laboratory-based science made their claim, the science of the early pioneers was frowned on as inferior, the mark of amateurs, a subject of ridicule. The Canadian James Fletcher, at an 1893 meeting of the American Association for the

Advancement of Science, overheard a conversation indicating that what William Henry Edwards did at his home in Coalburgh, West Virginia, "was not science"—a thought inconceivable in 1880, when John Lubbock, one of England's most respected naturalists and an amateur to boot, had singled out Edwards as one of the best natural scientists of the age. Fletcher, himself without professional training, considered the new entomology "stupid" and admired natural history's emphasis on "seeing" and "looking." "I retorted," he told Scudder of the AAAS conversation, "that they did not know what science was, and that my idea was simply, accurate observation."[106] Even Charles Darwin was being stigmatized as a mere amateur, some scorning him for doing science in his country house and not in a laboratory. They might have scorned August Weismann as well, a consummate scientist who, as a young man in 1860s, conducted his brilliant cold experiments on butterflies under conditions even more primitive than Edwards knew. "Unfortunately, I did not have a zoological institute then, not even a place where I could install an ice box," Weismann said. "I had to ask the proprietor of the Zähringerhof [a local hotel] whether I could place a case of the pupae in the hotel ice box, not a perfect arrangement, since I could not really record the temperature, much less manipulate it as I needed."[107] Weismann was an amateur of the highest order. So were "Wallace, Bates, Edwards, and Scudder," who "all pursued Natural History for the true love of it," wrote Fordyce Grinnell, a respected entomologist who, in Los Angeles in 1916, founded the Lorquin Society. "Thus," he added, "their collections have come to be called by some 'sentimental junk.' "[108]

Henry Bird, a young carpenter and part-time farmer from Rye, New York, and a specialist in the genus *Papaipema,* a group of pretty noctuid moths, learned his craft at the feet of the rivals Strecker and Grote. He felt warmly toward both and, in a 1911 letter to a friend, protested that "the relation among entomologists is changing from what it was some years ago. The numerous young men coming into the ranks through the various experiment stations are specially trained, have degrees, and naturally feel of a different caliber from the common collector—the fellow who gives a few hours outside business pursuits to the study. And they do not always clothe their contempt or opinions with much of a covering of gentility. So the young collector often feels there is little or nothing for him to do aside from the mere pleasure of making a collection, whereas there is everything to do, if one does not undertake to do too much."[109]

Meanwhile, Charles Valentine Riley and his successors were turning natural history into a killing machine, wrenching it away from its ecological and aesthetic-oriented roots.[110] They may have helped make America the greatest agricultural producer in the world, delivering more jobs to naturalists than, in the 1870s, they could have possibly imagined, but they transformed entomology into a bureaucratic business and validated for the young "schemes to slaughter insects wholesale and otherwise," to quote Henry Bird again. For the rest of his life, Bird would continue to complain of the economic entomologists who had persuaded the public to detest what he loved.[111]

If Bird had read Santayana's *Sense of Beauty,* he would have understood, should he have needed any proof, the way beauty—the beauty of the *Papaipema*—drew him to nature and to the work of understanding it. Recognition of beauty had probably done more than anything to deepen the human attachment to butterflies and to the nature they inhabited. It was in the interest of natural science, therefore, to remain aware of it, sensitive to it, steeped in it, whether as something to understand for its own sake (what is natural beauty?) or as a humanizing force or presence, a counterweight or antidote to an analytical science that reduced animals to pieces or parts or particles—a process, indeed, very often indisputably beneficial to human purposes but, on its own, degrading for humans and animals. And what did it mean for butterflies and moths, among the most stunning of nature's creatures, to be stripped of their finest glory? Grote, Edwards, Scudder, and even Strecker practiced analytical natural science, but they remained wedded to natural beauty, for what it meant and for how it embellished human experience. The growth of economic entomology and of laboratory science, and the increasing preoccupation with mere taxonomy, broke the connection between beauty and science that natural history had tied together. The result was to leave science to the professionals, and the beauty of nature to the amateurs—and to the commercial market, the bauble stores, and the spectacle theater.

The destruction of the family hybrid farm landscape, with its unplanned provision for butterflies and other living things, did not happen overnight; nor was industrial agriculture its only cause. Another was the formation of America's earliest suburbs. Right up to the early 1890s, hundreds of small market farms in Brooklyn fed the people of Manhattan; in this landscape, Brooklynites hunted butterflies and moths, and out of it arose the New York and Brooklyn Entomological Societies. Naturalists like George Hulst could collect moths, as late as the 1870s, along Flatbush's wooded paths, and in the area's glens and hidden recesses, Fred Tepper stumbled upon many beautiful catocalas.[112] By 1905, all the market farms were gone, not because they were unproductive (they weren't) but because their owners had sold them off to real estate developers, as many thousands of New Yorkers left crowded Manhattan in quest of better housing, an outward migration accelerated by the railroads and by Brooklyn's merger into New York City in 1898, which sent land values zooming, fostering sell-offs. The ecological result was the elimination of both hybrid and wild nature, forcing butterfly people to look for hunting grounds farther out, on the margins of eastern Brooklyn, and then into the Long Island prairie and then, even farther out, until—with the opening up and expansion of the highway system on Long Island, followed by aggressive suburbanization—there was no place else to look.[113]

For well over a hundred years, nature, for many Americans, had been almost synonymous with farming and country life, meadows, glens, pastures, and woodlands all shaped by human hands. To abandon family farming seemed to mean abandoning nature, because no other activity—however draining or exhausting to the human spirit—had proved so effective in bringing human beings, day and night, into relationship with living things. Anna Botsworth Comstock, a pioneer of the American nature-study movement, feared the impoverishment of the senses as people left farms, and she advocated nature study as a way to reawaken the bond and to revive an experience on the brink of dying. "Nature-study," she wrote in 1905, "is the effort to make the individual use his senses instead of losing them. Eyes open, ears open, and heart open is all that nature, the teacher, requires of her pupils, and in return she will reveal to them the marvels of life, the riches of the world, and the beauty of the universe."[114] See, touch, smell, and feel. "What moves in the grass I love—the dead will not lie still," the poet Theodore Roethke would later write. "Things throw light on things, and all the stones have wings."[115]

EIGHT

Death of the Butterfly People

O f the butterfly people who served William Henry Edwards, arthritic and leathery seventy-year-old David Bruce collapsed first, in the woods on a fall morning in 1903, after starting out apparently fit as a fiddle, on a hunting trip from Brockport to Lake Ontario. "We lost a fine butterfly collector then," Edwards lamented to William Wright, who would meet his own end at eighty-three in San Bernardino in 1912, six years after finishing *The Butterflies of the West Coast,* with his lively colored, if not always accurate, illustrations.[1] After the Canadian Railroad Company fired him from his job in 1899, Thomas Bean, self-sacrificing and shy and indifferent to wealth, headed back to Galena, Ohio, putting all forty-six boxes of his butterflies into a railcar and riding in the car himself, eating and sleeping on the floor, to protect his wards from harm.[2] He came home to care for his mother, and then for a sister who was "never entirely well." He discouraged visits from naturalists who sought him out in Ohio to learn about alpine butterflies, protesting that his "house is inconveniently small."[3] In 1930, he was working as a gardener for a local high school; two years later, at eighty-six, he was dead.

As of 1904, Mary Peart lived at 113 North Woodstock Street in Philadelphia and was still busy as an artist, "in need of all the money she can get," Edwards informed Holland, after suggesting that he buy Peart's larval drawings. ("They are too beautiful to be lost sight of," Edwards wrote Scudder in 1901.)[4] Holland refused. For thirty years Edwards had depended on the skill of Mary Peart to enhance his books. Often ill and sometimes down-and-out, she nevertheless stuck by him in hard times, determined to be among Edwards's "immortals." In the late 1880s and early 1890s, she and Edwards faced an unwelcome challenge. Both Lydia Bowen and her sister, Mrs. Leslie, after serving as Edwards's expert colorists for many years, succumbed to age, leaving Peart, especially, without their insight and companionship. But Edwards, too, was dispirited, for almost as much as he had feared losing Peart, Edwards had feared losing these women, who shared a flat in Philadelphia, having long outlived their husbands. They had no servants, and Mrs. Leslie did all the housework, something Edwards hated, since it often delayed the butterfly plates. As early as 1885, Bowen was strong enough only to give advice to her sister on how to make colors; before then she herself had done most of the coloring, an "expert artist," working hard all her adult life.[5] In 1888, nearing eighty, she died, and Mrs. Leslie, at seventy, inherited the full burden of the work. She is "my best colorist," Edwards said. "When I lose [her], I shall feel it very badly."[6] For a time after her death, Edwards turned to the Philadelphia firms Scudder had used for his coloring, and, almost against his will, he found himself reliant on the talents of a German-American, Edward Ketterer, who, though undisciplined and often drunk, did all the work "on the stone" for volume 3, save for two brilliantly rendered species by Mary Peart, the Rocky Mountain parnassian (*Parnassius smintheus*) and an arctic butterfly, *Chionobas varuna*.[7]

Peart, however, not only did some of the coloring herself but drew the early stages of *all* the species in volume 3 with the finest degree of accuracy, which Ketterer later lithographed. In 1892, Edwards wrote Scudder, "Mrs. Peart has access to [a] $1500 [solar] microscope in Philadelphia, and she is eyes for me. I would stop tomorrow if I had not her aid. *She is everything.*"[8] Her knowledge of the biology of butterflies grew in tandem, as Edwards noted repeatedly in his descriptions throughout volume 3. By 1898, she had completed more than twenty-five-hundred plates, "the beauty and precision of which," Edwards observed, "it has not been possible to copy on the lithographic stone." Karl Jordan and Walter Rothschild, vanguard butterfly men, agreed with Edwards: "We have no work in England or in Europe," they observed in 1906, "in which the life history of butterflies is so well-illustrated."[9] In gratitude Edwards named an arctic satyr after her, *Chionobas alberta peartiae,* the final species depicted in the volume, a flag raised by Edwards as proof to everyone of the magnitude of his debt. "I take pleasure for naming it for my associate, Mrs. Mary Peart, without whose cooperation from the beginning these volumes would not have been in existence," he wrote. Mary Peart died in 1917 in Philadelphia, the city of her birth. She was eighty years old.

As for William Holland, so festive was his eightieth birthday, in 1928, that it seemed the whole transatlantic world had come to shower him with honors. Two years later, he revised his *Butterfly Book* in hopes of making it "the last word as to the diurnal lepidoptera of North America." At the same time, he cut out some of his most caustic comments about competitors, amending, notably, his portrait of Herman Strecker, whose 1878 catalog, he had previously declared, "had no value to the beginner."[10] Holland struck out this sentence shortly before his death in 1932, perhaps in an attempt to leave this land as a kindly old eminence. He had outlasted all the leading first-generation butterfly people except Theodore Mead, who died in 1936.

Early in the 1890s, ill Berthold Neumoegen, only five feet tall, looked shrunken or "*geschrumpt*" (to quote his curator, Jacob Doll). Yet chronic tuberculosis had barely dented his upbeat disposition or his fervor for butterflies. In the summer of 1893 he had camped out in the Adirondacks for two months, sure it would "make a man of him again," and he was still stretching every line of credit to get more insects: "30,000 insects from Mexico, 10,000 from Honduras, 5000 from Brazil." "Keeps me busy and pleasantly excited," he told William Henry Edwards in 1894.[11] But he died a year later, at fifty-one, nearly penniless. "In his death," the New York Entomological Society resolved at its 1895 executive meeting, "the world has lost one of its most eminent students and collectors of Lepidoptera."[12] Still, museums did not rush to harvest his collection. It took another five years for Rebecca Neumoegen to sell his specimens; the Brooklyn Institute of Arts and Sciences (now the Brooklyn Museum) bought them in 1899; in turn, in the 1920s, the butterfly man William Schaus purchased them from that institute on behalf of the Smithsonian Institution, where they still reside.[13]

"I feel almost broken down," Henry Edwards, the impassioned ally of nearly all butterfly people, wrote to Scudder in April 1891, "the result of grip-pneumonia and rheumatism. I went with [my theater] Company to Washington and Baltimore, but I had to give up at the latter place and come home."[14] William Henry Edwards tried to take care of him, inviting him to West Virginia to "camp with me for a week or a month as you please and have the run of the house" (everyone else had gone to Florida to visit the Meads). "We will have High Jinks."[15] Henry didn't go. But toward mid-May, after more insisting from William, he took a train to Hunter, in quest of therapeutic relaxation in the Catskills. Within a week, just short of sixty-four, he passed into what he had called, in an earlier eulogy to a friend, "the impenetrable shadow of the world beyond."[16] Many, mostly from the theater world, paid their respects at his Manhattan home at 185 East 116th Street. William Winter, a leading theater critic of the *New York Herald Tribune,* delivered a short tribute, and, set among the flowers sent by the Bohemian Club of San Francisco, was a card with the words "The Curtain Falls." An agnostic, Henry had demanded brevity and simplicity in the service; he was cremated at Fresh Pond, on Long Island, and his ashes thrown into the sea.[17]

Soon after, William Beutenmuller, who had been the curator of Henry Edwards's collection for five or six years, set about selling it, with help from William Henry Edwards, who paid to advertise it in newspapers and encouraged William Holland to buy it after learning of Holland and Carnegie's plans for "a fine museum-of-the-future." "Of course I do not know" he noted, what kind of "collections you mean to have, but certainly you should have to look far before you could find another such general and cosmopolitan collection as that of Henry Edwards."[18] Henry's wife, Polly, often sick herself, had nothing to call her own except the household furniture and Henry's books and bugs, but she refused to part with the collection in "separated lots," willing to sell it only "as a whole," in accord with her husband's wishes.[19] Henry Edwards's more than 250,000 specimens, from all insect orders, reflected his democratic appetite, although preponderant among them were butterflies from the Pacific Coast region (many he had caught, named, and described himself) and from the tropics, most of which were carefully arranged and labeled, and many yet unnamed. "In a great collection like mine," he had written in 1885 to Arthur Butler, the butterfly curator at the British Museum, "there must be a great lot of things unnamed. During the past two years, I have some lovely species of Lepidoptera from Mexico, many of which are new."[20] With so many bugs (and so little time), he, like Neumoegen, had hired his own expert, William Beutenmuller, a young German-American butterfly man from Hoboken, New Jersey.[21] The collection traveled with Edwards to Australia in 1889 and back a year later, then big enough to fill an entire floor in his house, every available nook occupied by "boxes and cabinets of an infinite variety and shape."

William Holland bargained for the great collection, offering a low price, thinking Beutenmuller and Polly Edwards were on their knees. "Mr. Carnegie and I made them an offer," he curtly told Scudder, "which is not as great as they demand, but in view of all the facts seems to me to be about right."[22] Well, it was not to be, sadly for him. Beutenmuller persuaded Morris Jesup, the director of the American Museum of Natural History, to acquire the collection—which, ironically, Edwards himself had earlier tried to sell to Jesup. In the late 1880s, fearing that he would die leaving nothing of value to his wife, Edwards had urged Jesup to buy the insects. Jesup had dawdled, teased, then refused. Edwards had complained to Scudder, "They kept me 'on a string' with my collection for over three years, having made two distinct engagements to buy it, but it all fell through at last, and I have given it and them up. It appears that their idea of an entomological collection is to have a few cases of showy things from anywhere and without any system of arrangement for the admiration of 'jays' [simpletons] and nurse-girls. I am utterly disgusted with them." Henry Edwards abandoned his hope in the days of *Papilio* of making New York a magnet for scientific study. "If I were a millionaire I would found an entomological museum—per se—but not in New York—not in New York!"[23] Now, with Henry Edwards dead, Jesup consented, but on nearly humiliating terms: the museum would house the collection only *if somebody else paid for it.*

In response, Beutenmuller sent out numerous letters to potential contributors and set up a subscription fund, inviting gifts from affluent New Yorkers. Annie Trumbull Slosson, a fiction writer and a naturalist, who had begun to achieve a reputation as a lepidopterist, donated $200; it was her money that, in 1893, had revived Edwards's old society, the New York Entomological Club, expanding "Club" into "Society." Henry Edwards, on the other hand, had many more poor friends than rich ones—actors, actresses, stage hands, stage technicians who collectively contributed $10,000 out of the total of $15,000 demanded by Jesup. Indeed, Holland blamed them for "making it impossible for Mrs. Edwards to accept his own offer."[24] Jesup, now in control of an impressive trove of lepidoptera, permitted Beutenmuller, soon to become the museum's curator of insects, to put three thousand of Edwards's butterflies on public display.[25] His insects had found a home in a place that had not wanted them when he was alive.

William Henry Edwards greatly mourned Henry Edwards. He owed him a lot; years before, Henry had singled him out as the leading American butterfly naturalist, one who deserved all Henry could give him—bugs, data, insight, friendship, loyalty. William thanked Henry by quoting him in all three of his volumes, where Henry's voice can be heard describing freshly this species or that, noting the character of its life history, its locality, perhaps some high western peak or hidden valley in California or Colorado. "Harry was a most delightful man and warm friend," he wrote. "He had no enemy and never had, I venture to say. A great loss to our branch of science. I lament him exceedingly."[26]

William Henry Edwards was too generous: everyone has enemies, for whatever reason and however unjustified, and sometimes the most bitter ones are those who had once been close, who had belonged to the same group or sect, or shared intimacies and secrets. Herman Strecker had opened his soul to Henry Edwards in the late 1860s, and Henry helped usher him into tropical lands, in 1870 mailing the great "green Ornithoptera" Strecker had mooned over as a boy. Now, twenty years later, with Henry gone, Strecker called him a "great humbug" and "Old Buzzfuzz," and his collection "a mass of rubbish in such a condition that it matters little whether the insects are in boxes or in the gutter."[27] In the *Papilio* days Henry Edwards had excluded Strecker from the magazine and scolded him for his treatment of William Henry Edwards. Strecker had not forgotten. He had brooded, too, over what John Smith, in his analysis of American collections before the American Association for the Advancement of Science, had said about Henry Edwards's cabinet ("the best" all-around collection in the country, available to any who cared to see it) as against his own ("richer perhaps in exotics than in American fauna," but "tied up and inaccessible"). So the inflammable German-American stonecutter, who in earlier times had been served so ably by Henry Edwards, had turned ruthlessly against Edwards, who had never attempted to regain Strecker's confidence.

Strecker, in the old days, had stayed at the home of Berthold Neumoegen sometimes for many days at a time, drinking his wine freely and often borrowing or getting free lepidoptera. But he came to hate Neu-

Pub⁴ H.V.Rippon. del. et. lith. 1853.

M.& N. Hanhart imp.

O.(POMPEOPTERA) DOHERTYI, *Rippon. 1.2.♂; 3♂. var; 4,5,♀; 6, Subdorsal of Abdomen, ♂; 7. ibid. ♀*

25. Will Doherty caught this silky black butterfly (number 1, in the male form) on the island of Talaud in Malaysia. In 1903, the Canadian naturalist Robert Rippon named it *Troides dohertyi*, which still stands. *Courtesy of Craig Chesek, © American Museum of Natural History.*

SCHOENBERGIA PARADISEA, *Pagenstecher, and Staudgr., 1, 2, ♂; 1ᵃ, Neuration of ♂; 3, 4, ♀; 3ᵇ, 3ᶜ, Neuration of ♀; 1ᵇ, Abdominal margin of ♂; 5. Anterior Leg of ♀.*

26. In 1898, Robert Rippon drew and colored *Ornithoptera paradisea* for his giant catalog, *Icones Ornithopterorum*, a testimony to the beauty and diversity of tropical butterflies. It was first named by Otto Staudinger in 1891 and then renamed by Rippon *Schoenbergia paradisea*, an appellation still accepted. The butterfly appears as number 1, in the male form. *Courtesy of Craig Chesek, © American Museum of Natural History.*

1.2.MORPHO JUSTITIÆ.   3.M.GRANADENSIS.
4.5.M.OCTAVIA.

R.H.F. Rippon del. et lith.   M & N Hanhart imp.
13

27. This plate displays the forewings and hindwings of three Morpho species, the last, the blue *Morpho octavia*, a Mexican species. The other two are *Morpho justitiae* and *Morpho grandensis*, drawn and colored by Robert Rippon in *Biologia Centrali Americana*, vol. 3 (London, 1879–1901), a landmark multivolume natural history of the plants and animals of Mexico and Central America, edited by Frederick DuCane Godman and Osbert Salvin of the Natural History Museum, London. *Courtesy of Craig Chesek, © American Museum of Natural History.*

Lepidoptera. I. Papiliones. I. Nymphales, VII. Potamides, C. Conspicuæ, o .

1.

2.

Potamis conspicua Menelaus. M. 1. 2.

81.

28 and 29. In an effort to build up his butterfly collection, Herman Strecker mailed countless tropical blue Morpho butterflies to Americans around the country, hoping to inspire the recipients to exchange or to buy and sell butterflies with him. The strategy had its desired effect while at the same time exposing many Americans to the beauty of the natural world. This image of *Morpho menelaus,* named for the king of Sparta and husband of Helen of Troy, along with the images on plate 29 of two other Morpho species, was drawn and colored by the Austrian Jacob Hübner, one of the most influential nineteenth-century butterfly people, in his *Sammlung exotischer Schmetterlinge* (*Collection of Exotic Butterflies,* Augsburg, 1806). These images excited Strecker as a boy. *Courtesy of Craig Chesek,* © *American Museum of Natural History.*

Potamis conspicua Leonte. F. 3. 4.

80.

Potamis conspicua Achilles. M. 1. 2.

78.

PARNASSIUS.

I.

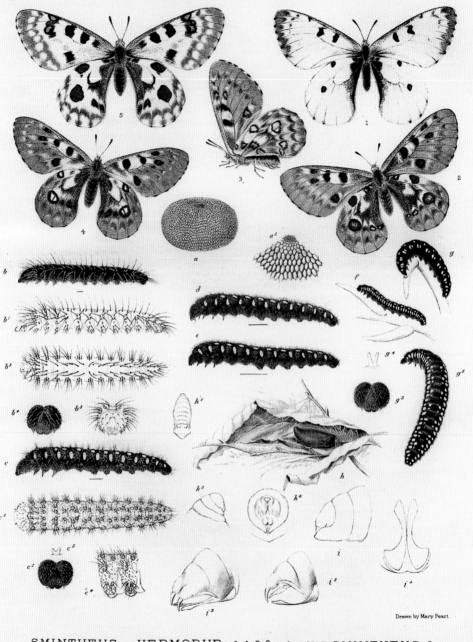

Drawn by Mary Peart.

SMINTHEUS—HERMODUR: 1 ♂. 2.3 ♀. 4. ♀ VAR. SMINTHEUS. 5 ♀.

a — a² Egg SMINTHEUS. ... *magnified*	*f. Larva, 4ᵗʰ mlt.- g adult nat. size.*	
b — b⁵ Larva, young "	*g² - g⁴* " " ... *magnified.*	
c — e " *1ˢᵗ to 3ʳᵈ moults* "	*h — h⁴. Chrysalis:*	*i - i² Periplast.*

30. Compare, for power and effectiveness, plates 30 and 31, one relying entirely on the artistry of Mary Peart, the other on color photography. In 1898 Peart ended her work for William Henry Edwards; she had done both the drawing and the coloring. This plate shows the Rocky Mountain parnassian, caught in the Rockies and found in Edwards's collection.

31. If Mary Peart showed expertly what could be achieved by the human hand, William Holland's *Butterfly Book* of 1898 demonstrated how machines could displace it. This plate, crowned by the beautiful Diana fritillary (1 and 2) and photographed from specimens sold to Holland by William Henry Edwards, relied on the most advanced technology of the nineteenth centruy. For beauty and accuracy, such images challenged at the core the great artisan tradition of the handcrafted that had reached something of a climax in the work of Mary Peart.

Drawn by Mary Peart.　　　　　　　　　　L. Bowen, lith. Phila.

NOKOMIS, 1,2 ♂,3,4 ♀.

32. The Nokomis fritillary, or "daughter of the moon," named by William Henry Edwards in 1862 after Hiawatha's grandmother in Longfellow's *The Song of Hiawatha*. Drawn by Mary Peart and colored by Lydia Bowen for Edwards's *Butterflies of North America*.

moegen as much as he hated Henry Edwards, again mostly because—in his mind—Neumoegen had betrayed him in the early 1880s by siding with Augustus Grote and Edwards in the creation of the journal *Papilio*. Making matters worse was Neumoegen's incessant bragging to him, and to others, of the number and rareness of his own insects, telling the *New York Times,* in 1882, for instance, that he possessed the biggest collection in the country. So in 1892, when Strecker learned from Staudinger that Neumoegen could not meet his debts for specimens bought over months on credit, Strecker took his revenge, informing Staudinger that Neumoegen owned nothing himself and lived in the house of his father-in-law, who controlled all the "property" in it, including the "Schmetterlings," the butterflies. Besides, he was "a Jew, and you know how they are." So here it was, the final card slammed down against a man who believed that he could "keep friends with everybody" and "hurt no one" and who resembled Strecker in all essential ways, to say nothing of having gotten him out of many fixes.[28] Strecker suggested that Staudinger find some way of "seizing" Neumoegen's "seat in the Stock Exchange." But fortunately for Neumoegen, he began delivering payments to Staudinger, without ever suspecting Strecker's duplicity.

Yet for all his mean-spirited, hateful resentments, Herman Strecker's aptitude for friendship, and for generosity, actually began to strengthen as he became less driven as a dealer, a shift aided by an ever-swelling number of "professional collectors," who all together, as a class, helped to drive down the prices of insects. As Will Doherty would discover, these men made it harder to earn a living from selling butterflies. And as Strecker himself wrote a dealer in 1897, "once anything would sell, but now through many dealers, the naturalists have been particular and nothing but specimens in almost perfect order will command any sale."[29] But Strecker had other reasons besides a saturated market for turning once again to natural science.

Around 1900, he renewed contact with an old childhood buddy and fellow collector, Russell Robinson, whose father had built the railroad from Reading to Philadelphia and who, unlike Strecker, had left butterflies behind for a quite different career—a decision, he told Strecker, he now, at age fifty-seven, much regretted: "I wish to god I'd stuck to butterflies. If I had, I would have been a happier man." Still, there were compensations. He had begun butterfly hunting again, this time "in the

clover fields and gardens" around his house in Virginia, and he had a
son, Wirt, a young army captain, who had just come back from Margarita Island, off the coast of Venezuela, where he had discovered an undescribed dwarf variety of the Thoas swallowtail (*Papilio thoas*), a tropical
species occasionally found in southern Texas. The father had passed on
to the son the love of nature Strecker had imbued in him. Wirt would
teach chemistry and natural history at West Point and bequeath to other
young men the same interest; a gravestone to his memory with a small
butterfly inscribed on it can be visited on the academy grounds. Russell
Robinson reminded his old friend "of the radiant brightness connected
with our young life" and invited him to spend a month at his Virginia
home, "when the butterflies are moving and you can do so." "I'll feed
you well, hear your prayers, and take you away from women and brick
and mortar, where the Good Lord is if you want to see him."[30]

Contact with Robinson must have set Strecker's scientific juices flowing, inspiring him to return to fresh work on butterflies, as he had long
wanted to do. Many of his old friends, such as a now aged John Morris, the author of America's first butterfly catalog, had backed him up
in this. Shortly before his death at ninety-two, in 1895, Morris told
Strecker that "I just today picked up your capital book on Butterflies"
and "got so deeply interested in what I had read before over and over, as
to spend I do not know how much time, in poring over it."[31] He longed
to have Strecker as a guest again at his country home in Lutherville,
Maryland, with its spreading oaks, pear trees, lanes and meadows, and
butterflies without number.

Strecker also had many new friends as well as faithful ones from
the past, all naturalists with few commercial entanglements—Harrison
Dyar, John Smith, William Beutenmuller, Henry Skinner, and Henry
Bird, among them. Bird wrote several articles on the *Papaipema* genus
of moths, which delighted Strecker, so he got himself invited to Bird's
home in Rye, New York, and sat for hours in his kitchen, drinking all
the medicinal Scotch Bird's mother had stored in the cupboard.[32] The
memory of his supposed thievery of insects at the American Museum of
Natural History, given credence by Grote thirty years before and never
decisively disproven, had by now become a subject of jokes by friends,
not to malign but to get closer to him.[33] "If I were Dr. Skinner," John
Smith told Strecker in 1900, "I would never under any circumstance

allow you to see another insect under my charge; would never allow you to enter my door again, and would give peremptory orders that at the American Entomological Society all keys to the collections should be refused you."[34]

Strecker managed to write and publish two supplements to his 1878 catalog, but both lacked, of all things, colored pictures, to the dismay of friends who knew how crucial they had been to him and how gifted he was as a lithographic artist. John Smith, who admired him as much as the others, reminded Strecker of Strecker's own motto, that "a description without a figure was good for nothing." Smith's gentle rebuke hurt; after all, Strecker could trace his whole butterfly life back to the pictures he had devoured in the Philadelphia academy as a boy.[35]

Henry Skinner tried to persuade Strecker that color photographs had great promise and pointed to Holland's *Butterfly Book* as proof. "Have you seen Holland's new book?" he asked; "if not I think it will open your eyes to the possibilities of the 3 color half-tone process. . . . The Hesperiidae [skipper butterflies] came out wonderfully."[36] Strecker might even make money if he followed Holland's lead, Skinner suggested, and Strecker responded not by ridiculing the idea (Holland as a model?) but by meekly observing that he had no desire to "make money," only to ensure that "the things in my collection will have a status. I don't care if any one buys it or not, it fixes the species, and that is all I care for." Yet "if I could get a decent plate cheaply I would not hesitate," he admitted. "Yes, I wish the color-type business would get cheaper then I could figure my monstrosities and other things of interest."[37] Strecker had ceased to imagine that he might rely on his own craft, rooted in an earlier artistic tradition reaching back hundreds of years in Germany, on the grounds that, as he said in 1900, "to draw on stone as I did in former years is too great an undertaking with the limited time at my disposal."[38]

Things might have been different had Strecker not been chained to business, whether in butterfly selling or monument carving for the dead. He did both right up to the end of his life. "I hate business," he wrote a friend in 1895, regarding the butterflies, "and hate it to be intruded into science beyond the mere unavoidable mention of getting boxes made. I never had any predilection for business. It makes me mad when I am forced into rules or set phrases."[39] "It is hard to get away from business," he told Harrison Dyar, one of his new naturalist admirers (here he meant the gravestone business). "It is my bread and butter, but I hate it. If only I could give all my time to science. I am getting near the end and have accomplished nothing."[40] "I have been worked fearfully getting gravestones in place for Decoration Day," he wrote Henry Skinner in 1900. Staudinger's death that year at seventy made Strecker feel even closer to the grave. How *much* these two men had in common: an almost innate love for all kinds of butterflies, the same longing to be known as lepidopterists rather than as collectors, the same business phobia. But unlike Strecker, Staudinger, at fifty, had dropped his business in the lap of his son-in-law, freeing him to write his big catalogs, and when he died, he had reached commanding stature, in part because of his relationship with the German kaiser and the Royal Museum of Natural History in Berlin. "He was the best and greatest of all," Strecker told a friend, without a hint of envy, "and his place will not be filled again perhaps for ages."[41]

In early 1901, William Kearfott, a moth specialist in Manhattan, asked Strecker, then sixty-five, to join him and "a dozen or more members of the New York Entomological Society, all friends of yours," at his house in New York. "We will have a love feast and I will not take no for an answer."[42] Of course, the butterfly maniac could not resist and relished every minute, but over the next five or six months, he suffered from crippling influenza and then some kind of stomach ailment; in late November, he had a stroke as he was getting on a train in downtown Reading, and he never recovered.[43] His friends and enemies assumed he died well-off, but he surprised them all by leaving behind nothing but his collection, his house, and $9,000.[44] Eveline Strecker hastened to offer her husband's collection for sale, for $20,000, but though both Holland of the Carnegie and Skinner of Philadelphia's Academy of Natural Sciences wanted it badly, each balked at the price. "Strecker's wife has an exaggerated notion of the value of her husband's collection," Holland complained to George Hampson, a former tea planter and moth authority at the British Museum. "It was Strecker's fate to make a great many new species on the basis of worn or rubbed specimens, sometimes on the basis of merely abnormal varieties. The poor old fellow had but small resources and thought he knew a great deal more than he did."[45] Year after year, as Eveline grew poorer, she lowered her price. In 1905, her son Paul borrowed $2,500 from her, her only savings, and fled town.

The lepidoptera and the great archive of letters remained, confined on the third floor, until 1908, when a native of Reading, William Gerhard, a collector of Bolivian insects for a drugstore mogul in the mid-1890s and later the first curator of insects at the Field Museum in Chicago, took a train to Reading, his heart set on the collection. Gerhard arranged to have the glass cases and drawers full of Strecker's precious cargo, plus the sixty thousand letters and books Strecker had saved during his nearly fifty years as a butterfly man, lowered tenderly out the third-floor window by block and tackle and packed off in a freight car to Chicago. The whole thing weighed more than fourteen thousand pounds and cost Gerhard $15,000.[46] Until the 1980s, without any decree from any imperial authority, as in Staudinger's case, the Field Museum housed Strecker's insects in a separate place within its collections, keeping both Strecker's ordering of species and his hand-printed labels.[47]

Will Doherty died in 1901, in the same year as Strecker. Three years before, in 1898, he had come home again to get over another tropical sickness, so ill that his doctor had forbade him to write letters, because they exhausted him. His mother took care of him until the following November, when he felt ready to go back to London. There was no other course, in any case; he needed the money and was determined to keep his place as one of the great collectors.[48] In London, Doherty charted his way back to Malaysia and then to a new place for him, East Africa. But right up to the brink of departure, he had doubts, especially about the African trip. Perhaps he wouldn't find good men to collect for him. Maybe there would be no profit in it. "I feel rather discouraged by Africa, and doubt if I can make it pay," he wrote. He considered going instead to the Comoro Islands, near Madagascar, where Christopher Aurivillius, an admired Swedish butterfly man, had obtained many insects, but he rejected it as too costly.[49] On a visit to Emily Sharpe's little store in London near the British Museum, around Christmastime, he chatted with Sharpe, a specialist in African butterflies who had named a few herself (for instance, *Papilio jacksoni* E. Sharpe). They talked of seasonal concentrations, a matter of great concern to Doherty, since he had blundered in New Guinea for such reasons. They may have discussed, too, the lion attacks along the new Uganda Railway, then being built by the British, a tremendous technical feat, erected in part with the aid of American engineers. Theodore Roosevelt himself took that railway in 1909, as his entry point to Uganda for a hunting expedition to kill big mammals for museums in Washington and New York.[50]

"How far is [the Uganda railroad] open to traffic?" Doherty asked Walsingham's personal secretary. "I have designs on Victoria Nyanza [the old name for Lake Victoria]." Collectors all along the railroad had been killed, several buried in a cemetery set aside for Westerners in Nairobi, then a busy railroad depot, later the capital of Kenya.[51] At a London dinner party, both Rothschild and Henley Gross-Smith, a lawyer eager for tropical rarities, warned Doherty not to go to Uganda. George Hampson at the British Museum "never loses a chance to discourage me," Doherty reported in a letter to Ernst Hartert, "and Elwes at dinner Monday was very down on it." On the other hand, he learned that William Holland was exuberant and appealed to him to go.[52]

Holland's influence may have been decisive, tipping Doherty into making a fatal decision. By the late 1890s, Holland, empowered by Carnegie, stood for an enormous coterie of institutions, not just natural history museums but botanical gardens, zoos, aquariums, department stores, and specialty shops, all forming a new commercial market that pressed in upon collectors, enticing them with money to take risks throughout the world. Holland had men everywhere, above all in South America and Africa. One such was José Steinbach, a multilingual German immigrant in Bolivia who, beginning in 1900, delivered ten thousand Bolivian birds to the museum, in defiance of the spirit (if not the letter) of the 1900 Lacey Act and with the full knowledge, of course, of Holland. Steinbach sent probably triple that number of butterflies and moths to Holland himself, sometimes traveling for months by ox carriage, with his wife and one of his children in tow, jammed into a "tiny hole" in the carriage to make way for storage.[53] Holland paid relatively well but treated Steinbach with contempt, suspicious that even the most accomplished collectors might scheme to cheat him. When Tyler Townshend, a skilled and reliable naturalist, sent Holland his "entire lot" of rare lepidoptera caught in Mexico's Sierra Madres, Holland dismissed them all as "a lot of rubbish that he absolutely did not care for"; "it looked like a lot gathered in carelessly with a drag net."[54] In 1902, Hans Fruhstorfer, the "fiery" man so admired by Doherty, mailed Holland numerous Burmese butterflies, only to be told that "taken as a whole this is the most abominable lot of rubbish that has ever been offered to me by anyone with whom I have ever had entomological dealings."[55]

In 1898, Holland dragooned America's greatest all-round tropical collectors, Herbert and Daisy Smith, to get butterflies and birds in South America. Holland knew a good deal about the Smiths after Herbert contacted him, looking for work at the museum. Poor and raising a young son, he told Holland, who was well aware of Smith's competence, that he would be happy to "accept a small salary," while his wife would work for nothing. "I can truly say that I am actuated by no mercenary motive," he wrote.[56] Briefly, both Smiths worked in the museum itself, Daisy, of course, without pay, but Holland wanted them back in the jungles to get him insects, and, fortunately or unfortunately for selfless Smith, that was exactly what he wanted, "crazy to go back to South America to collect insects."[57] A profoundly dedicated naturalist, he hoped, in his "spare" time, to launch a "Biologia Meridionali Americana," a biological study of a South American region covering an immense amount of fauna and modeled after Frederick Godman and Osbert Salvin's efforts in Central America.[58] Late in 1898, Smith took Daisy, their son, and two assistants to Santa Marta, Colombia, to collect birds for the Carnegie museum. Privately, he contracted with Holland to supply "all the Lycaenidae, Hesperiidae, and Lemoniidae, which said Smith may take during his visit, to a number not to exceed thirty specimens of any one species," along with "at least one pair of all other species of diurnal lepidoptera taken by said Smith." Holland committed to pay ten to twenty cents per specimen.[59]

But Holland had his sights on more than South America. His greatest dream was to be seen as a worldwide authority on African butterflies, and by African, he meant *all* of sub-Saharan Africa. As early as 1888 he had told Strecker, "I should like to do for Africa what Marshall or De Nicéville are doing for India."[60] He even *paid* to get articles he had written on African insects into strategic periodicals, footing the bill for publication in Scudder's *Psyche* and Henry Skinner's *Entomological News,* where he wanted to present his "600 African species new to science," collected in the field by somebody other than Holland. "Next to going to Africa," Holland informed his doting, always unquestioning parents, "is the pleasure of studying the beautiful creatures which come from that far-off land and of *naming them*."[61] "I desire to give them to the world," he wrote Skinner.[62] And, with help from people like Will Doherty, everybody's premier collector, who *knew* what he was observing, Holland might nail a sure place in entomological history. In February 1900 he wrote Doherty to get "all the Hesperiidae and Lycaenidae which you can take" from East Africa, the "entire catch of these things," and the "moths of Africa," too, "from any given locality or well-marked region, say, for instance, Kilimanjaro." And get me "all the butterflies which you can take from Madagascar or all the moths you can take from Madagascar," he added. He wanted *omnium* because "I like to work up the insects of a group or of a faunal region at my entire leisure, without feeling that because of the possession of material possibly made by the same collector my labors will be nullified unless I rush into print." He promised to pay thousands of dollars for the butterflies and moths.[63]

In March 1900, Doherty came down with a mild attack of malarial fever and couldn't work or eat; Hartert offered to nurse him, but he revived sufficiently to take a last stroll through London and pick up the cashier's checks Rothschild had generously provided, "pulling" him "out of hole."[64] He paid a sad farewell to Hartert: "So good bye, old fellow, in case we don't meet again. Possibly I may make Trinidad or some other place my headquarters, but not as a married man. You are luckier."[65] Doherty left for Paris in late March to meet with Charles Oberthür (also enthusiastic about Africa) to get help with passports to the French colonies.[66] Sometime in late May, thin and sickly, he took a steamer from Paris to Mombasa, an old Arab port city under British control, and began hunting butterflies, notably for Holland.

Between the fall of 1900 and the early spring of 1901, Doherty and Holland had a nasty exchange of letters. The Mombasa butterflies were "very common," "exceedingly small," and "hopelessly familiar," Holland spewed forth. Mombasa is "already an overworked and hopeless locality," he declared. "I could do nothing in the way of adding to the literature on the subject except to publish a list of species."[67] He tried to check the negatives, knowing how much Rothschild looked after Doherty's well-being. He paid about $450 for the collection, a "good deal more than it is worth," he had to add, but he didn't dare treat Doherty as he had Fruhstorfer, who got nothing for his pains.[68] Holland had no desire "at all" to "find fault," only to "stimulate to best effort." "I think that if you do the kind of work that I believe you are capable of in this region, science is going to be wonderfully benefitted," he wrote at the end of a January 11, 1901, letter, "but the first haul by your nets in the suburbs of Mombasa I fear has yielded very little."[69]

Weeks later, talking into his Dictaphone, he "found fault" with Doherty's notion that he should be paid in advance for his demanding collecting work. He would, of course, make allowances for Doherty's own "fault-finding" and the "querulous tone" in his letters, in light of Doherty's "discomforts" caused by the extremes of cold and heat near Mount Kilimanjaro, but, "the trouble with you my dear Doherty," he said,

is that you are not definite and business-like in your dealings with us. I have written a number of letters to which I received no reply whatever. . . . It is not customary for institutions to pay for collections in advance of their receipt, especially when not a single intimation is given by letter as to the number of specimens or anything else which is necessary to know in order to make a settlement. . . . If you expect us to be prompt and explicit it is not asking too much to require of you that you will also take the time to write down what you understand and what is wished. Our business contracts with collectors and others are generally made by letter. Your communications have been few and infrequent, and by no means explicit in reference to money matters that it is highly important that we should know.

Doherty's letters to Holland did not survive (perhaps Holland, wisely, destroyed them), but one can infer from the words "fault-finding" and "querulous tone" that Doherty took nothing sitting down. And surely, the bureaucratic lecturing by Holland, after years of his shabby business behavior, must have seemed rich indeed. And what, possibly, could Doherty have thought of Holland's final words: "As I look out the window while dictating my letter, my eyes are greeted by the sight of whirling snowflakes and hillsides in ermine. Winter is upon us with genuine force, and though you speak of enduring lapland nights where you are in the tropics, I am sure you would enjoy the refreshing exhilaration which we feel in view of the wintry storm which surrounds us."[70]

At the same time, clear across the Atlantic Ocean, Herbert and Daisy

Smith, in Colombia, were struggling to supply Holland with his butter-flies. Collecting along the coast of Santa Marta, alone and without assis-tants, Herbert was infected so badly by sand flies that the skin peeled off his palms, and Daisy caught malaria, the deadliest kind. Herbert paddled her and their son by canoe through driving rain back to Santa Marta, to seek care from the local "Sisters of Charity." Daisy suffered greatly.[71] Yet despite this misery, Smith mailed off hundreds of butterfly species to Holland, apologizing for lacking time "to work in the moun-tains." Holland's response was to criticize him for the lack of "novel-ties." Staudinger, he said, had sold him these things years ago! "Have some mercy on a man!" Smith replied. "After you have ransacked the world to get together the finest collection in North America, you can hardly expect to turn out many novelties for so small a lot as mine."[72]

Both Smiths were experienced amateur naturalists who loved South America and ached to do a survey of its fauna. They were quite happy to work for rich "aristocrats," who were often as dedicated as themselves to expanding for everyone the knowledge of the natural world. But there was no symbiotic relationship between them and Holland. Smith, so decent himself, seemed never able to admit that Holland was not decent, that he was no honorable boss mason, that he was, in fact, a new kind of man in a new institutional age—a parasitoid on other people's talents and achievements. ("The best place to collect butterflies," Hol-land told an Iowan naturalist in 1903, "is in another man's collection who has been at it for many years.")[73] Ironically, Holland, too, was an amateur, but by the late 1890s, he had come to despise amateurs as too uppity and romantic, as losers, in fact, even though he habitually robbed them blind of their own gifts.

In 1900, Holland, dressed like a gentleman in the dirty summer heat, traveled to a fossil bed in Wyoming as Carnegie's obedient servant, to break the state university's claim on a giant dinosaur Carnegie wanted for his own museum.[74] "This is his fancy," Holland explained, "and we must please him." He hired a lawyer to buy all the land around the claim and seduced the man who had staked the claim—W. H. Reed, a competent amateur fossil expert—with a three-year contract at the museum.[75] He warned the Wyoming legislature not to "antagonize the gracious kindness of Mr. Carnegie" and twice tried to bribe its mem-bers.[76] Anyone who interfered with his mission or authority, he fired.

He fired the coachman who drove him around after the man refused to take his advice on how to drive.[77] He fired J. L. Wortman, the head of the museum's fossil operations, after Wortman told him to "go to hell" for suggesting that he get a couple of "Italians" from the nearby "workmen's agency" to help him dig up the dinosaur. He fired the original claim staker, W. H. Reed, too, for siding with Wortman and for failing, as he wrote Reed, "to submit yourself, as others must submit themselves to the judgement of those who are over them." Reed, Holland claimed to a colleague, was a "self-made man who had an exaggerated idea of his own importance and the value of his attainments." "All things being equal, I prefer the man who had the benefit of a liberal course of training in a literary institution. Such men are far less apt to be troublesome than those who have not had such advantages"; they have "docility, and the disposition of Soldiers, who obey orders." In a breath, Holland had rid himself of self-made amateurs who had made America's early nature science (including butterfly science) what it was. "We wish no more Reeds, no more Wortmans," he declared—and, he might have added, no more Smiths.[78]

On his return home from Colombia, Herbert Smith felt totally demoralized, unable to work or to think or speak clearly. It took him months to recover. In 1902, Holland, in a letter to a fellow naturalist, bad-mouthed Smith as "the best example of a scientific tramp whom I have ever known—unstable as water" and "irreconcilably discontented with all existing conditions. I fully predict it will not be three months until he will be back again, confessing himself an ass—as he has done before—and asking for a job."[79] Scorned by Holland, a year later, Herbert and Daisy left Pittsburgh forever, to take up a new career at a natural history museum in Tuscaloosa, Alabama, specializing in the freshwater shells of the region (they became world-class experts). Herbert Smith was hit by a railroad car and died in March 1919. By then totally deaf, he did not hear the train coming as he crossed the tracks. Daisy stayed on at the museum, appointed to her husband's job.[80]

Doherty never allowed such a man as Holland to dictate his activities. Nevertheless, he was not as lucky as the Smiths, in the end trapped more than they by the exigencies of the commercial market. In 1900, he traveled on the new railway to Nairobi, seeking butterflies, probably in June and July, but he missed everything he went out to acquire, "except in microlepidoptera, which don't pay."[81] Then he reached the Kikuyu section of the Escarpment, a "magnificent" elevated landmass of ancient origins. "We are on the slope of the Sattima Range," he told Hartert, "and just at our feet is the broad yellow desert of the Rift Valley. We are cut off from Kilimanjaro by nearly 200 miles of naked desert or grassy plains." In August, at the temporary terminus of the Uganda Railway, northwest of Nairobi, he put up a small tent, "well-armed" but not as secure as the waterproof hut he constructed for his men, "much safer as regards lions and natives, but too dark for me."[82] The men hunted birds, beetles, mammals (from rats to antelopes), cicadas, wasps, and unusual mosquitoes, and, net in hand, Doherty searched for butterflies, all at between 6,500 and 9,000 feet.[83] He wrote his mother and Hartert about his entire catch. "We have got some dozen of the King of African butterflies," one of the high fliers, or *Papilio rex,* a beautiful big, tailless dark brown swallowtail, with white dots on all surfaces and shades of rufous red on the forewing near the thorax, a great mimicker, too, reflecting in every detail a toxic species in the Danaidae family, as Doherty observed to Hartert. For years, Westerners had known of only two specimens of *Papilio rex,* until Oberthür formally described it in 1886; after that, collectors caught more, slowly forcing down the price.[84] To Hartert, Doherty listed seven other swallowtail species caught on the Escarpment, all "in great numbers, including both sexes." He had got some Palearctic forms, too, and one *Acraea* species closely resembling *Acraea excelsior,* which Emily Sharpe had identified and described in 1891.

Despite this bounty, he was disappointed that there were no new species. The king of African butterflies had been dethroned, he reported, "and is now as cheap as the nobility in Georgia [in Russia]." "I'm afraid Mr. Jackson got most of the conspicuous butterflies I have got here," he told Hartert. His prospects might improve, he suggested: "Last week I reached the $60 limit and for ten days or more I have averaged over $40 a day."[85] In late September, Doherty rode the Uganda Railway four hundred miles back to Mombasa, on the coast, to mail off his "low country catch." When he returned to his camp in early October, the rainy season had begun, canceling out any chance for butterflies or moths. Animal attacks had started, too. In a deep ravine, holding only a butterfly net, Doherty and another man, who had a rifle, fended off lions. Again and again, there were "marauding" tribes, and a drenching cold wetness

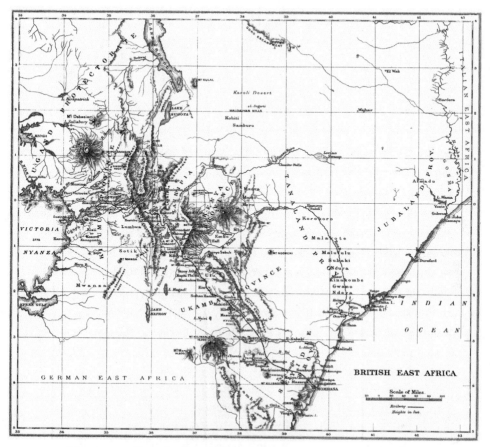

This map, published in Sir Charles Eliot's The East Africa Protectorate
*(London: Frank Cass, 1905), shows the extent of the Uganda Railway,
an imperialist road begun by the British in 1896 to move goods, men, and
military supplies. In 1905, four years after Doherty died, it extended from
Mombassa on the east coast to Lake Nyanza (Lake Victoria) in the interior.
The road virtually created Uganda as a national entity.*

so "intense" that he had to sleep "under a sheet, three blankets, my
clothes, and a heavy overcoat, and my men even more, having thicker
blankets." By January 1901, Doherty had contracted scurvy, "the curse
of the mountains," and was ravaged by dysentery. Four months later,
he could no longer walk, and his men carried him to the railway hospi-
tal in Nairobi. On May 25, he died. He was only forty-four years old.
Doherty's parents asked Hartert to help them "settle" Will's estate and
urged him to "find something among Will's effects to retain in remem-

brance of him." At their request, Hartert sent back Will's traps, his gold watch, a few coins, his manuscripts, and, above all, his "notebooks," his "Black Marias," as he called them, written in detailed, almost inde-cipherable script, more priceless, perhaps, than any swallowtail, all now lost.[86] Butterflies flew over his grave in Nairobi.

Two years after the death of both Doherty and Strecker, Augustus Rad-cliffe Grote died in Hildesheim, Germany. Unlike Strecker, Grote had harnessed his demons in time to make strides in the scientific study of lepidoptera. In the beginning, he had no interest in the insects of Hildesheim or of Europe, only in "the odd ones that reminded me of home," but as his productive energy rekindled, he began to study seri-ously German diurnals even more than the moths, and drew strength from American butterfly men, whom he felt certain were the best ento-mologists in the world.[87] For every new friend Grote made across the Atlantic, he seemed to need a new enemy, especially "those he consid-ered his rivals," and he carried on some feuds in public as ugly and fruitless as the one he had waged with Strecker years before.[88] One such person was John Smith, a moth man and Strecker's admirer, who arro-gantly scoffed at all ways to identify lepidoptera *except* by their genita-lia, a view that Edwards—with his loyalty to larval characters—rejected, although Grote and Scudder took genitalia seriously as identifiers, as have naturalists into our own time.[89] Smith needled Grote repeatedly in print, and Grote called him, off and on, a fool, a liar, an uneducated illiterate, clown who could barely write English, and, perhaps worst of all, "a classificator without ideas." "Before long he will be at the bottom of the bay," he pronounced to Dyar.[90]

Grote warmly befriended three other Americans—Henry Bird, Har-rison Dyar, and John Comstock. Bird and Dyar, young and student-like, were, ironically, as close to Strecker as to Grote, and Comstock, a mature scientist, was an inspiration to Grote himself. Bird collected moths in shallow swamps full of cardinal flowers near his house in Rye or on horseback along the banks of the Hudson, and studied Grote's favorite family, the Noctuidae, especially the genus *Gortyna*—small, often exquisite, moths that, through painstaking field investigation of their life stages, beginning in 1895, he came to know better than anyone else.[91] Grote wisely told Bird "not to scatter your work too much, but to stick to the Gortyna in which you have made so praiseworthy a first effort."[92] Dyar, a highly educated New Yorker who inherited a fortune, shared Grote's love of music and philosophy and practiced the same kind of evolutionary entomology.[93] Grote, turning fifty-four in 1895, assured Dyar that "there is no one I care more about or wish so well." "I get lonely here with no one to talk to and discuss a subject with," signing his letters, "with much love."[94] Dyar's original research on butterfly caterpillars, enlisted to establish phylogenetic relationships among butterfly families, much influenced Grote's own analysis.

By the late 1890s, John Comstock of Cornell University had become one of the country's leading naturalists, beloved of countless students, his work bearing all the earmarks of both natural history and Darwinian biology as practiced by Grote, Edwards, Scudder, and others, from storytelling, collecting, and respect for beauty to systematic analysis and anthropomorphic ecology. In 1904 Comstock cowrote with his wife, Anna Botsford Comstock, *How to Know the Butterflies: A Manual of the Butterflies of the Eastern United States,* published by D. Appleton of New York. The manual went through many editions and perhaps did as much as Holland's to inspire amateurs.[95] The Comstocks, childless, had enormous hearts, caring for each other as they did for the students at Cornell and for the natural splendor of the surrounding environment. They so thoroughly mixed art and science in their own lives as to make the two seem one, just as they did in their book. The Comstocks' manual was superior to anything before it, blessed with splendid color photographs and a text enlivened by stories taken from their own experiences, and as sardonic as anything ever done by Strecker. The color photographs may have been the only departure for them from the natural history tradition, with its reliance on the handcrafted, which Anna Comstock herself so ably exemplified. But they were such *good* photographs, crisper, brighter, the butterflies for the most part more artfully and simply arranged, than anything in Holland.

The Comstocks sketched analogies between humans and butterflies, believing them alike "in many important particulars," and painted sensual verbal pictures of such insects as the clouded sulphurs, which hold "banquets around the mud puddles in the road in August" and look like "shining yellow blotches" that "scatter when approached into a hundred yellow butterfly fragments." In a portrait of the "silvery blue," a gossamer-winged butterfly whose full life history in their day was still uncharted, they took pleasure in disregarding the scientific call to demystify all natural phenomena. "There are several things in this world that it were better to know nothing about, such as a perfect passage of music or a bit of exquisite color. Both were meant to appeal to the soul through the senses, and knowledge about them is superfluous and a distracting factor. Therefore we feel a certain satisfaction in not being able to give any facts about the life history of the silvery blue. All that we know is that it bears on its wings a blue found nowhere else in the world except in the pearly spectrum of the sea-shell."

In this book and others, John Comstock systematically analyzed butterflies, focusing on the wings, unlike Dyar and Edwards, who preferred the caterpillars, or Smith, who studied the genitalia. But not *just* the wings, rather the wing *veins;* many others in the past—beginning perhaps with Moses Harris in England and Gottlieb Herrick-Schaeffer in Germany—had relied on the veins to establish species, but Comstock saw them in terms of evolution, an unprecedented way to get at insect identity. It set him apart from all other previous wing analysts. For him, the wings bore nearly the whole weight and stress of evolution, and on them appeared "the record of the action of natural selection as upon a broad page."[96] By studying how the veins shifted, crisscrossed, converged, or narrowed to indecipherable points, one might trace descent, Comstock believed, from one group to another, and thereby classify and "represent" species by natural relationships and in "constantly branching lines." He fashioned a system of lettering and numbering for the veins that allowed naturalists to see these relationships clearly; it became *the* standard system for entomologists deep into the twentieth century.[97]

Grote himself had long advocated the study of wing venation. "I published the first diagrams and plates of veins in scientific periodicals in America," he told Dyar.[98] Along with Scudder and Edwards, he had also urged the incorporation of evolutionary change into insect classification, at the same time aware of how hard it was to describe or identify organisms in this way. John Comstock's work, bringing veins and evolution together, fortified Grote and brought him back to life, and he did everything he could to popularize it in Germany.[99] Working eight-hour days nonstop, he bred insects again and, like a detective after a crook, plastered maps on the walls of his museum room, illustrating the phylogeny and ontogeny of butterflies (Comstock had done the same thing). By 1900, Grote had produced enough long articles to fill two books on the wing venation of both living and fossil butterflies, two of which displayed the first photographs of *living* lepidoptera ever published. Most naturalists viewed butterflies as evolving from a common stem, but that was wrong, according to Grote. Instead, two supergroups had emerged, the first, the Papilionidae, made up of the swallowtails, the parnassians, and something Grote called the Teinopalpidae, and the second, the Hesperiidae and *all the rest of the butterflies* (skippers, blues, and fritillaries). Each supergroup revealed different veins, indicating evolution along independent parallel lines and proving, Grote believed, that the classification of Henry Bates and Samuel Scudder, which insisted that the skippers came from the swallowtails and, therefore, should be classified as a "higher" group on the evolutionary ladder, must be rejected.[100] Making use of Comstockian terms, Grote told Dyar "never to forget" that the Papilionidae are "more specialized than *any* of the *Hesperids,* in having lost the 2nd internal vein of secondaries."[101] "After all," he wrote, "butterflies have wings, and these wings constitute a record."[102]

While Scudder hated the reenthronement of the swallowtails and believed that Grote's "innovations showed the length to which one may go in discussing classification from a single standpoint," Edwards was happy to see the swallowtails back on top, holding their noses before the nasty skippers, the end of the line.[103] Staudinger observed, in the preface to his newest catalog, that "for the retention of the Papilionidae at the beginning of the Rhopalocera, and for the arrangement of this group altogether, Grote's recent phylogenetic studies are authoritative."[104] Dyar, ignoring Grote's residence in Hildesheim, Germany, called him "the best lepidopterist of America, living or dead."[105]

And yet for all of his seeming success, Grote remained, most of the time, victim to his never-dying resentments toward his living American detractors, wishing them all agonizing ends and wondering, "Why does the all-suffering Deity suffer John B. Smith to live?"[106] Grote was blessed by a good wife and children, but he feared losing "attachments." In 1896 he complained to Dyar about Edward Graef, his old collecting companion from his early Brooklyn days, because Graef had failed to respond promptly to a request for some moth cocoons *with their attach-*

ments still on—that is, with those parts that the insects used to tether themselves in the winter to a tree branch, a door, or an eave of a house. This "attachment," in an obvious free association, mattered a good deal to him. "You see what I want is *the attachment*. This is seldom preserved (also in social life, but I hope we will be exceptional!). It is shown in the Philosamia you sent me. The ideal pleasure these cocoons give me is great, for I fancy myself searching for them! I still want a promethea *with attachment* (in love as it were), the twig cut off."[107] Homesickness plagued Grote. "I want to get back home, but there seems no opening for me," he wrote on the Fourth of July to William Henry Edwards. It "is very bitter to me," he told Charles Fernald of Massachusetts, that "I will never return." "I wish to Heaven I could get back," Grote confessed to Dyar. "I get very blue sometimes." But to Dyar he also reflected, in January 1896: "I never thought I would live so long on this beastly planet, only relieved by beautiful butterflies and a friendship such as yours."[108] And three years later, he reported how a living cocoon of the pale green luna moth someone had sent him from America had fulfilled its purpose: "A beautiful luna came out today, and I have been watching it for hours and imagining I was back on Staten Island."[109] His heart failed in Hildesheim in 1903.

The two Yankees died two years apart, William Henry Edwards in April 1909, at eighty-seven, after years of near perfect health, and Samuel Scudder in May 1911, at seventy-four, after years of suffering that began in 1896 with the death of his only child and best companion, Gardiner Scudder. In his mid-twenties, Gardiner had just graduated at the top of his class at Harvard Medical School when he fell seriously ill with a high fever. The doctors first diagnosed typhoid, but then decided on tuberculosis. In late September Scudder informed a young colleague that he would be unable to examine the insects he had sent, writing, "I should have stopped your sending them, if I had known what was in store for me. My son is dangerously sick and the doctors give us practically no hope of his life." In November, in a letter to a close friend, Scudder observed that the disease was "terribly ravaging all the organs." Gardiner died the day after Christmas. Upon hearing the news, William Henry Edwards sympathized "with all my heart," noting, "It is cruel and hard to lose a child at any time, and just when he is old enough to enjoy life and be a comfort to his friends."[110] Father and son had been inseparable, in summer, year after year, canoeing and camping with each other for weeks at a time in the Maine woods, and, in winter, trekking as high as the alpine peak of Mount Washington in New Hampshire, when it was foolhardy.

"If I may but bear with courage what God gives me to endure, as my dear boy bore his illness," Scudder told Lawrence Bruner, a naturalist at the University of Nebraska, "then all will be well."[111] But the emptiness was nearly unbearable, just as the loss of his wife had been twenty-five years before. The experience may have exacerbated a nerve disorder (probably Parkinson's disease), which, over time, would shut down his mobility. "For the last year," Scudder told Bruner in March, "I have been losing control over the muscles of my left leg, so that I halt in walking and cannot walk any long distances. So my collecting days are about over. I should only be a drag on a companion. The doctor says I am growing old in a one sided fashion, that is all! I shall be sixty next month, and I think my left leg has the growing old all to itself!" A year later his feet hurt so badly that he had trouble walking the few blocks from his home on Brattle Street to the Museum of Comparative Zoology, and a few years after that he lost sensation in his left hand and arm, making it impossible for him to pack and mail insect boxes. "My right hand is also beginning to tremble and I must soon give over handling fragile specimens," he told Bruner.[112] He took a medicine called torinol to sleep.[113]

Scudder's affliction was beyond words. He lost his job as chief paleontologist in Washington, his son, and much of his ability to walk all almost at once, but if he ever pitied himself or thought God was punishing him, there is no record of him having said so. The purpose he found in natural history helped, for a time, to save him. You will do "better work for the struggle to get it," he reassured Lawrence Bruner, a young man who, at one time, had doubted he had any ability.[114] And so Scudder must have told himself, stalwart, passionate Protestant that he was. Throughout the remainder of the century, with his power to move about slowly ebbing away, he opened his home to naturalists, who often met in a big laboratory behind his house that had a door knocker shaped like a grasshopper. Heated by an enormous open fireplace and filled with books, this scientific sanctuary struck one visitor as a "mecca for every entomologist, resident or migrant, native or foreign," and another as an "enchanted chamber wherein Scudder's kindly spirit dominated."[115]

Scudder was not only a pioneering butterfly man but the country's principal expert on grasshoppers, and while his physical powers lasted, he researched forty articles on grasshoppers, among them essays on cave crickets and pink grasshoppers and an account of the prayers Mexicans relied on to fend off insect plagues. A month after his son's death, he began work on a "general guide" to the grasshoppers of North America, and shortly after that, he started a 150-page catalog on fossil beetles. At the same time he was pounding out one guidebook after another on American butterflies, derived from his magnum opus.[116] In the late summer of 1900, on vacation at Prouts Neck, in Maine, he thought about the monarch butterfly, his name for that soaring orange-and-black vagrant, its life history still shrouded in mystery, and, feeling better than usual, took a short walk "on Sunday morning," as he wrote his friend Samuel Henshaw, the director of the Museum of Comparative Zoology, and "came across a spot on the sunny side of a fir grove where the boughs of trees, and especially the dead ones, were covered with Plexippus, which occasionally flew up and then settled again. There were perhaps thousands of them." "On coming home just before reaching it, I came across another lot of some hundreds in another sunny nook. Noticing yesterday that the butterflies were rather scarce about the house, I went out this morning at the same hour as Sunday and to the same spots, with the result that I saw only three there, though others were flying about, but singly." He told Henshaw to keep a look out and perhaps he might see the same butterflies on their way southward from Prout's Neck. "Observe whether they take any definite general direction."[117]

For five or six years, Scudder could barely move his body, unable to work, or even to think clearly for long. Yet he responded warmly to his friends, who treated him with much affection. "I am sure it was you who sent me those lovely flowers," Scudder wrote Henshaw in April 1907, "which came this afternoon without any name of sender. You place me under new and too friendly obligation all the time and make me feel I can never repay you. I am not worthy of such kindness as you always show me, but what I should do without you I don't know."[118] In warm weather, he sat on the veranda of his Cambridge home, listening to his sister-in-law read from the newspapers or books. For nearly ten months, she exhausted every page of Dr. Johnson's *Dictionary,* enunciating and often defining each word with the kind of studious care Scudder himself used for his own words, as living and lovely to him as butterflies. "It is one of the most pathetic facts in the history of science," Theodore Cockerell, a Colorado naturalist, wrote soon after Scudder's death, in May 1911, "that for seven years this great naturalist remained paralyzed and helpless, with so much of the work he had planned to do still unfinished."[119]

The finest American butterfly man of them all, William Henry Edwards, kept a news clipping in his entomological diary about Jules Barthélemy-Saint-Hilaire, an admired French linguist and statesman who, in his ninetieth year, wrote: "I understand nothing but work, and if anybody wishes to attain my age, he must never cease working, and abandon the idea of even retiring from labor. Work! Work! That should be the motto of everybody who desires the welfare of his country and himself."[120] Edwards himself was in his mid-seventies when he delivered up volume 3 of his *Butterflies of North America* to Houghton, Mifflin, and soon after he felt the itch again for yet another volume, despite what would be its inevitable financial burden. He fantasized that some magnate would throw money at his feet for a "supplementary volume," and wrote Wright that if he could get his hands on "a guaranteed fund of $1100 I would begin again. I was 75 a short time ago, but my health is good, and I have a good chance of 5 years more."[121] Some friends, in fact, led by Henry Lyman in Canada, had already started a campaign to find him funding.[122] He told Wright to go out and "get the eggs of *Ivallda*" (an alpine satyr) "that the species might be done on plate."[123]

But nothing happened; no major groundswell of support materialized. Even his family appears to have lost interest in backing him in a new venture. His wife, Catherine, was nearly an invalid; Theodore Mead, up to his eyeballs in orchids and caladiums and following a route that Edwards's son had helped chart, seldom talked of butterflies anymore; and Willie himself had a far greater desire to dig up oil and gas than chase insects. In 1905 Edwards wrote Wright, "It occurs to me that it might be well if I returned to you all your letters to me. When I depart, these will be unheeded by my heirs. Whereas they would remind you of much you have now forgotten." He mentioned that he "had all the letters Bruce ever wrote me, as big as yours. I wish I could put them in a safe place."[124] But he never did mail back the many hundreds of Wright's letters; nor did he keep Bruce's safe. So, as he'd predicted—since he took no legal steps to prevent it (but why, one might ask, did he fail to act?)—his heirs threw most of the letters away, keeping only the choicest, with an eye to those written by such luminaries as Darwin and Wallace, along with a tiny "representative" selection from the many thousands Edwards had received over nearly seventy-five years. Had his family kept them all, they would have retained a precious archive, one surpassing in size both Scudder's and Strecker's.

Still, Edwards hadn't stopped working—how could he? He got up a long, if tedious, genealogy of his family, which anyone with a "lesser" lineage might not have bothered to finish. It opened with the epigram "The glory of children are their fathers," something we might well dispute, while still recognizing the historical truth behind it: that those born with much often get much.[125] William chose to relate Jonathan Edwards's biography entirely in the words of others, having probably no desire to disparage his own relative in print. But why did he leave so much paternal "glory" unreflected on? Why did he refuse to examine his phylogenetic tree the way he had done the butterflies, in pursuit of a core thread in his identity? Had he ever read his relative's "Beauty of the World" with any objectivity? Had he read it at all? By 1900 he had finished writing, of all things, a biography of William Shakespeare, which he called *Shaksper Not Shakespeare*. He'd been researching the book since the late 1880s, partly out of attachment to Henry Edwards, and meant to prove that William Shakespeare had not written the plays. (Mark Twain thought the same, and scholars debate it today.) His thesis rested on a mountain of poorly digested secondary reading and had little of the elegance of Edwards's prose on the life and death of butterflies. Still, Edwards's own voice unexpectedly breaks in, as when he observes, in a passage critical of Shakespeare's supposed love of money over art, that "the passion for money-making is antagonistic to the passion for study. The two cannot exist in the same mind. A man may become rich as a result of his passion for literature, but he cannot become learned by study, or distinguished in literature, when money-making has been the first object."[126]

In April 1901, Edwards told Scudder, "I have let the butterflies fly since the end of my Volume." But it would be a mistake to take these words too literally.[127] He never forsook the thing he had lived with and suffered for throughout much of the nineteenth century, lost in it as much as Jonathan Edwards had been absorbed in the mysteries of God. In 1899, a year after his volume 3 had appeared, he had received some "superb" satyr larvae from David Bruce and had set to thinking about them. "You know I have the larvae of that up to three molts inclusive," he wrote Wright in November. "I have so many species of rare larvae to figure" and "could go to sixty plates completely, if I had the money." He dwelled on the Nokomis fritillary—or, in his preferred Latin, *Argynnis nokomis*—one of America's most handsome dimorphic insects and almost equal in its beauty to Diana, Nokomis meaning "daughter of the moon" and named by Edwards in 1862, after Hiawatha's grandmother in Longfellow's poem *The Song of Hiawatha*. Flying in the American West rather than in the East where Diana flies, the female Nokomis is black with white-yellow toward the margins of both wings, the male consumed almost entirely by orange. The 1871 plate by Peart and Bowen of the butterfly, male and female, in volume 1 of *The Butterflies of North America* still glows radiantly on the page. In 1904, Edwards reported to Wright on an article he'd read in *Scribner's* on exploration in southern Utah: "Years ago, Neumoegen had a large invoice of Argynnis Nokomis sent him, he said, by a Mormon from there. From the description of the region in this Magazine, I am sure the specimens came from those wild mountains." And again, three weeks later, and five years before he died, "just where did Neumoegen's Argynnis Nokomis come from? If the district could be located, I might send some man to this region." "I wish it were in *my* power to go."[128] Edwards carried within him the tension at the heart of the American experience, from the colonial period onward, between extraction and adoration, artifactual beauty and natural beauty, commerce and science, and here the weight of the tension seemed to fall—as it did for nearly all of America's leading butterfly people—toward adoration, natural beauty, and science. At the end of the railroad, such wonderful creatures, nurtured within the purposeless, chaotic density of nature, flying freely and wildly.

Acknowledgments

More people assisted me on this book than on any of my previous books. The lepidopterists were generous with me, forgiving, I like to think, of someone with no formal instruction in insects whom they probably at first believed knew more about butterflies than I actually did. My first encounter with a modern-day butterfly man occurred in early spring of 1999 when I interviewed the late Nicholas Shoumatoff at his home in Bedford, New York. Shoumatoff was an authority on Palearctic butterflies and a nephew of Andrey Avinoff, émigré butterfly man from czarist Russia, a talented painter, and William Holland's choice to succeed him as curator of lepidoptera at the Carnegie Museum of Natural History in Pittsburgh. Exemplary of the blend of art and science, Avinoff's sensuous watercolors were hung all over the walls of Shoumatoff's household. Shoumatoff wasn't put off by my ignorance of Lord Rothschild or of Rothschild's estate, Tring (which I had never heard of), nor by my uninformed judgment that all the influential American butterfly people must have come from upper-class backgrounds. He showed me the remnants of his small 1940s collection, with species from the Bedford area, crowned by one of the last regal fritillaries caught in the region, its wet habitat demolished by development. It was the regal's demise in the East that persuaded Shoumatoff to trade in his butterfly net for binoculars. I owe a good deal to him, not least his advice that I visit the Carnegie Museum of Natural History and meet John Rawlins, curator of invertebrate zoology, advice that led to research that would last for another twelve years.

Born on an isolated sheep ranch in eastern Oregon, and blessed with tremendous energy and a childlike devotion to the natural world, John Rawlins initiated me into the wonders of natural history and of butterfly systematics. I could not have wished for a more inspiring guide. After Rawlins, I had the good fortune to meet, in one way or another, many fine naturalists and entomologists—David Grimaldi, Kurt Johnson, Mark Epstein, Carla Penz, and Philip DeVries. DeVries, an impressive butterfly man with feet squarely planted in the natural history tradition, graciously guided me, in particular, down the right roads, never failing to answer my queries in a timely and complete way; I learned a lot directly from him and from his eloquent two-volume field guide, *The Butterflies of Costa Rica and Their Natural History*. Early on in my work at the American Museum of Natural History, I came to know Eric Quinter, moth-obsessed lepidopterist at the museum. Born in eastern Pennsylvania near Strecker's Reading, in the midst of what was a flourishing community of family farms, Quinter took pleasure in a world of insects that seemed to bloom all around him; at frequent lunches in a nearby Chinese restaurant, he answered my questions and educated me about the mysteries of the life cycle of moths, with special and original understanding of the impact of artificial light on their behavior. As acting curator, he got me passage into the rich (but cramped) entomological library on the museum's fourth floor, now absorbed into the museum's special collections.

In 2005, on a research trip to London's grand natural history museum, I met in the entomological library Bernard d'Abrera, world-renowned expert on tropical butterflies and a man of visceral intensity. He helped me discard a claim made by Will Doherty that Otto Staudinger was Jewish. "How could he have been Jewish?" reacted d'Breara with dismay. "If he had been, he wouldn't have been buried in a Prussian cemetery." Known for his contempt for political correctness as well as for a fearless disdain of Darwinian thinking ("all the evidence," he has written, "points to the fixism or stasis of species in their morphology and behavior, as long as their gene pool and environment have not been destroyed"), d'Abrera made an unforgettable impact on me, one enlivened by a wonderful e-mail he later sent me summarizing the history of natural history. I regret seeing him only briefly.

I would like to thank Carol Sheppard, entomology professor at Washington State University, who happily shared her knowledge of Benjamin Walsh; Jason Weintraub, insect collections manager at the Academy of Natural Sciences, Philadelphia, who told me about Titian Peale's early collection boxes and about the diversity of bird-wing butterflies; Krishnamegh Kunte, scientist at the National Center for Biological Sciences, Bengaluru, India, who sent me in-depth e-mails about the traditional Indian approach to butterflies and about how fundamentally it differed from the Western approach; and Andrew Rindsburg, conchologist and archivist at the Alabama Museum of Natural History in Tuscaloosa, who helped me investigate the papers of Herbert Huntingon Smith and, later, thoughtfully e-mailed me much information about Smith. I am grateful to entomologist Louis Sorkin, at the American Museum of Natural History, for giving me access to the papers of the New York Entomological Society, housed at the museum; to Jackie Miller and the late Lee Miller, for allowing me to do research at the Allyn Museum in Sarasota, Florida; and to John V. Calhoun of Palm Harbor, Florida, for kindly sharing with me an unpublished diary of Theodore Mead and for spotting numerous errors in the text.

I would like to acknowledge the late Tom Allen. Allen was a research biologist for the West Virginia Division of Natural Resources, author of a fine guidebook, *The Butterflies of West Virginia and Their Caterpillars* (1997), and a skilled visual artist. In the summer of 2002, he drove me around in his jeep to many of William Henry Edwards's old collecting grounds, most unchanged except for the trails often disgraced by garbage or, even worse, the top of a mountain near Edwards's home decapitated by strip mining (the kind of mining Edwards did not do). My visit was made memorable by Allen's acid commentary and by watching him search for the tiny eggs of zebra swallowtails from beneath the leaves of the pawpaw bushes and rejoice at capturing a rare golden-banded skipper as we walked along a creek bed.

In the summer of 2003 I had my first significant interview with the late Charles Remington, co-founder of the modern Lepidopterist Society in 1948, exponent of evolutionary ecology, and a much-admired teacher of entomology at Yale University. Remington expressed his lifetime admiration for Samuel Scudder and William Henry Edwards (whose volumes stared at us from across the room as we spoke) as well as for the whole natural history tradition, which, as he well knew, formed the bedrock of all modern work on butterflies. In 1947, he helped to create (and to edit) *Lepidopterist News*, still the central organ for butterfly people today; he insisted from the start that it contain lively little biographies of early lepidopterists as a reminder to modern ones of what they owed the past. Remington encouraged "American amateur lepidopterists," especially, to "pursue a line of work right at hand, which does not require an extensive library or high-priced equipment. This line of work is life history studies."

Remington spoke to me glowingly about his students, such as Larry Gall, curatorial affiliate in entomology at the Yale Peabody Museum of Natural History, and one of the two lepidopterists I asked to read this book in manuscript form. I met Gall thirteen years ago at the Carnegie Museum of Natural History, when both of us were doing research, he on the noctuid moths of Augustus Radcliffe Grote, many of which the museum owned. John Rawlins introduced us, and, at lunch, Gall told us of how he climbed up lampposts or "beat the bushes" to capture specimens. Later he became a president of the Lepidopterist Society. I thank him for his splendid effort, above all for uncovering a legion of errors in nomenclature.

The other expert I relied on was Arthur Shapiro, professor of evolution and ecology at the University of Calfornia, Davis. Like Eric Quinter, Shapiro grew up in Pennsylvania, scouring as a boy the woodlots and fields northwest of Philadelphia or in that region of the Piedmont Plateau marked by dairy and general farming throughout. But he also collected in railroad yards, along railroad tracks, and in the marshes near the Philadelphia International Airport. "I was immensely fortunate to have grown up where the city ended and the country began," he told me. I first came to know of him after reading his lyrical little book, *New York City's Last Frontier* (1972), on the natural history of Staten Island, still a fresh and moving depiction of an ecology suffering irreversible decline; later, after migrating to California, Shapiro wrote beautifully on the butterflies of that state, studying especially those conditions that had caused the gradual decimation of many species. His 2007 field guide, distinguished by his well-crafted prose and by the original handmade illustrations by Timothy Manolis, mixes science with art in such a way as to recall the finest natural history of the past. Shapiro delivered to me fifteen single-spaced pages on everything from the tiniest Latinate blunders to mistakes of knowledge and judgment. His nearly pitiless directness sometimes winded me. But I still return to those pages, as I do to many of his other e-mails to me, for their warnings and insights, and for their assured guidance regarding new lines of inquiry.

Shapiro also introduced me to the work of Adolph Portmann, a mid-twentieth-century Swiss biologist whose eloquent discussion of the diversity of natural organisms, their forms, shapes, colors, sizes, and patterns, articulated in his 1952 book, *Animal Forms and Patterns,* helped me understand better the impact of such diversity on Americans after the Civil War. His views also made clearer to me the historical significance of Samuel Scudder's ideas. Portmann took a position very similar to Scudder's. He accepted completely the explanatory validity of natural selection, but, at the same time, he believed that "developments in the production" of animal form go "far beyond anything which can be understood as adaptations to special conditions of life, such as better equipment in the struggle for existence." "The scales of butterflies give the appearance of metal, of gold and blue textures, whose sense surpasses any merely adaptive ability." "All around us are forms of life, small or large, in which have been realized other possibilities of existence than those found in our own lives."

To Arthur Shapiro—and to all the other lepidopterists who helped me—I can only apologize for the mistakes of fact and interpretation that doubtless still remain in the book.

Archivists from across the country and Europe also gave me generous assistance in my research. Pamela Henson, director of the Institutional Division of the Smithsonian Institution Archives in Washington, D.C., stands out as an invaluable guide of much distinction, an expert on the history of natural history, and an authority on the intellectual careers of John and Anna Comstock. Henson helped me find whatever of value the Smithsonian contained relating to the history of butterfly people; at lunches or on coffee breaks over many years, she conveyed the ins and outs of research in Washington and urged me to participate in the academic life of the natural sciences. I would like to thank as well the archivists at the following places: the California Academy of Sciences, San Francisco, California; the Museum of Comparative Zoology, Cambridge, Massachusetts; the Boston Museum of Science, Boston, Massachusetts; the University of Connecticut, Storrs, Connecticut; the Natural History Museum, London, England; the Hildesheim Museum, Hildesheim, Germany; the Museum für Naturkunde, at Humboldt University, Berlin, Germany; the Academy of Natural Sciences, Philadelphia, Pennsylvania; the Cincinnati Museum Center at Union Terminal, Cincinnati, Ohio; the Cincinnati Historical Society, Cincinnati, Ohio; Southern Illinois University, Carbondale, Illinois; the Mathematics and Science Library, Columbia University, New York, New York; the Putnam Museum of History and Natural Science, Davenport, Iowa; the Spencer Research Library, University of Kansas, Lawrence, Kansas; the Special Collections and Archives, W. E. B. Du Bois Library, University of Massachusetts, Amherst; Morris County Historical Society, Morristown, New Jersey; the Bancroft Library, University of California, Berkeley; the Buffalo Museum of Science, Buffalo, New York; Rare Books and Manuscripts Collections, Cornell University Library, Ithaca, New York; University Archives, Baylor University, Waco, Texas; Department of Archives and Special Collections, Rollins College, Winter Park, Florida; and the Staten Island Institute of Arts and Sciences, Staten Island, New York.

Over the years I have spent many rewarding hours at the American Museum of Natural History, a truly fabulous treasury of materials bearing on the history in this book, without which there would have been no such book. During this time, Barbara Mathe, head of special collections, and Tom Baione, director of the library, gave me indispensable help for which I thank them, as did Mai Qaramen, a more recent member of the museum staff, with many of the color plates. In 2007, on my research trip to Germany, I relied on the kindness of Editha Schubert, a young archivist at the Deutsches Entomologisches Institut, Müncheberg, Germany, who brought me a cartload of documents relating to the German/American treatment of butterflies in the nineteenth century; she left me alone in her office to go through everything, and later had copied electronically a trove of letters and diary fragments drawn from this unusual, and still largely unexplored, entomological archive. Another estimable archivist was Ben Williams, who served as director of the library at the Field Museum of Natural History for nearly all the years I did research there, watching over his ward like a brooding hen. The library contains the Herman Strecker collection, unmatched in the United States as a source for studying the butterfly people of America as well as of Europe and England; Williams recognized its value and did all he could to ensure that I made the best use of it.

And then there is Bernadette G. Callery, sophisticated sleuth at the Carnegie Museum of Natural History, in Pittsburgh, Pennsylvania, the best archivist I have ever known. From the first day I arrived at the museum in 1999 to the last time I spent with her in the summer of 2006, Callery did what the best archivists do: listened attentively to my requests, did everything within her power to understand and satisfy my research needs, and then, when I was not around, wrote me long e-mails full of useful material. Together, we came to grips with the peculiarly flawed character of William Holland, the former director of the great museum whose archives she so beautifully superintends.

I owe Robert Michael Pyle a great deal. A brilliant and generous naturalist, Pyle read the manuscript in the *very* last stages of production, finding all the errors my other reads failed to see. It was too late for me to thank him in the hardcover, so I do so now, with heartfelt gratitude.

Several living direct descendants of Augustus Radcliffe Grote and William Henry Edwards generously aided me in my research, as did friends, colleagues, and fellow historians. Peter and Marianne Gaethgens of Berlin, Germany, the latter the great-grandson of Grote, happily supplied me with all the relevant documents they could gather on Grote, best of all, sheet music to a march composed by Grote to celebrate the election of Grover Cleveland as mayor of Buffalo; I thank them especially for correcting a mistake I made regarding the way Grote died. The family of the late John A. Willis, descendant of William Henry Edwards, allowed me to visit Edwards's home in Coalburgh, still standing, and still occupied to this day by Willis kin. The family was gracious, especially Douglas Willis, who, months before my journey, instructed me about archival sources on Edwards; when I arrived in Coalburgh, he and his brother, Tom, walked with me from Edwards's house down a path across a ridge studded by pawpaw bushes to Edwards's old 1865 coal mine, the remains of which still stood, along with the rail tracks that once carried coal by trolley down the mountainside to the tipple on the Kanawha River below.

I am grateful to Victoria Cain, Ian Miller, Barbara Fields, Lynn Nyhart, and Marsha Wright, all of whom, in one form or another, helped me complete this book. Five historians accepted my request to read the entire manuscript, and deserve my particular, and deeply felt, gratitude. Peter Dimock's and Mary Furner's warmly enthusiastic readings had the effect—as William Henry Edwards said of the impact of a long-awaited letter from Henry Edwards—of giving me "a sense of pleasure like that of sunshine on a bank of flowers." Donald Worster persuaded me to make major changes in the introduction in order to create a more integrated analysis throughout the book, freeing it from the burden of ideas raised in the beginning but later never well developed; Robert Richardson urged me to rid the book of needless technical verbiage, and I hope I managed to do that; and the insightful Polly Winsor compelled me, in an extensive, marvelous e-mail and two later phone calls, to understand better the complexities of systematics and the character of natural history tradition.

I would like to thank the people at Pantheon—Jill Verrillo, for taking it on the chin when she didn't really deserve it and for always acting professionally (even if I didn't); Ellen Feldman, who superbly oversaw the copyediting of the book, not once, but twice, thus making the book twice-born, as it were, in the language of the American philosopher William James; Bonnie Thompson, who did awesome copyediting; and, above all, Dan Frank, my editor at Pantheon, who read whatever I put before him thoughtfully and with penetration. A wonderfully tolerant and civilized individual, Dan supported this book—and me—in a way I will never forget.

Elizabeth Blackmar and the late Jeannette Hopkins were my best allies in this often difficult, prolonged experience. Betsy, my dear partner in life, read the manuscript so many times as to nearly commit it to memory and to make it her own; to have done that was to have taken responsibility for the manuscript, and to have taken responsibility for it amounted to a refusal to accept failure, her failure, as well as mine.

Jeannette Hopkins died in August 2011, and I will miss her for the rest of my life. I relied on her to edit three of my books, written over a period of thirty years; she immersed herself in them inside and out. The first was an immense thrill for me, given what I knew about her reputation. She wielded almost total control over the manuscript, and I adopted almost everything she proposed; the manuscript sits in my attic, refugee from her relentless pen, every page covered in red ink. The second book was a harder experience, and the third the hardest of all, since she fought with me, tooth and nail, over the "gestalt" of the book, as she called it, the book's governing concept or meaning. My struggle with her was painful, and, at one point I wanted to end the relationship, but I did not. I cannot say Jeannette had total victory, but she won where it mattered. The book is dedicated to her.

Notes

Abbreviations of Frequently Used Sources

BEUSC Samuel Scudder, *The Butterflies of the Eastern United States and Canada*

BNA William Henry Edwards, *The Butterflies of North America*

CE *Canadian Entomologist*

DEI Deutschen Entomologischen Institut, Müncheberg, Germany

DP Duncan Putnam Papers, Putnam Museum of History and Natural Science, Davenport, Iowa

EML Ernst Mayer Library of the Museum of Comparative Zoology, Harvard University, Cambridge, Massachusetts

EN *Entomological News*

GF George Hazen French Papers, Special Collections Research Center, Southern Illinois University, Carbondale

HD Harrison Gray Dyar Papers, 1882–1927, Smithsonian Institution Archives, Washington, DC

HE Henry Edwards Correspondence, 1882–1891, E39, American Museum of Natural History, New York, New York

HS-ANS Henry Skinner Papers, 1879–1925, Collection 920, Academy of Natural Science Archives, Philadelphia, Pennsylvania

HS-FM Herman Strecker Papers, Field Museum Library and Archives, Chicago, Illinois

JL Joseph Lintner Papers, Manuscripts and Special Collections, New York State Library, Albany

JMH James Morgan Hart Papers, #14-18-65, Division of Rare and Manuscript Collections, Cornell University Library, Ithaca, New York

LB Lawrence Bruner Papers, California Academy of Sciences, San Francisco

LRH Herman Strecker, *Lepidoptera: Rhopaloceres et Heteroceres*

MNK Museum für Naturkunde, Archives, Berlin, Germany

NHM-LONDON Natural History Museum Archives, London, England

SS-BMS Samuel Scudder Papers, Boston Museum of Science Archives, Boston, Massachusetts

TB Thomas Bean Correspondence, 1877–1879, American Museum of Natural History, New York, New York

TM Theodore Mead Papers, Department of College Archives and Special Collections, Olin Library, Rollins College, Winter Park, Florida

WGW William Greenwood Wright Papers, Special Collections, California Academy of Sciences, San Francisco

WH-CM, William J. Holland Papers, 1896–1925, CMNH, 1988-3, Carnegie Museum of Natural History Archives, 2007-5, Pittsburgh, Pennsylvania

WHE-SA William H. Edwards Collection, West Virginia State Archives Manuscript Collection, Charleston

WH-HSWP William Holland Family Papers, 1747–1933, Historical Society of Western Pennsylvania, Senator John Heinz History Center, Pittsburgh

Introduction

1. Augustus Grote, "On the Geographical Distribution of North American Lepi-doptera," *CE* (September 1886): 165; for the Jonathan Edwards quote, see his sermon "Beauty of the World" (1725), in *A Jonathan Edwards Reader,* ed. John E. Smith et al. (New Haven, CT: Yale University Press, 2003).

2. W. L. Devereaux to Strecker, March 26, 1892, HS-FM. A farmer from upstate New York, Devereaux reflected in this letter on his youth in the 1870s, when he collected moths and butterflies.

3. On railway imperialism around the world in the nineteenth century, see Clarence B. Davis and Kenneth E. Wilburn Jr., with Ronald E. Robinson, *Railway Imperialism* (NewYork: Greenwood, 1991).

4. I do not wish to romanticize the farming landscape, especially in its extensive, monocultural forms, which, whenever they appeared, led to the eradication of diverse habitats, not to their creation, and, thereby, to the extinction or near extinction of many butterflies (such as the regal fritillary, baltimore checkerspot, and atlantis fritillary). Small farms had adverse impacts, too, but their effects have also been very positive. On the negatives, see Michael Gochfeld and Joanna Burger, *Butterflies of New Jersey* (New Brunswick, NJ: Rutgers University Press, 1997), 180–82, 188–89.

5. This analysis is drawn largely from David R. Foster, *Thoreau's Country: Journey Through a Transformed Landscape* (Cambridge, MA: Harvard University Press, 1999). This book, mixing analysis with full quotations from Thoreau's journals, shows wonderfully how much Thoreau's notion of nature and "wildness" owed to the hybrid farm landscape. Foster's work belongs to a new literature that has emerged over the past ten or so years and that builds, in part, on the views of Aldo Leopold. Leopold contended that certain kinds of farming (but *not* monoculture) can enrich the diversity of life, rather than reduce or destroy it. For Leopold's position, see "The Farmer as Conservationist," a 1939 essay republished in *The River of God and Other Essays,* ed. Susan L. Flader and J. Baird Callicott (Madison: University of Wisconsin Press, 1991), 255–65. For recent assessments, see in addition to Foster, Brian Donahue, *The Great Meadow: Farmers and the Land in Colonial Concord* (New Haven, CT: Yale University Press, 2004), 1–23,155–233. For a fine history of the evolution of farming practices in America, see Christopher Clark, *The Roots of Rural Capitalism: Western Massachusetts, 1780–1860* (Ithaca, NY: Cornell University Press, 1990).

6. On pasture species, see Scott L. Ellis, "Biogeography," in *Butterflies of the Rocky Mountain States,* ed. Clifford D. Ferris and F. Martin Brown (Norman: University of Oklahoma Press, 1980), 17. On the monarch, see Paul Opler to Dr. Marilyn T. Vassallo, September 12, 1978, Paul A. Opler Papers, Archives, California Academy of Sciences, San Francisco. Opler wrote that monarchs actually became "more abundant after the arrival of European Man on the continent since many milkweeds, the monarch's caterpillar food, thrived in the disturbed habitats that are the products of man's environmental manipulations." See also Lincoln P. Brower, "Understanding and Misunderstanding the Migration of the Monarch Butterfly (Nymphalidae) in North America, 1857–1995," *Journal of the Lepidopterists' Society* 49, no. 4 (1995): 304–85. On the black

swallowtail and the meadow fritillary, see Gochfeld and Burger, *The Butterflies of New Jersey,* 122, 183; Samuel Scudder, *BEUSC* (Cambridge, 1889), vol. 1, p. 608; and Gary Noel Ross, "Butterflies of the Wah'Kon-Tah Prairie," *Holarctic Lepidoptera* (March–September 2005): 1–30.

7. "Winged wanderers on clover sweet" appears in the poem "A Butterfly in Wall Street," quoted by Samuel Scudder in *BEUSC,* vol. 1, p. 249. Scudder cites only "Sherman" as the author of the poem. For an extended description of collecting in clover, see the following from John Byrkit of Indianapolis, Indiana, to Herman Strecker, November 18, 1874, HS-FM:

> The Argynnis was taken on 11th June 1871 about two miles from south of the city on a clover field about half a mile from a low swampy meadow. The surface of the field was rolling, gravelly piece of ground situated near a wood heavy set with fine red clover in a rich mass of bloom. *Argynnis Cybele* in numbers very fine were abundant. *Colias Philodice* and *Pieris Protodice* were also plenty. Took the same day one specimen *L. Bachmanii* and one specimen *Thecla Halesus.* I have worked hard many a day over the same ground to find more examples of the *Argynnis* that I might establish what I supposed to be a new species but have so far failed.

On the butterfly abundance in wilderness regions, see John Thomas Powell and Walter and Irja Knight, "A Vegetation Survey of the Butterfly Botanical Area, California," *Wasmann Journal of Biology* 28, no. 1 (1970).

8. Higginson, quoted in Scudder, *BEUSC,* vol. 1, p. 569; Edward Doubleday, *The Genera of Diurnal Lepidoptera,* vol. 1 (London, 1846), 225; and Crève-coeur, selections from *Letters From an American Farmer* (London, 1782), in Robert Finch and John Elder, eds., *The Norton Book of Nature Writing* (New York: Norton, 2002), 53–66. The historian Willis Conner Sorensen, in his fine, groundbreaking book *Brethren of the Net: American Entomology, 1840–1880* (Tuscaloosa: University of Alabama Press, 1995), argues that most entomologists in the nineteenth century were not farmers and that farming, therefore, could not be listed as an explanation for why they became naturalists or lepidopterists. But Sorensen ignores the existence of the ubiquitous hybrid agrarian landscape, which all the leaders knew very well from childhood. The matter is important, given that, after 1950, this landscape was dismantled and everywhere suburbanized. See Sorensen, *Brethren of the Net,* chapter 8, "Profile of the American Entomological Community. About 1870," 150–96.

9. Walt Whitman, *Specimen Days and Collect* (1883; rpt. New York: Dover Publications, 1995), 83, 121.

10. Herman Strecker, *Butterflies and Moths of North America: Complete Synonymical Catalogue of Macrolepidoptera* (Reading, PA: B. F. Owen, 1878), 9.

11. On European and English natural history, see David Elliston Allen, *The Naturalist in Britain* (Princeton, NJ: Princeton University Press, [1976] 1994); and the introduction to *From Natural History to the History of Nature: Readings from Buffon and His Critics,* ed. John Lyon and Philip Sloan (Notre Dame, IN: University of Notre Dame Press, 1981), 1–2. See also Keith Thomas, *Man and the Natural World* (Oxford: Oxford University Press, 1983); N. Jardine et al., eds., *Cultures of Natural History* (Cambridge: Cambridge University Press,

1996); Stephen T. Asma, *Stuffed Animals and Pickled Heads: The Culture and Evolution of Natural History Museums* (Oxford: Oxford University Press, 2001); Harriet Ritvo, *The Platypus and the Mermaid: And Other Figments of the Classifying Imagination* (Cambridge, MA: Harvard University Press, 1997); David Freedberg, *The Eye of the Lynx: Galileo, His Friends, and the Beginnings of Modern Natural History* (Chicago: University of Chicago Press, 2003); and Brian W. Ogilvie, *The Science of Describing: Natural History in Renaissance Europe* (Chicago: University of Chicago Press, 2006).

12. Lyon and Sloan, *From Natural History*, 1–2.

13. On Linnaeus, see Wilfred Blount, *Linnaeus: The Compleat Naturalist* (Princeton, NJ: Princeton University Press, 2001), and Lisbet Koerner, "Carl Linnaeus in His Time and Place," in *Cultures of Natural History*, ed. N. Jardine et al., 145–62; on Humboldt, see Jason Wilson, introduction to Humboldt's *Personal Narrative of a Journey to the Equinoctial Region of the New Continent*, abridged and trans. Jason Wilson (New York: Penguin, 1995), xliii; Douglas Botting, *Humboldt and the Cosmos* (New York: Harper & Row, 1973); and Gerard Helferich, *Humboldt's Cosmos* (New York: Gotham, 2004). On Buffon, see Jacques Roger, *Buffon: A Life in Natural History*, trans. Sarah Lucille Bonnefoi (Ithaca, NY: Cornell University Press, 1997); and on Darwin, see John Bowlby, *Charles Darwin: A New Life* (New York: Norton, 1991). Carol Kaesuk Yoon has many insightful things to say on the history of systematics in her penetrating book *Naming Nature: The Clash Between Instinct and Science* (New York: Norton, 2009). Unfortunately, I read Yoon's book too late to make use of it here.

14. See Mary P. Winsor, "Linnaeus Was Not an Essentialist," *Annals of the Missouri Botanical Garden* 93, no. 1 (2006): 2–7; Mary P. Winsor, "Non-essentialist Methods in Pre-Darwinian Taxonomy," *Biology and Philosophy* 18 (2003): 387–400; and Staffan Muller-Wille, "Collection and Collation: Theory and Practice in Linnaean Botany," *Studies in History and Philosophy of Biological and Biomedical Sciences* 38, no. 3 (2007): 541–62. I would like to thank Professor Winsor for her indispensable guidance on this matter.

15. Lyon and Sloan, *From Natural History*, 1–2; and "Premier Discours," in *From Natural History*, 101.

16. Quoted in Roger, *Buffon*, 329–30. See also Buffon, "Premier Discours," in Lyon and Sloan, *From Natural History*, 105–23; and Roger, *Buffon*, 320. For a comparison of Buffon and Linnaeus, see Asma, *Stuffed Animals and Pickled Heads*, 114–53; and Roger, *Buffon*, 71–92, 311–13.

17. On this evolution, see Lynn K. Nyhart, "Natural History and the 'New' Biology," in *Cultures of Natural History*, ed. Jardine et al., 426–43; and Muller-Wille, "Collection and Collation."

18. Linnaeus, *Philosophica Botanica*, trans. Stephen Freer (1751; repr., New York: Oxford University Press, 2003), 332.

19. Buffon, quoted in Roger, *Buffon*, 83, 91. This discussion of Buffon and Linnaeus rests on my reading of Lisbet Koerner, *Linnaeus: Nature and Nation* (Cambridge, MA: Harvard University Press, 1999); Roger, *Buffon*, 71–73, 279, 288–89, 309–12; and Winsor, "Linnaeus Was Not an Essentialist."

20. Gerardo Lamas et al., *Atlas of Neotropical Lepidoptera*, vol. 124 of *Bibliography of Butterflies* (Gainesville, FL: Scientific Publishers, 1995), 207; Edward

Whymper, *Travels Amongst the Great Andes of the Equator*, vol. I (New York: C. Scribner's Sons, 1892), 365; Alexander von Humboldt, *Views of Nature, or, Contemplations on the Sublime Phenomena of Creation* (London, 1850), 232–33; and Botting, *Humboldt and the Cosmos*, 155.

21. Quoted in Robert Richards, *The Romantic Conception of Life: Science and Philosophy in the Age of Goethe* (Chicago: University of Chicago Press, 2003), 34.

22. David Freedberg, *The Eye of the Lynx*; and Richards, *The Romantic Conception of Life*.

23. Pamela Smith, *The Body of the Artisan: Art and Experience in the Scientific Revolution* (Chicago: University of Chicago Press, 2004).

24. Albertus Seba, *Cabinet of Curiosities,* a facsimile copy based on an original in the Koninklijke Bibliotheek (1734–56; repr., New York: Taschen, n.d.), 452, 464, 476.

25. On the strengths and weaknesses of these writers, see Walter Rothschild and Karl Jordan, "A Revision of the American Papilios," *Novitates Zoologicae* 13 (August 1906): 415–16.

26. Lyon and Sloan, "Premier Discours," in *From Natural History,* 110–11.

27. Alexander von Humboldt, *Aspects of Nature*, vol. 1, trans. Elizabeth Sabine (London, 1849), 208; Humboldt, *Personal Narrative*.

28. Quoted in Helferich, *Humboldt's Cosmos*, 6; and Malcolm Nicolson, "Historical Introduction" to Humboldt, *Personal Narrative,* xix–xx.

29. Ann Shelby Blum, *Picturing Nature: American Nineteenth-Century Zoological Illustration* (Princeton, NJ: Princeton University Press, 1993).

30. This democratic situation is laid out beautifully in Paul Starr's *The Creation of the Media: Political Origins of Modern Communications* (New York: Basic Books, 2004), especially in the first one hundred pages. On the diffusion of wealth, see Joyce Appleby, *Capitalism and the New Social Order: The Republican Vision of the 1790s* (New York: New York University Press, 1984), and *Relentless Revolution: A History of Capitalism* (New York: Norton, 2010).

31. In his biography of Ernst Haeckel, Robert Richards writes that only academics made such "ritual"exchanges of photos, but the practice was widespread, embracing many individuals from many walks of life. See Richards, *The Tragic Sense of Life: Ernst Haeckel and the Struggle over Evolutionary Thought* (Chicago: University of Chicago Press, 2008), 114.

32. Benjamin Walsh to Hermann Hagen, December 7, 1863, EML. For an account of this development earlier in England, see Allen, *The Naturalist in Britain,* chapters 2–6.

33. I would like to thank Jeannette Hopkins for persuading me to make this argument.

34. George Hulst to Herman Strecker, February 9 and 28, 1876, HS-FM.

1. Yankee Butterfly People

1. For an early account of the Linnaean legacy, see George Louis Leclerc Buffon, *Natural History of Birds, Fish, Insects, and Reptiles*, vol. 5 (London, 1793), 135–37.

2. On this weakness in Fabricius and Linnaeus, see Walter Rothschild and Karl Jordan, "A Revision of the American Papilios," *Novitates Zoologicae* 13 (August 1906): 412–15.

3. Jean Baptiste Boisduval, *Histoire générale et iconographie des lépidoptères et des chenilles de l'Amérique septentrionale* [A General History and Iconography of the Butterflies and Caterpillars of North America] (Paris: Librarie Encyclopedique de Roret, 1833).

4. On Lorquin, see E. O. Essig, *A History of Entomology* (1931; repr., New York: Hafner, 1965), 694–97; and on Boisduval, see ibid., 559–61, and Jean Gouillard, *Histoire des entomologistes français, 1750–1950* (Paris: Société Nouvelles des Éditions Boubée, 2004), 58–59. For the two volumes on American butterflies, see Boisduval, *Histoire générale,* and Boisduval, "Lépidoptères de la Californie," in *Annales de la Société Entomologique de Belgique* (Brussels, 1868–69).

5. For a complete definition of systematics, see Ernst Mayr, *The Growth of Biological Thought* (Cambridge, MA: Harvard University Press, 1982), 145–46.

6. Edward Doubleday, "Communications on the Natural History of North America," *Entomological Magazine* (March 1838): 26–27, 31.

7. *Entomological Magazine* (September 1838): 199–200; and, in the same issue, "Proceedings of the Entomological Club of London," 206–7.

8. Edward Doubleday, *The Genera of Diurnal Lepidoptera* (London, 1846–50), vol. 1, pp. 226, 210–11, and 226.

9. *Entomological Magazine* (September 1838), 203.

10. Edward Doubleday to Thaddeus Harris, July 12, 1839, draft of letter, in "Harris Manuscript II, Lepidoptera I," in the section "Arrangement of the Bombyces, Sept. 1839," no pagination, Thaddeus William Harris Papers, 1823–1855, EML.

11. This was the considered judgment of Karl Jordan and Walter Rothschild in their authoritative "A Revision of the American Papilios," 411–37. This article offers an informative overview of the changes in insect iconography, as well as a discussion of why images matter in the first place.

12. On Abbot, see Pamela Gilbert, *John Abbot: Birds, Butterflies, and Other Wonders* (London: Natural History Museum, 1998). For a recent assessment of Abbot's pictorial legacy, see John V. Calhoun's informative essays in the *Journal of the Lepidopterists' Society,* published over a three-year period, 2004–7, vols. 58–61.

13. Jessie Poesch, *Titian Ramsey Peale* (Philadelphia: American Philosophical Society, 1961), 4.

14. Howard Ensign Evans, *The Natural History of the Long Expedition to the Rocky Mountains, 1819–1820* (New York: Oxford University Press, 1997), 14.

15. Quoted in Poesch, *Titian Ramsey Peale,* 53–54.

16. Doubleday, "Communications on the Natural History of North America," 199.

17. Titian Peale, "Proposals for Publishing by Subscription, a Work to Be Entitled *Lepidoptera Americana*" (1833), original in special collections, American Museum of Natural History, New York; Poesch, *Titian Ramsey Peale,* 60–61; Thomas Say, *American Entomology: A Description of the Insects of North America with Illustrations Drawn and Colored After Nature* (1815), unpagi-

nated; and Titian Peale, "Lepidoptera: Larva, Foodplant, Etc." (1833, 1877, 1879), with handwritten introduction, entries, and drawings, Rare Book Collection, American Museum of Natural History. For more on Peale's plates, see Patricia Tyson Strand, *Thomas Say: New World Naturalist* (Philadelphia: University of Pennsylvania Press, 1992).

18. Michael Salmon, in his history of English butterfly people, claims that "the twentieth century" was "the age for the study of living insects," pioneered especially by an Englishman, Frederick Frohawk, who "reared every British species from the egg." But this is incorrect, since American entomologists had led the way on this matter in the nineteenth century. See Salmon, *The Aurelian Legacy: British Butterflies and Their Collectors* (Berkeley: University of California Press, 2001), 370.

19. See William Henry Edwards to Philip Zeller, November 22, 1873, Philipp Zeller Papers, Manuscripts Collection, Entomological Library, NHM-LONDON, in which he describes the *Canadian Entomologist* as "the only organ American Entomologists have for making known their observations." On clubs and societies, see Sally Kohlstedt, "The Nineteenth-Century Amateur Tradition: The Case of the Boston Society of Natural History," in *Science and Its Public,* ed. Gerald Holton and W. A. Blanpied (Dordrecht: D. Reidal, 1976); and Sally Kohlstedt, ed., *The Origins of Natural Science in America: The Essays of George Brown Goode* (Washington, DC: Smithsonian Institution Press, 1991).

20. The dedication for Samuel Scudder's *The Life of a Butterfly* (New York: Holt, 1893).

21. "Beauty of the World" (1725), published in *A Jonathan Edwards Reader,* ed. John E. Smith et al. (New Haven, CT: Yale University Press, 2003), 14–15.

22. I have relied here on George Marsden's biography *Jonathan Edwards: A Life* (New Haven, CT: Yale University Press, 2003), especially chapter 4, "The Harmony of All Knowledge," pp. 58–81. Marsden's discussion of Edwards's religious views, on pp. 433–505, is excellent, the concluding pages profound.

23. Quoted in Marsden, *Jonathan Edwards: A Life,* 65.

24. William Henry Edwards, *Autobiographical Notes* (privately printed, 1901), 14–16, WHE-SA.

25. William Henry Edwards to Samuel Scudder, October 21, 1893, SS-BMS. In another letter to Scudder, Edwards wrote, "Mr. Hopkins was a wise old fox, well fitted to govern such an institution. I was twice fined three dollars by him. . . . I went to him one Sunday, when *I thought the sun was down,* to ask if I might go to Troy for a celebration the next day. He glanced at Round Top Mountain, and saw the last rays shining on its summit, and told me to go and come back *in half an hour,* which I did, and then he told me I could go to Troy!" (May 2, 1892, SS-BMS).

26. Edwards, *Autobiographical Notes,* 66–67, 64, 63, and 72.

27. William Henry Edwards, "Journal of Natural History," November 4, 1844, and February 22 and May 7, 1845, WHE-SA.

28. Edwards, "Journal of Natural History," 103–4.

29. Edwards, *Autobiographical Notes,* 72. See Edwards to William Greenwood Wright, January 25, 1891, WGW. In his sunniness Edwards even presented a sanguine view of slavery. "Brazilian slavery, as it is," he wrote, "is little more

than slavery in name. Prejudice against color is scarcely known, and no white thinks less of his wife because her ancestors came from over the water. Half the offices in the government and of the army, are mingled blood; and padres, and lawyers, and doctors of the intensest hue, are none the less esteemed. The educated blacks are just as talented, and just as gentlemanly as the whites, and in repeated instances we received favors from them, which we were happy to acknowledge." Whether fully accurate or not, this reflection is nevertheless remarkable, not only for what it says about Brazil, but for what it reveals about Edwards, whose family members were at the time practicing abolitionists.

30. William Henry Edwards, "Preface to My London Diary of 1848," July 8, 1906, and March 25, 1848, WHE-SA.

31. Spence later reported glowingly to his colleagues, after Edwards had returned home, about Edwards own "stories" in his *Voyage*. See "Proceedings of Entomological Society of London," May 1, 1848, in *Transactions of the Entomological Society of London*, vol. 5 (1847–49): xxxviii; and Edwards, "London Diary," April 8, 1848. Also William Kirby and William Spence, *Introduction to Entomology* (London, 1819), 69, 72; and William Spence to Ernst Germar (an entomologist in Halle, Germany), July 21, 1817, DEI.

32. David Allen, *The Naturalist in Britain* (Princeton, NJ: Princeton University Press, [1976] 1994), 89.

33. Edwards, "London Diary," April 8, 10, and 28, 1848.

34. Ibid., May 10, 1848, and April 28, 1848.

35. On Edwards and the early coal business, see Otis K. Rice, "Coal Mining in the Kanawha Valley to 1861: A View of Industrialization in the Old South," *Journal of Southern History* 31, no. 4 (1965): 393–416; and George W. Atkinson, *History of Kanawha County* (Charleston, 1876), 216–20. "West Virginia," Atkinson wrote, "contains more than one-fourth of the coal of this country, and Kanawha County is the great center of this immensity of natural wealth."

36. For this history, see Wilma A. Dunaway, *The First American Frontier: Transition to Capitalism in Southern Appalachia, 1700–1860* (Chapel Hill: University of North Carolina Press, 1996), 56–86; Ronald L. Lewis, *Transforming the Appalachian Countryside: Railroads, Deforestation, and Social Change in West Virginia, 1880–1920* (Chapel Hill: University of North Carolina Press, 1998), 85–89; and Barbara Rasmussen, *Absentee Landowning and Exploitation of West Virginia, 1760–1920* (Lexington: University of Kentucky Press, 1994), 93–179.

37. Edwards, *Autobiographical Notes*, 109–11; photocopy courtesy of Douglas Willis, direct descendant of William Henry Edwards.

38. William H. Edwards, "History of the Kanawha and Ohio Coal Company" (1902), 1–7, LOHE-SA; *West Virginia Geological Survey* (Wheeling, WV, 1914), 33; "Records of the Kanawha and Ohio Coal Company Store," June 1864–January 30, 1865, WHE-SA; "History of the Kanawha and Ohio Coal Company," 2–3.

39. Ten years earlier, in 1854, a Virginian, R. H. Maury, opened a mine using slaves on the same mining property, after buying the land from Edwards, which Edwards, in turn, bought back from him in 1864. His miners and their families lived on the bottomlands along the river, very near the store, a local monopoly selling not only dry goods, meat, and groceries but also such "personals" as

"looking glasses," "hair oil," "cologne," something called "1 B pain killer," and laudanum (according to the records for 1864–65, one man, Peter M. McCourt, bought all the laudanum in stock, and we can only imagine for what purpose). "History of the Kanawha and Ohio Coal Company," 2–3; *West Virginia Geological Survey,* 33; "Records of the Kanawha and Ohio Coal Company Store."

40. On Edwards's debt to Weidemeyer, see Edwards to Scudder, December 31, 1874, SS-BMS.

41. On Akhurst, see Edward Graef, "Some Early Brooklyn Entomologists," *Bulletin of the Brooklyn Entomological Society* 9, no. 3 (June 1914): 49–50. And on Edwards's debt to Weidemeyer, see Edwards to Theodore Mead, December 15, 1876, TM: "When I began collecting, Weidemeyer was in the vigor of collector-ship, and I got a great deal of information from him." Weidemeyer later moved to Paterson, New Jersey, where Edwards and others lost touch with him (perhaps he either ceased his butterfly work or died). See his *Catalogue of North American Butterflies* (Philadelphia, 1864), taken from the "Proceedings of the Entomological Society of Philadelphia." Edwards's butterfly collecting began along the Hudson River in New Hamburg, New York, where he lived in a rented cottage between 1857 and 1858, shortly before moving to nearby Newburgh. "New Hamburgh, where the house was in a bit of a forest," Edwards recalled, "started me in the butterfly path. I found many sorts of caterpillars, and my wife made colored drawings of them. The next summer we were at Lenox, Massachusetts, and I used my net to much advantage." See Edwards, *Autobiographical Notes,* 201–2.

42. Edwards to Scudder, December 20, 1870, SS-BMS.

43. Edwards to William Greenwood Wright, January 28, 1896, WGW. In April 1891, Edwards wrote to the naturalist Henry Skinner about this discovery. "I saw Mr. Peale's collection in Washington in 1861, I should think. Was taken to his house by Robert Kennicott. Noticed his style of putting up his insects, and straightway went home and had a lot of cases made, about 16in X 10, glass on both sides, tin foil over the inside frame, and fastened corks to the bottom. Glass just as Peale showed, and that case I have used to this day. Mine are not sealed. But I have scarcely ever lost an insect by dermestes" (Edwards to Skinner, April 5, 1891, HS-ANS). Peale laid out his preserving methods in an article prepared for the Smithsonian Institution, "Method of Preserving Lepidoptera," in the *Annual Report of the Board of Regents of the Smithsonian Institution* (Washington, DC, 1864), 404–6. Edwards brought fifty of Peale's boxes home with him to Newburgh for his own butterflies, and would use them for another forty years. Edwards to Joseph Lintner, June 29, 1897, JL.

44. Edwards, *Autobiographical Notes,* 202–7.

45. For a map of the counties in West Virginia, see Lewis, *Transforming the Appalachian Countryside,* 24–25.

46. Edwards recounts this history in a letter to Lintner, August 2, 1879, JL.

47. William Henry Edwards to Henry Edwards, May 4 and May 29, 1879, HE.

48. William Henry Edwards, Entomological Diary, July 1876, WHE-SA.

49. Edwards regretted that there were not more clover fields around the house. He wished he'd had "fields of it," to study better the member of *Argynnis* (fritillaries). See Edwards to Thomas Bean, January 13, 1876, TB.

50. Edwards to William Greenwood Wright, August 2, 1896, WGW.

51. On Edwards's sighting of the Diana fritillary and on its impact on his decision to write a book, see Edwards's autobiography, *Autobiographical Notes*, 207–8.

52. "Argynnis I," *BNA* (Philadelphia: American Entomological Society, 1868–72), vol. 1, unpaginated.

53. Letter to the editor, *CE* (July 1878): 140; and letter to Scudder, June 14, 1888, SS-BMS, in which he recounts this spectacular sight. The other citations in this paragraph can be found in Edwards "Entomological Journal," April 14, 1867; April 16, 1869; May 28, 1869; and May 8, 1870, WHE-SA.

54. On Thoreau, see David R. Foster, *Thoreau's Country: Journey Through a Transformed Landscape* (Cambridge, MA: Harvard University Press, 1999); on Marsh, see George Perkins Marsh, *Man and Nature* (1864; repr., Cambridge, MA: Harvard University Press, 1965).

55. Thaddeus Harris to Thomas Wentworth Higginson, c. June 1851, in *The Entomological Correspondence of Thaddeus W. Harris*, ed. Samuel Scudder (Boston, 1869), 263–64. This letter appears at the very end of the volume.

56. August Weismann, *Studies in the Theory of Descent* (London, 1882), 651. This quotation comes from an essay in this volume, "On the Mechanical Conception of Nature," written sometime in the 1870s, probably in 1877.

57. Alexander Humboldt, *Personal Narrative*, quoted by Richard H. Grove, in his *Green Imperialism: Colonial Expansion, Tropical Island Edens, and the Origins of Environmentalism, 1600–1860* (Cambridge: Cambridge University Press, 1995), 367.

58. On early coal mining in this region, see John Alexander Williams, *West Virginia* (New York: Norton, 1976), especially chapter 5, "Paint Creek," pp. 130–58.

59. Philipp Zeller to William Henry Edwards, May 27, 1879 (this letter was appended to another letter to Edwards by Zeller, with the same date), Philipp Zeller Papers, NHM-LONDON.

60. Henry Edwards, "Iron and Its Relation to Civilization," published in his *A Mingled Yarn: Sketches on Various Subjects* (New York: G. P. Putnam's Sons, 1883), 65–85. I know of only one butterfly man or miner who did have doubts about mining. Thomas Belt, a brilliant English naturalist who joined the Australian gold rush in the early 1850s and later managed the gold mines of Chontales, Nicaragua, noted briefly in his book *The Naturalist in Nicaragua* how a river passing through Chontales was "woefully polluted by the gold-mining on its banks, and flows, a dark muddy stream, through the village of Santo Domingo." In 1872 he went to Colorado to mine gold; three years later, he died there. See Belt, *The Naturalist in Nicaragua* (1874; repr., London, 1888), 106–8.

61. *Walt Whitman: Complete Poetry and Collected Prose* (New York: Library of America, 1982), 986; and Ralph Waldo Emerson, "Nature" (1836), in *Selections from Ralph Waldo Emerson*, ed. Stephen E. Whicher (Boston: Houghton Mifflin, 1960), 22.

62. On Philadelphia as center of printing and lithography, see Nicholas B. Wainwright, *Philadelphia in the Romantic Age of Lithography* (Philadelphia: Historical Society of Pennsylvania, 1958).

63. Preface to *BNA*, unpaginated.

64. William Henry Edwards to Hermann Hagen, May 16, 1871, Hagen Correspondence, EML.

65. Dru Drury, *Illustrations of Exotic Entomology*, vol. 1, ed. J. O. Westwood (1770; repr., London, 1837), xviii–xix.

66. In this volume, however, he tried to prove the existence of several distinct species of blues, a position he later discarded but that current systematics confirms. See "The Azures," in Michael Gochfeld and Joanna Burger, *Butterflies of New Jersey* (New Brunswick, NJ: Rutgers University Press, 1997), 165–66; and "Spring Azure," in Alexander B. Klots, *A Field Guide to the Butterflies* (Boston: Houghton Mifflin, 1951), 169.

67. William Henry Edwards to Philipp Zeller, May 27, 1879, and Zeller to Edwards (copy of letter contained in Edwards's letter to Zeller, same date), Philipp Zeller Papers, NHM-LONDON. Zeller wrote that "it is certain that Europeans have been far from executing your method, which seems the only correct one, in order to obtain the truth." Edwards did not invent this new method. William Saunders, a Canadian who corresponded with him from the early 1860s on, had been confining females in this way for many years. Saunders's letters to Edwards were destroyed, but see Saunders to Samuel Scudder, January 4, 1864, and March 26, 1864, SS-BMS.

68. In 1876 he would write Philipp Zeller, a German naturalist, "I do little more now but breed from eggs" (Edwards to Zeller, 1876, Philipp Zeller Papers, NHM-LONDON). On the ice house and the greenhouse, see Edwards to William Holland, April 21, 1886, WH-CM. On butterflies in the dressing room, and so on, see Edwards, Entomological Diary, September 25, 1873, WHE-SA, and Edwards to Henry Edwards, May 28, 1875, HE.

69. "Whole story" appears throughout Edwards's letters, but see his letter to Scudder, July 11, 1871, SS-BMS. The term comes out of natural history, and for insects was best expressed in *The Introduction to Entomology* by William Kirby and William Spence (London, 1819), a book Edwards knew quite well. Viewing nature as a series of "stories" sounds unscientific, but in Edwards's rendering, all living things have stories, people and butterflies alike.

70. William Henry Edwards to Henry Edwards, November 15, 1875, and June 23, 1876, HE.

71. William Henry Edwards to Scudder, December 20, 1870, and January 24, 1871, SS-BMS. On the impact of Darwin's views on other naturalists, see Mark V. Barrow Jr., *A Passion for Birds* (Princeton, NJ: Princeton University Press, 1998), 27; and Edward Larson, *Evolution's Workshop* (New York: Basic Books, 2001), 102–9.

72. Edwards to Scudder, December 1, 1875, SS-BMS.

73. Edwards to Henry Edwards, November 15, 1875, and June 23, 1876, HE.

74. See Benjamin Walsh to Hermann Hagen, May 11, 1860, and August 4, 1861, EML.

75. See, on Walsh, Charles Valentine Riley, "In Memoriam," *American Entomologist* 2, no. 3 (1869–70); and Carol Sheppard, "Benjamin Dann Walsh: Pioneer Entomologist and Proponent of Darwinian Theory," *Annual Review of Entomology* 49 (January 2004): 1–25. See also Edna Tucker, "Benjamin D. Walsh—First State Entomologist," *Transactions of the Illinois State Historical Society* (Springfield, IL, May 14, 1920), 54–61; and Walsh's biographical sketch in *American National Biography*, ed. John Garraty and Mark Carnes (New York: Oxford University Press, 1999).

76. By 1869, Edwards was sending so many letters to Walsh that he could not answer them all. See Walsh to Edwards, January 18, 1869, WHE-SA. On the historical significance of Walsh, see Carol Sheppard, "Benjamin Dann Walsh."
77. Walsh to Hagen, January 14, 1869, EML.
78. "In every department of nature," Wallace wrote in 1865, "there occur instances of the instability of specific form, which increase of materials aggravates rather than diminishes. . . . It is for such reasons that naturalists now look upon the study of varieties as more important than that of well-fixed species. It is in the former that we see nature at work." Quoted in *Infinite Tropics: An Alfred Russel Wallace Anthology,* ed. Andrew Berry (London: Verso, 2002), 97–98.
79. See Wallace's essay in Berry, ed., *Infinite Tropics,* 88, plus Berry's commentary.
80. "Papilio III," in *BNA,* vol. 1, unpaginated.
81. Edwards's immaturity about the life history of butterflies may have shaped his failure to see how surface coal mining on any scale might have threatened butterflies. In the late 1860s, despite his desire to focus on *living* things, he still did not have much *feel* for how they interacted with one another or with the larger environment around them. Thaddeus Harris, at the end of his life, had achieved this ecological state of mind, but Edwards was only beginning to understand how every stage of a butterfly's life history, from egg to butterfly, could open up one's insight into larger environmental conditions.
82. For an excellent and extensive assessment of this breakthrough, see W. Conner Sorensen, *Brethren of the Net: American Entomology, 1840–1880* (Tuscaloosa: University of Alabama Press, 1995), 219–22.
83. Ibid., 222–24.
84. Edwards to Scudder, October 20, 1870, SS-BMS; Edwards, *BNA,* vol. 1, Grapta IV, unpaginated.
85. Edwards to Hagen, May 16, 1871, EML.
86. Edwards to Scudder, August 1871, SS-BMS.
87. "What Volume I cost I never knew," he wrote Samuel Scudder in 1890. "I burned all the bills so that no one should see how extravagant I had been." Edwards to Scudder, September 25, 1890, SS-BMS.
88. Edwards, Entomological Diary, July 5, 1868.
89. Ibid., September 2, 1868.
90. William Henry Edwards, "Habits of Melitaea Phaeton," *CE* (February 15, 1869): 59; and "Melitaea I," in *BNA* (Boston, 1884), vol. 2, unpaginated. In both places, Edwards refers to Frank Fraser as "one of my young friends." See also Entomological Diary, June 6, 1871; August 21, 1869; and May 20, 1869.
91. Edwards, Entomological Diary, June 24, 1870; Edwards to Theodore Mead, June 10, 1871, TM; and Edwards to Henry Edwards, July 14, 1873, HE. On Edwards's sister in Florida, see Edwards to Hermann Hagen, December 29, 1878, Letters to Hermann Hagen, EML; and on his other sister, see Edwards to Joseph Lintner, September 20, 1884, JL.
92. David Bruce to Henry Skinner, March 17, 1896, HS-PAS; and William Henry Edwards to Charles F. McGlashan, June 11, 1891, Charles McGlashan Papers, Bancroft Library, Berkeley, CA. In a letter to William Wright, William Henry Edwards mentions Henry's agnosticism (see June 15, 1891, WGW). Arnold Mallis has Edwards's birth date incorrectly at 1830, but the May brothers saw Edwards's birth certificate. See Andrew Brown-May and Tom W. May, " 'A

Mingled Yarn': Henry Edwards, Thespian and Naturalist, in the Austral Land of Plenty, 1853–1866," *Historical Records of Australian Science* 11, no. 3 (1997): 407–8; and Arnold Mallis, *American Entomologists* (New Brunswick, NJ: Rutgers University Press, 1971), 292.

93. For Anna Dickinson, see Mead's 1871 journal, January 10, 1871, in private possession of John V. Calhoun, who graciously let me read this document. See also his Letterbook, 1870–74, especially Mead to Arthur Whittemore, November 21, 1872; Mead to Willie Edwards, February 26, 1872; Mead to Willie Edwards, January 11, 1874; and Mead to his father and mother, July 12, 1886, TM. As a very young man, Mead argued for more (not less) Chinese immigration (the Chinese being "honest, sober, industrious, and clean"), defended the rights of Jews to join established clubs in New York City, and viewed blacks as just as intellectually able as whites. His only prejudice (or the only one he would admit to) was against what he called the "savage" Catholic Irish, "incapable of reasoning and destitute of any sense of justice or personal moral responsibility." When he heard that William Henry Edwards considered American Irish the most "efficient foremen"—John Burke, the man who managed his coal miners on Paint Creek, was one such—and believed that they had an "unexcelled genius for government," he could not believe his ears.

94. Mead to Willie Edwards, May 15, 1871; September 10 and November 3, 1872; May 15, 1871; and February 9, 1873, TM.

95. Mead to Willie Edwards, May 19, 1872; Mead to William Henry Edwards, February 12, 1873, and May 29, 1872; and Mead to James Behrens, June 19, 1872, TM.

96. Mead to Herbert Morrison, February 12, 1873, TM.

97. Mead to William Henry Edwards, December 7, 1870, TM.

98. Mead, "Notes upon Some Butterfly Eggs and Larvae," *CE*, 7, no. 9 (1875): 161–63.

99. Edwards to Scudder, September 13, 1873, SS-BMS; and William Henry Edwards to Henry Edwards, October 5, 1871, HE.

100. William Henry Edwards to Henry Edwards, October 5, 1871; December 5, 1871; and October 16, 1876, HE.

101. Edwards to Mead, July 4 (or 7), 1878, TM.

102. Theodore L. Mead, "Naturalist, Entomologist, and Plantsman: An Autobiography," published in *The Yearbook of the Amaryllis Society* (1935), 2, 3–14, TM.

103. Edward Doubleday, *The Genera of Diurnal Lepidoptera*, vol. 1 (London, 1846), 46. Among other earlier pioneers to the Rockies were the Americans William Wood, Winslow Howard, and James Ridings, who visited in the 1850s and '60s. See F. Martin Brown's biographical sketches, especially "Two Early Collectors of Colorado Insects," draft, in the F. Martin Brown Papers, American Museum of Natural History, New York.

104. Mead to William Henry Edwards, June 27, 1871, in *Chasing Butterflies in the Colorado Rockies with Theodore Mead*, ed. Grace H. Brown and annotated by F. Martin Brown, Bulletin No. 3 (Florissant, CO: Pike's Peak Research Station, Colorado Outdoor Research Center, 1996), 20; Mead to Will Scott, November 1, 1871, TM; and Mead, "Notes upon Some Butterfly Eggs and Larvae," 161–63.

105. Edwards, Entomological Diary, July 4, 1871; July 11, 1871; and December 5, 1871.

106. Mead to William Henry Edwards, October 27, 1871, TM; and William Henry Edwards to Henry Edwards, July 5, 1878, HE.

107. William Henry Edwards to Henry Edwards, July 5, 1878, HE.

108. Mead to William Henry Edwards, October 20, 1871, in Brown, ed., *Chasing Butterflies,* 70.

109. Henry Edwards to Scudder, June 2, 1870, SS-BMS; and Charles F. McGlashan to Edwards, June 7, 1887, HE.

110. Obituary, "Henry Edwards," *EN* 2, no. 7 (1891): 7–8; and Andrew Brown-May and Tom W. May, " 'A Mingled Yarn,' " 407–9.

111. At the same time, Edwards did "battle" (his word) against Australian clerics who were convinced that theatrical "amusement" promoted prostitution. In an open letter to a leading ministerial opponent, Edwards observed that actors and actresses have faults, to be sure, but "more virtue," not less, than most other human beings; besides, "the Great Author of divine things himself" has, for centuries, "permitted amusement to exist," because he "regards" it as "a natural craving of the human heart" and because it has "aided and directed the progress of the human mind." Edwards later published this letter in the United States in his collection *A Mingled Yarn,* 90–91.

112. On Brooke and Polly, see William John Lawrence, *The Life of Gustavus Vaughan Brooke: Tragedian* (London, 1893), 160–64, 190–95, 200–2. I thank Andrew Brown-May for supplying me with the name of Brooke's wife (Marianne Bray).

113. Letter cited by William Henry Edwards in *BNA,* vol. 1, in the description "Pieris I, Pieris Beckerii," n.p.; and Brown-May and May, " 'A Mingled Yarn,' " 411.

114. Brown-May and May, " 'A Mingled Yarn,' " 411.

115. Henry Edwards to J. Angelo Ferrari, Vienna, May 24, 1862, DEI.

116. On the California theater in the days of gold rush and beyond, see Constance Rourke, *Troupers of the Gold Coast* (New York: Harcourt, 1928), and Richard A. Van Orman, "The Bard in the West," *Western Historical Quarterly* 5 (January 1974): 29–38.

117. For these words, see an essay by Edwards published in *High Jinks,* a San Francisco magazine, in 1873 and republished in *A Mingled Yarn,* 90–91.

118. Essig, *A History of Entomology,* 611–12.

119. Edwards, *A Mingled Yarn,* 1–63.

120. Obituary of Henry Edwards, appended by William Henry Edwards in his Entomological Diary, 1890.

121. Arthur Shapiro to the author, e-mail dated November 17, 2010; Henry Edwards to Charles McGlashan, July 25, 1885, Charles McGlashan Papers; and Samuel Scudder, *BEUSC,* vol. 2, p. 845.

122. Donald Worster, *A Passion for Nature: The Life of John Muir* (New York: Oxford University Press, 2008), 144–45, 136–80.

123. On Muir's conversion to the "religion of nature" in the late 1860s, see Worster, *A Passion for Nature,* 141–64; and Muir, quoted by J. S. Wade in "Vignettes of Henry Edwards and John Muir," *Scientific Monthly* 30 (March 1930): 240–50.

124. Arthur Shapiro to the author, e-mail dated November 17, 2010; Arthur Shapiro, *Field Guide to Butterflies of the San Francisco Bay and Sacramento Valley*

Regions (Berkeley: University of California Press, 2007), 137–38; and Wade, "Vignettes of Henry Edwards and John Muir."

125. Henry Edwards to Herman Strecker, October 11, 1873, HS-FM.

126. Henry Edwards to Charles McGlashan, July 25, 1885, NHM-LONDON.

127. Edwards to Lintner, October 13, 1877, JL.

128. William Henry Edwards to Henry Edwards, February 12 and 22, 1876, HE. The butterfly man David Bruce used the same metaphor to express Henry's impact on him: "A letter from you is like a gleam of sunshine on a dull day." See Bruce to Henry Edwards, November 4, 1884, HE.

129. Augustus Grote, *The Hawk Moths of North America* (Bremen, 1886), 61.

130. *Entomologists' Magazine* (London; July 1868): 50. Peart's art was later displayed in the Women's Pavilion of the 1876 Centennial Exposition in Philadelphia, as an example of what women could do. See George Dimmock, "Entomology at the Centennial Exhibition," *Psyche* 1 (October 1876): 201–5. "Six sample plates from Edwards' *Butterflies of North America*," observed Dimmock, "testify to Miss Peart's extraordinary ability in figuring insects on stone."

131. Butler was quoted by Edwards in a letter to Scudder, April 23, 1874, SS-BMS.

132. Jean Baptiste Boisduval to Edwards, June 1, 1873, in Brown, ed., *Chasing Butterflies*, 198. Ironically, Edwards himself detested Hübner *and* his system. On Hübner, see Essig, *A History of Entomology*, 664–66.

133. *American Entomologist* 1, no. 1 (1868).

134. On Agassiz's geological creationism, see Martin J. S. Rudwick, *Worlds Before Adam* (Chicago: University of Chicago Press, 2008), 517–33.

135. Horace Scudder, *Life and Letters of David Coit Scudder, Missionary in Southern India* (New York: Hurd and Houghton, 1864), 3–8, 239, 380.

136. Mark Hopkins, "An Address Delivered in Boston, May 26, 1852, Before the Society for the Promotion of Collegiate and Theological Education in the West" (Boston, 1852), 9. For Charles Scudder's admiration of Hopkins and his hope that Samuel "might come under his intellectual guidance," see Alfred Goldsborough Mayor, "Samuel Scudder, 1837–1911," in *Memoirs of the National Academy of Sciences* 17, no. 3 (1924): 82.

137. Samuel Scudder, "How I Served My Apprenticeship as a Naturalist," *Youth's Companion* (November 5, 1896): 593.

138. Scudder, *BEUSC*, vol. 2, p. 821; and Scudder, *Every-day Butterflies: A Group of Biographies* (Boston: Houghton Mifflin, 1899), 58, 225–30.

139. On Chadbourne's influence on Scudder and other students, see J. Walter Wilson's biography of Alpheus Packard (one of Scudder's student friends), unpublished manuscript, p. 8, Alpheus Spring Packard Papers, George J. Mitchell Department of Special Collections & Archives, Bowdoin College Library, Brunswick, ME. In the early 1860s, Packard was a student with Scudder at the Museum of Comparative Zoology, studying under Agassiz. Chadbourne introduced the two earlier, however, when he taught briefly at Bowdoin College in Maine, where Packard was an undergraduate. He told Packard that "he had many young men working under him at Williams College who were studying to make Natural History their profession. In particular he suggested that he write to Scudder to initiate a correspondence in regard to entomology" (8–9).

140. Wilson biography of Packard, 282–83, 299.

141. Paul Ansel Chadbourne, *Instinct: Its Office in the Animal Kingdom, and Its Relation to the Higher Powers of Man* (New York: Putnam's, 1883), originally given in the 1871 Lowell Lecture series, which in turn drew on "the Author's *Natural Theology* published in 1867," p. 13. Chadbourne served as president of Williams in the 1860s, and of the University of Wisconsin shortly after. He had a political career as well.

142. Scudder to R. Ostensacker, July 24, 1858, EML.

143. Mayor, "Samuel Scudder, 1837–1911," 83.

144. I have based this paragraph entirely on information from Edward Lurie's biography of Agassiz, *Louis Agassiz: A Life in Science* (Baltimore: Johns Hopkins University Press, 1988), which though first published in 1960, remains a fresh assessment of Agassiz and of a major transformation in American culture.

145. Ibid., 194–95. For Agassiz's debt to Humboldt, see his tribute to Humboldt, "Address Delivered on the Centennial Anniversary of the Birth of Alexander Humboldt" (Boston: Boston Society of Natural History, 1869).

146. Louis Agassiz, *Essay on Classification* (1857; repr., Baltimore: Johns Hopkins University Press, 1962), 177–78. Edward Lurie's introduction informs this analysis.

147. Lurie, *Louis Agassiz,* 186–87, 196–97, 212–17.

148. Ibid., 82–83.

149. The society published a sophisticated journal, held an extensive array of meetings and classes, and employed a knowledgeable staff; it nearly equaled in influence the existing museums, Philadelphia's Academy of Natural Sciences, the Smithsonian, and the Museum of Comparative Zoology. For historical material on the Boston Society of Natural History, see Mary P. Winsor, *Reading the Shape of Nature: Comparative Zoology at the Agassiz Museum* (Chicago: University of Chicago Press, 1991), 10–11, 129–30; and Samuel Scudder, presidential speech, *Proceedings of the Boston Society of Natural History* 21 (1885): 15–16. "The Boston Society," Scudder observed, "was more largely endowed than any similar institution in the country."

150. Lurie, *Louis Agassiz,* 252–302; but I have taken from the whole book.

151. See Winsor, *Reading the Shape of Nature,* 37–38. Winsor presents a caustic, idiosyncratic, and brilliant assessment of both the Museum of Comparative Zoology and Agassiz, the best ever published. But it, too, relies heavily in places on the work of Edward Lurie. See, also, for different approaches to Agassiz, Philip J. Pauly, *Biologists and the Promise of American Life* (Princeton, NJ: Princeton University Press, 2000), 33–51; Edward J. Larson, *Evolution's Workshop* (New York: Basic Books, 2001), 95–105, 122–25, and 218–19; and Rudwick, *Worlds Before Adam,* 437–49, 518–38. For a discussion of Agassiz's influential predecessors—above all, Cuvier—see Stephen T. Asma, *Stuffed Animals and Pickled Heads: The Culture and Evolution of Natural History Museums* (Oxford: Oxford University Press, 2001), 114–53.

152. Lurie, *Louis Agassiz,* 295–97.

153. Winsor, *Reading the Shape of Nature,* 25.

154. Lurie, *Louis Agassiz,* 57.

155. Here again, I owe this interpretation to Lurie's excellent discussion in chapter 7 of *Louis Agassiz.*

156. Winsor, *Reading the Shape of Nature,* 31–37.

157. Ibid., 12–15.

158. Lurie, *Louis Agassiz,* 303–17.

159. Alexander Humboldt, *Personal Narrative of a Journey to the Equinoctial Regions of the New Continent* (1804; repr., New York: Penguin, 1995), 126–30.

160. Scudder, "Perambulations in Search of an Eclipse!," handwritten in 1860, 44–45, Samuel Scudder Papers, Houghton Library, Harvard University. Scudder discusses his asthma on pp. 17–19. There were three versions of this trip, which took him through many Indian villages, one by Commander C. H. Davis and two by Scudder. See Davis, "Journal on the Astronomical Expedition Sent Out by Commander C. H. Davis (1860)," handwritten, Houghton Library, Harvard University (Film 03-1642). The first Scudder version was "Perambulations," later published in a bowdlerized form (cutting out Scudder's references to pretty Indian girls, for instance, as well as the fact that he went to church whenever he could find one, mornings and afternoons, sometimes in moccasins, to suit the Indian custom) as *The Winnipeg Country; or, Roughing It with an Eclipse Party* (Boston, 1886). *The Winnipeg Country* lacks the life and detail of the earlier version. Moreover, there is no indication that Scudder wrote it (the author is listed as "A Rochester Fellow") or was even on the trip. Why it was published in this form is a mystery.

161. Walsh to Scudder, December 17, 1864, SS-BMS.

162. Scudder, *BEUSC,* vol. 1, pp. 299–300.

163. See Scudder, "Annual Meeting of the Society," *Proceedings of the Boston Society of Natural History* 9 (May 20, 1863): 230–32.

164. Quoted by Clark A. Elliott in his *Thaddeus William Harris (1795–1856)* (Bethlehem, PA: Lehigh University Press, 2008), 174–75. See also Thomas Wentworth Higginson, "Memoir of T. W. Harris," in Samuel Scudder, ed., *The Entomological Correspondence of Thaddeus W. Harris* (Boston, 1869), xxvi; and Harris to Edward Doubleday, August 31, 1840, in Scudder, ed., *Entomological Correspondence,* 147.

165. Scudder to Herman Hagen, January 22, 1866, EML.

166. For Scudder's announcement, see *American Naturalist,* vol. 3, nos. 3 and 4 (May and June 1869).

167. Scudder summarized this emphasis on biology in a public address, "Recent Progress of Entomology in America," *Psyche* (January-February 1878): 97–116.

168. Agassiz set the precedent, although, curiously, Scudder himself would later credit his student colleagues with the idea of focusing on the American scene. See the preface to vol. 1, *BEUSC.*

169. *American Naturalist* (June 1869): 213.

170. Scudder, *BEUSC,* vol 1, p. 468, and vol. 2, pp. 1007–8.

171. Gene Stratton-Porter, *Moths of the Limberlost* (New York: Doubleday Page, 1912), 369.

172. Scudder to Henry Edwards, April 9, 1869, HE.

173. William Henry Edwards to Scudder, October 20, 1870 (Alexandria), and August 7, 1871 (Montreux), SS-BMS.

174. Scudder to William Henry Edwards, November 5, 1871, WHE-SA; and Scudder, "Fossil Butterflies," in *Memoirs of the American Association for the*

Advancement of Science, vol. 1 (Salem, MA, 1875); for an account of his European experience, see p. 43. Scudder discussed his alpine trip in his *The Geology of New Hampshire,* vol. 1 (Concord, 1874), 343.

175. The quotes come from an essay by Augustus Grote, "On Genera and the Law of Priority," *CE* (March 1876): 36–37. For definitions of lumping and splitting, see Kurt Johnson and Steve Coates, *Nabokov's Blues* (Cambridge, MA: Zoland, 1999), 54, 101–2, 285–86.

176. Scudder to William Henry Edwards, November 5, 1871, WHE-SA.

177. Scudder, *Historical Sketch of the Generic Names Proposed for Butterflies: A Contribution to Systematic Nomenclature* (Salem, 1875), 91–96; and Scudder, *Systematic Revision of Some of the American Butterflies, with Brief Notes on Those Known to Occur in Essex County, Massachusetts* (from a Report of the Peabody Academy of Science, for 1871), 3–4.

178. On Hübner's biography, see William Henry Edwards, *CE* 8, no. 3 (1876): 43.

179. Scudder, *Butterflies: Their Structure, Changes, and Life-Histories* (New York: Holt, 1881), 6–7, 54–59.

180. Scudder to William Henry Edwards, November 5, 1871, WHE-SA.

181. William Henry Edwards to Scudder, April 28, 1872, and December 20, 1870, SS-BMS.

182. Edwards to Scudder, October 6, 1871, and April 28, 1872, SS-BMS.

183. William Henry Edwards to Scudder, August 27, 1872, SS-BMS.

184. William Henry Edwards to Scudder, October 6, 1871, SS-BMS. "Nothing shows more clearly the absurdity of looking at specific differences in the imago alone," he wrote, "than the late discoveries of Ajax and Interrog. A lot of butterflies—say from Africa—come before a lepidopterist, and he forthwith, strictly according to received practice, separates them into as many species as he can from little differences, in color, in spots. When actually it is all guess work and breeding from egg would show a dozen species in one."

185. William Henry Edwards to Scudder, September 3, 1871, SS-BMS.

186. Scudder to William Henry Edwards, November 5, 1871, HE.

187. William Henry Edwards to Scudder March 3, 1873, HE.

188. William Henry Edwards to Scudder, January 24, 1871, and January 19 and March 3, 1873, SS-BMS.

189. William Henry Edwards to Scudder, December 24, 1870, and March 22 and April 14, 1871, SS-BMS.

190. William Henry Edwards to Scudder, June 15 and July 11, 1871, SS-BMS.

191. *BNA,* vol. 1, section "Argynnis XIV," for this quotation. The volume itself is unpaginated.

192. Scudder to William Henry Edwards, November 5, 1871, HE.

193. Scudder to Strecker, announcing his coming to Reading, November 23, 1873, HS-FM.

194. On Mount Washington, club members roughed it for weeks on end, in search of crepuscular and nocturnal insects, and gathered a great deal of biological information on moths and butterflies. See the account by Joseph Lintner, in his "List of Lepidoptera," in the *Thirtieth Report of the State Museum of Natural History* (Albany, 1879): 141.

195. Scudder, "English Names of Butterflies," *Psyche* 1 (May 1874): 1.

196. J. H. Behrens, "Vernacular Names for Butterflies," *CE* (July 1874): 9–10; S. H.

Peabody, "Mr. Scudder's Butterflies," *CE* (December 1881): 246–50; Scudder, "English Names of Butterflies."

197. See Worster, *A Passion for Nature*, 185–90.

198. Scudder, quoted in "Tenth Anniversary of the Club," *Appalachia* (1886): 366.

199. Scudder, "The Distribution of Insects in New Hampshire," a chapter in *The First Volume of the Final Report upon the Geology of New Hampshire* (Concord, MA, 1874), 331–62.

2. The German-American Romantics

1. For an excellent overview of this immigration, see Kathleen Neils Conzen, "Germans," in *Harvard Encyclopedia of American Ethnic Groups,* ed. Stephen Thernstrom (Cambridge, MA: Belknap Press of Harvard University Press, 1980), 405–25.

2. On beer and entomology, see Edward L. Graef, "Some Early Brooklyn Entomologists," *Bulletin of the Brooklyn Entomological Society* 9, no. 3 (1914): 55; Charles W. Leng, "Memories of Fifty Years Ago," *Bulletin of the Brooklyn Entomological Society* 18 (February 1923): 1–12; and George P. Engelhardt, "The Brooklyn and New York Entomological Societies, Past and Present," *Annals of the Entomological Society of America* 22, no. 3 (1929): 392–400. See also William Bather, "Another Reminiscence of Early Days," *Bulletin of the Brooklyn Entomological Society,* 18 (1923): 56–57.

3. Zimmerman, well known and admired by other American naturalists of that time especially, collected and studied American beetles. See John Morris to Thaddeus Harris, February 13, 1841, Harris Papers, EML: "Zimmerman is a strange genius with many most excellent traits of character. He is quite wealthy, and of a family rather of the higher order in Europe. I mean that he is not a plebian." Zimmerman's only fault, according to Morris, was that he tended to fall in love too readily, and suffered because of it, especially in relation to one young woman he met "while boarding" in a "house with the girl who refused his suit." This experience possibly helped "send" him to South America.

4. Morris to Harris, July 22, 1839, Harris Papers, EML.

5. John Morris, *Synopsis of the Described Lepidoptera of North America* (Washington, DC: Smithsonian Institution, 1862), vii–xvi. For a biography of Morris (but with little discussion of his butterfly work), see Michael J. Kurtz, *John Gottlieb Morris: Man of God, Man of Science* (Baltimore: Maryland Historical Society, 1997), 74–76, 86–91.

6. William Henry Edwards, *Autobiographical Notes* (privately printed, 1901), 203; and Kurtz, *John Gottlieb Morris,* 91.

7. I would like to thank Lawrence F. Gall of the Peabody Museum of Natural History, New Haven, Connecticut, for this information. Gall is the recognized authority on Grote's entomological achievement.

8. Herman Strecker, *LRH* (Reading, PA, 1872–78), 78; and William Hewitson to Strecker, September 1873, and October 9, 1873, HS-FM.

9. Grote mentions "the progress of aesthetic entomology" in a letter to Scudder, observing at the same time that "in Germany a few study this well." See Grote to Scudder, October 30, 1869, SS-BMS.

10. Augustus Grote, *The Hawk Moths of North America,* 5; and Grote, "Historical Sketches of *Gortyna* and Allied Genera," *CE* (May 4, 1900), 70–81.

11. Augustus Grote, "Moths and Moth-Catchers: Part I," *Popular Science Monthly* (June 1885): 246–52; and *Hawk Moths of North America* (Bremen, 1886), 5.

12. For its English application, see Corbin Scott Carnell, *Bright Shadow of Reality: Spiritual Longing in C. S. Lewis* (1974; repr., Cambridge, MA: Eerdmans, 1999), especially chapter 1, "Sehnsucht," pp. 13–29.

13. See Nicholas B. Wainwright, *Philadelphia in the Romantic Age of Lithography* (Philadelphia: Historical Society of Pennsylvania, 1958); and Jessie Poesch, *Titian Ramsey Peale* (Philadelphia: American Philosophical Society, 1961).

14. On the neighbors, see George Meiser, *Echoes of Scholla* (Reading, PA: Berksiana Foundation, 1976), 73; newspaper clipping, HS-FM. For Strecker's self-description, see his close Reading friend, Fred Spang, to Strecker, September 22, 1869, HS-FM: "Speaking of the Devil, I believe you always considered yourself one of his favorite children, a point which I am not ready to dispute with you." Spang invited Strecker to visit San Francisco, where Spang had recently migrated, because "you will be nearer Hell than any other place I know of."

15. Woldemar Geffcken to Strecker, March 7 and 18, 1874, HS-FM. In a letter from James Behrens, a German lepidopterist and an immigrant to San Francisco, to Strecker, Behrens details the classic work of an exemplary artist-entomologist, Carl Julius Milde, of Lübeck, thereby giving insight into this general European practice. See Behrens to Strecker, November 19, 1875, HS-FM.

16. This information appeared in many early accounts of Strecker's life and comes directly from his own account, curiously almost buried in a concluding section of his 1878 catalog called "List Localities." The locality mentioned here was Kern County of California, possibly named by Frémont for the three brothers. Strecker apparently never boasted about his relation to the Kern family, whom he barely knew. See Herman Strecker, *Butterflies and Moths of North America: Complete Synonymical Catalogue of Macrolepidoptera* (Reading, PA: B. F. Owen, 1878), 53. On Ferdinand's hometown and Uncle Wilhelm, see Woldemar Geffcken (of Stuttgart) to Strecker, March 18, 1874, and March 29, 1878, HS-FM. On the Kern brothers and their Romantic zeal for nature and exploration, see David J. Weber, *Richard H. Kern: Expeditionary Artist in the Far Southwest, 1848–1853* (Albquerque: University of New Mexico Press, 1985); Robert V. Hine, *Edward Kern and American Expansion* (New Haven, CT: Yale University Press, 1962), vii–ix, 1–55; H. A. Spindt, "Notes on Life of Edward M. Kern," *Kern Historical Society* (November 1939): 5–20; William Heffernan, *Edward M. Kern: The Travels of an Artist-Explorer* (Bakersfield, CA: Kern Historical Society, 1953). In the late 1840s, Joseph Leidy hired Richard to do the illustrations for the annual reports of the Philadelphia's Academy of Natural Sciences, which may explain why the young Strecker got easy access to the pictures in the basement library. See Weber, *Richard H. Kern,* 23–24.

17. Strecker to Duncan Putnam, June 2, 1876, DP.

18. E. L. Hettinger and Milton W. Hamilton, "Dr. Herman Strecker—Artist and Scientist," *Historical Review of Berks County* (July 1946): 98–102. For his own business, see letterhead, Strecker to Henry Edwards, November 9, 1870, HE.

19. Strecker to Duncan Putnam, June 10, 1876, DP.
20. Reakirt to Strecker, June 23, 1869, HS-FM: "I can thoroughly appreciate the weight of moving a marble yard, and I am glad you didn't smash your feet again." On his gravestones, see, for example, John H. Kendall (husband of Mary) to Strecker, February 13, 1869, HS-FM.
21. Theodore Mead to Strecker, June 5, 1873, HS-FM.
22. Joseph Drexel to Strecker, August 29 and October 18, 1867, and March 23 and September 9, 1868, HS-FM. In the 1880s, Drexel moved to New York, where he ran his banking business and served as a trustee to the American Museum of Natural History, donating his lepidoptera to that institution. See Frank Lutz, "Amateur Entomologists and the Museum," *Natural History* 24, no. 3 (May–June 1924): 337.
23. On Charles Wood, see Wirt Robinson to Henry Skinner, April 20, 1908, HS-ANS. Robinson's grandfather knew Strecker when both were boys in Reading. Wirt recalled Wood in his letter, describing the role he played. See also Charles Wood to Herman Strecker, December 14, 18, 23, and 28, 1866, HS-FM. On his father's beating, see the interview with Strecker in Arthur Fuller's column in *Rural New York*, "Daily Rural Life," January 11, 1875, HS-FM.
24. Herman Strecker, *LRH*, 99. *C. Amatrix* is described in William Holland, *The Moth Book: A Guide to the Moths of North America* (1903; repr., New York: Dover, 1968), 263.
25. Strecker to Henry Edwards, December 11, 1870, HE.
26. Strecker, preface, *Butterflies and Moths of North America*, 2, and Alexander Humboldt, *Cosmos*, vol. 1 (1850; Baltimore: Johns Hopkins University Press, 1997), 82–83.
27. Strecker to Henry Edwards, August 23, 1871, and January 21, 1874, HE.
28. Paul A. Druzba, *Neversink: Reading's Other Mountain* (Reading, PA: Exeter House, 2003), 12–14.
29. On the relationship between Strecker and Robinson, see Wirt Robinson to Henry Skinner, March 30 and April 20, 1908, HS-ANS.
30. Sallie Goodfellow to Louisa Roy, December 18, 1856, HS-FM; and Herman Strecker to Louisa Roy, November 11, 1856, HS-FM.
31. Strecker to Henry Edwards, September 11, 1870, HE.
32. I learned of the death and disease only after reading countless letters to him from others, and of the death of his wife and mother only from a letter of a friend, who had heard the news from someone else.
33. George Hulst to Strecker, May 28, 1878, HS-FM. In his letter, Hulst expresses his condolences regarding the death of another Strecker child.
34. The draft of a letter by Strecker to Edward Owen, April 12, 1882, HS-FM.
35. Strecker, *Butterflies and Moths of North America*, 2. See also Strecker to Henry Edwards, April 4, 1871, HE. He refers to the "immense amount of labor my collection entails on me in the shape of correspondence, exchanges, slacking, identifying, on and on, which has all to be done after 8 or 9 o'clock in the evening."
36. Strecker to Henry Edwards, August 22 and November 7, 1872, and August 24, 1874, HE.
37. Fred Tepper to Strecker, HS-FM.

38. Strecker to McGlashan, June 18, 1885, Charles F. McGlashan Papers, Bancroft Library, Berkeley, California.

39. "The notes in the latter part" of the notebook "are, I think, partly by my father, and partly by Strecker—or perhaps entirely by my Father, as he was only fifteen," Wirt Robinson wrote Henry Skinner, March 30, 1908, HS-ANS. Toward the end of his life, Strecker gave this notebook to Russell Robinson as a gift, and Robinson's son, in turn, gave it to Philadelphia's Academy of Natural Sciences, after his father died, which was the purpose of these letters to Skinner.

40. Strecker's 1856 notebook, Entomological Records, Academy of Sciences, Philadelphia, Pennsylvania, presented to the academy by W. Russell Robinson, pp. 8, 19, 33, 64–65. For the current number of swallowtails, see James A. Scott, *The Butterflies of North America* (Stanford, CA: Stanford University Press, 1986), 161.

41. He still visited the fields, however, writing Henry Edwards in 1871, "Our collecting season has opened here, and I have been out several times and had fair luck" (May 12, 1871, HS-FM).

42. Graef, "Some Early Brooklyn Entomologists," 52.

43. George Hulst to Strecker, October 11 (or 16), 1876, HS-FM.

44. Strecker, *Butterflies and Moths of North America,* 32.

45. H. Landis to Strecker, September 24, 1878, HS-FM.

46. On conflicts naturalists had over prices and money payments, see Jim Endersby, *Imperial Nature: Joseph Hooker and the Practices of Victorian Science* (Chicago: University of Chicago Press, 2008), 84–111.

47. For Reakirt, see "Descriptions of Some New Species of Diurnal Lepidoptera," *Proceedings of the Academy of Natural Sciences* (1866): 238–49, 331–42; and F. Martin Brown, "Tryon Reakirt," *Journal of the Lepidopterists' Society* 18, no. 4 (1964): 211–14. For Drexel, see Drexel to Strecker, August 29 and October 18, 1867, and March 23 and September 9, 1868, HS-FM.

48. Henry Edwards to Strecker, February 10, 1870, HS-FM.

49. Strecker to Henry Edwards, September 4, 1869, and May 7, 1870, HE; and Henry Edwards to Strecker, August 23, 1869, HS-FM.

50. These words and phrases can be found in Strecker's letters to Henry Edwards, March 2, 1870; September 11, 1870; August 26, 1873; October 22, 1873; and August 24, 1874, HE.

51. Strecker to Henry Edwards, March 3, May 23, June 10, August 5, and November 11, 1872, HE.

52. Strecker to Henry Edwards, February 19, 1871, HE. Later, Edwards sent him a pair of specimens of *Ornithoptera richmondia,* discovered in 1852; his joy was equally undisguised. Henry Edwards to Strecker, June 15, 1871, HE.

53. Tepper to Strecker, July 24, 1874; November 6, 1874; and September 3, 1876, HE.

54. Tepper became an expert on *Catocala* moths (or underwings, in the vernacular) and amassed one of the finest collections in the world.

55. Fuller to Strecker, undated, c. 1877, HE.

56. See "New Illustrated Works on American Lepidoptera," *CE* 48 (August 1872): 158–59. The journal noted: "The Lepidopterist of the present day . . . possesses vastly improved advantages over his predecessor of even ten years ago in the

accurate and artistic drawings that are being so copiously issued from the press. There are now [two] works in the course of publication, whose chief object is to afford faithful coloured illustrations of Butterflies and Moths." The works were by William Henry Edwards and Strecker.

57. Strecker to Mead, March 24, 1874, TM.

58. See William Hewitson, *Bolivian Butterflies* (London, 1874); and his *Illustrations of Diurnal Lepidoptera* (London, 1862–78), vols. 1 and 2. See also L. G. Higgins, preface to *Hewitson on Butterflies, 1867–1877* (Hampton: E. W. Classey, 1972).

59. Hewitson to Strecker, September 1873; October 9, 1873, HS-FM. For biographical information on Hewitson, see the obituary in *Entomologist's Monthly* (July 1878): 44–45.

60. Hewitson to Strecker, August 10 and October 9, 1873, HS-FM. Hewitson said in the same letter that "the cold indifference of most of those who call themselves naturalists towards our dearly loved pets show that all the warmth of feeling they have is centered in self." On Hewitson's "creationism," see Richard I. Vane-Wright, "A Portrait of Clarence Buckley, Zoologist," *Linnean* 7, no. 3 (1991): 30.

61. Hewitson, "Review," *Entomologist's Monthly* (July 1875): 141.

62. See Strecker's 1878 catalog, *LRH,* 66.

63. William L. Devereaux, April 13, 1874, and April 4, 1898, HS-FM. Devereaux married a woman who also loved entomology. They named their first son Linné, after Linnaeus.

64. John Akhurst to Strecker, April 6, 1869, HS-FM; on Akhurst's assistance to Edwards, see, for example, Edwards's portrait of *Papilio turnus,* in vol. 2 of *BNA,* unpaginated.

65. James Angus to Strecker, February 28, 1871; Henry Schonborn to Strecker, October 18, 1874; and Adrian Latimer to Strecker, September 22, 1879, HS-FM.

66. Hulst to Strecker, June 9, 1876, and December 4, 1878, HS-FM.

67. Having grown up in France, once in America, Neumoegen married Rebecca Livingston, the daughter of a German Jewish immigrant and owner of a wholesale clothing business on Manhattan's Lower East Side. Neumoegen to Strecker, April 4, 1876, HS-FM. Livingston owned a summer estate in Morris County, New Jersey, near the estates of other Livingstons, though unrelated. I would like to thank Sarah Dockray and Carie Levin of the Joint Free Public Library of Morris and Morris Township, New Jersey, for tracking down this information for me, which they located by examining the 1870 census records at the North Jersey Genealogy Center at the JFPL. Why Neumoegen's father-in-law, Lewis Livingston, took the name of Livingston (if that was, in fact, what he did) and had a summer estate near the native-born Livingstons, I do not know. That Livingston was Jewish I discovered in a draft letter from Strecker to Otto Staudinger, dated July 1894: "They are all Hebrews you know," he wrote of the Livingston family.

68. Neumoegen to Strecker, February 28, 1876, HS-FM. ·

69. Neumoegen to Strecker, March 14, 1876, HS-FM. For biographical information, see April 4, 1876, and March 21, 1876, HS-FM. A brief sketch of Neumoegen appears in Harry Weiss, "Journal of New York Entomological Society,

1893–1942," *Journal of the New York Entomological Society* 1 (December 1943): 285–94.

70. Julius Meyer to Strecker, May 1885, HS-FM.

71. Titian Peale, letter to the editor of *Papilio* 4, nos. 7–8 (1884). Throughout the 1870s, near Red Bank, New Jersey, he did drawings from life of many butterflies and moths, fleshing out his portraits with "full descriptions and habits of each insect," although these studies never reached beyond drafts. See Peale, "Lepidoptera: Larva, Foodplants, Etc." (1833, 1877, 1879), handwritten, in possession of the American Museum of Natural History, New York.

72. Peale to Strecker, December 30, 1873, and April 26, 1875, HS-FM.

73. John Morris to Strecker, January 6, 1875, HS-FM.

74. Harrison Dyar, preface to *Insecutor Inscitiae Menstruus* 1, no. 1 (1913), a journal Dyar edited. For Henry Bird, see his unpublished study, "Epic of Papapeima," 52, 425, Research Library, American Museum of Natural History, New York. Henry Skinner, the influential editor of the *Entomological News,* called Grote "one of the greatest students of American Lepidopterology," high praise from Skinner, since he was no fan of Grote's. See Skinner, "Augustus Radcliffe Grote," *EN* 14, no. 9 (1903): 277–78.

75. William T. Davis, obituary of Grote, *Proceedings of the Natural Science Association of Staten Island* 3 (November 10, 1903). For other biographical information, see Charles Bethune, "Professor Augustus Radcliffe Grote," *The 34th Report of the Entomological Society of Ontario* (1903): 109–12. For a remembrance of the collecting ventures of these boys, see Graef, "Some Early Brooklyn Entomologists," 47–56; and Grote, "Moths and Moth-Catchers,"; part 1 appeared in June 1885, pp. 246–52, and part 2 in July 1885, pp. 377–89. On the rural character of the area in the 1850s, see Marc Linder and Lawrence S. Zacharias, *Of Cabbages and Kings County: Agriculture and the Formation of Modern Brooklyn* (Iowa City: University of Iowa Press, 1999), 19–51, 131.

76. Graef, "Some Early Brooklyn Entomologists," 47.

77. Johann Meigen, *Handbuch für Schmetterlingsliebhaber: Besonders für Anfänger im Sammeln* (Aachen, 1827), 1 and 9 (my translation).

78. Charles E. Lang and William T. Davis, *Staten Island and Its People: A History, 1609–1929* (New York, 1930), 231.

79. Quoted ibid., 246.

80. Grote, "Notes on Staten Island Noctuidae," *CE* (May 1886): 95.

81. Grote, "Notes on Staten Island Noctuidae"; and Grote, *Checklist of North American Moths* (1882), 5.

82. H. Frederik Nihout, *The Development and Evolution of Butterfly Wing Patterns* (Washington, DC: Smithsonian Institution Press, 1991), 1.

83. Grote, "Die Saturiniiden," *Mittheilungen aus dem Roemer-Museum, Hildesheim* 6 (June 1896): 1.

84. Grote, "Moths and Moth-Catchers: Part I," 246–52; and Grote, *Hawk Moths of North America,* 5.

85. Grote, *Hawk Moths of North America,* 5–12; and Grote, "On the Geographical Distribution of North American Lepidoptera," *CE* (September 9, 1886): 162–74.

86. Obituary, *CE* 4, no. 6 (1872).

87. Grote to Cajetan Felder, c. 1862–63, Felder Correspondence, NHM-LONDON.

88. Grote to Philipp Zeller, December 9, 1867, Zeller Papers, NHM-LONDON.
89. This information comes from a letter Grote wrote John L. Le Conte, February 18, 1876, John L. LeConte Papers, 1812–1897, American Philosophical Society, Philadelphia, Pennsylvania. Grote refers to "my mother's plantation" and says that he has "planted cotton practically for five years."
90. Grote to Scudder, October 6, 1869, and December 20, 1869, SS-BMS.
91. See his letter to Le Conte, February 18, 1876; Grote says, "My wife was a granddaughter of Judge Johnson of Charleston, South Carolina and relative of Reverby Johnson of Maryland and General Edward Johnson of Richmond, Va." Julia Blair's name appears on the Grote family *Stammbaum* (family tree), courtesy Peter and Marianne Gaehtgens.
92. Grote to Scudder, May 18, 1872, and September 2, 1872, SS-BMS.
93. In his 1876 letter to Le Conte, Grote says, "my step mother and children still live near the plantation at Demopolis, Alabama. I have been anxious to rejoin them for some time, but my means have not allowed it."
94. He also wrote such long and interesting poems as *Rip Van Winkle: A Sun Myth,* which integrated the story of Rip Van Winkle into musings on life and death, youth and old age. Spare and grim, reflecting the burden of his personal history, part 1 ends:

> Ere the passing of breath
> Comes the luring hour of death;
> Weariness restrains the feet,
> And the dying bed is sweet
> To the dying. All in vain
> Frets the busy working brain—
> It will slumber presently,
> It must sleep from toiling free.
> Love will vanish from the heart,
> Hate forget to play its part,
> And the harvest of the mind
> From the book, the field, the wind,
> Housed within us, will be freed:
> Death for Life must sow the seed.
> Time is Life in larger sweep,
> Death is but a longer sleep.

Grote, *Rip Van Winkle: A Sun Myth* (London: Kegan Paul, French, 1881).
95. John G. Milburn, "Recollections of A. R. Grote," *EN* 24 (April 1913), 182–83.
96. Editorial, "Why a New Weekly," *Evolution* (January 6, 1877): 1; and Grote, "The Laborer in Politics," *Evolution* (December 1877): 308–9. The people who wrote for the *Evolution*—Henry Edger, Annie Besant, Alexander Wilder, Augusta Cooper Bristol, Sara Underwood, Caroline Dall, Steven Pearl Andrews, and so on—often wrote for other similar papers (including *Woodhull and Claflin's Weekly*). Many were anarchists, some positivists or theosophists. All followed the banner of evolutionary thinking. Grote did not agree with them on all things, but he shared their commitment to equality and to challenging modern conditions and priorities. See, on these people, William Leach, *True Love*

and Perfect Union: The Feminist Reform of Sex and Society (Middletown, CT: Wesleyan University Press, 1989).

97. Grote, "Protestantism and Science: Part I," *Evolution* (July 1877): 204–5; "Protestantism and Science: Part II," *Evolution* (January 1878): 8–9; and *New Infidelity* (London, 1883). Grote wrote the lead article for the last issue of the *Evolution,* "Does Humanity Need a New Revelation?" No, Grote argued. "The true conclusion is, that there may and must be new revelations, but not supernatural ones, to Humanity," which "will be brought out by continued observation and experience," he wrote. "The only way to 'reconcile' the Bible with Science is to make anything out of anything" or "to twist words until they lose their meaning. Those who think such a course profitable will continue to indulge in it. They will write books to prove that the Biblical chronology holds true for the white race and except the blacks. But it may be said of such workers in any rank that it would have been better for them never to have been born." See *Evolution* 2, no. 2 (1878): 193–95.

98. "Minutes of the Executive Committee," March 26, 1873, Buffalo Museum of Natural Sciences.

99. For contributions by women to the *North American Entomologist,* see, for example, Emily Smith, "Natural History of *Eurra Salicicola,*" the leading piece in vol. 1, no. 6 (1879): 41–42; and "Biological and Other Notes on *Pseudococcus aceris,*" also a leading piece, vol. 1 no. 10 (1880): 73–77.

100. Grote, "On the Species of *Helicopis* Inhabiting the Valley of the Amazon," *Bulletin of the Buffalo Museum of Natural Sciences* 1 (April 1873–March 1874): 106; and for Grote's German work, see, in the same issue of the *Bulletin,* his "Kleiner Beitrag zur Kenntniss einiger Nordamerikanischer Lepidoptera" (A Small Contribution to Knowledge of a Few North American Lepidoptera), 168–70. Strecker described Merian's work as "wonderful" in his *Butterflies and Moths of North America,* 2.

101. *Practical Entomologist* (December 25, 1865): 17; Grote, "A Few Remarks on Silk-Producing Lepidoptera," *Practical Entomologist* (November 27, 1865): 13–14, and (February 26, 1866): 38–39; and Benjamin Walsh, *Practical Entomologist* "Salutatory" (October 1866): 25.

102. Walsh to William Henry Edwards, March 27, 1867, WHE-SA.

103. Grote to Scudder, May 18, 1872, and October 30 and December 20, 1869, SS-BMS.

104. See Grote's essays "Our Genera and the Law of Priority," *CE* (March 1876): 36–37, and "On Jacob Hübner and His Works on Butterflies and Moths," *CE* (July 1876): 131–35.

105. See, on this wide acceptance, Peter Bowler's *Charles Darwin: The Man and His Influence* (Cambridge: Cambridge University Press, 1990).

106. *Evolution* 1, no. 1 (1877): 9.

107. *Evolution* (December 1878): 194; and Grote, *New Infidelity,* 101.

108. Grote, *Checklist of Noctuidae* (1876), 32.

109. See Grote, "The Effect of the Glacial Epoch upon the Distribution of Insects in North America," *CE* (September 1875): 164–67; Grote "On the Insect Fauna of the White Mountains," *Psyche* 1, no. 1 (1875): 76–77. "I am the original propagator of the view of the colonization of Insects on Mountains through the Glacial Epoch," he later explained to a friend (Grote to John Comstock, Octo-

ber 29, 1898, John Henry and Anna Botsford Comstock Papers, 1833–1955, #21-23-25, Division of Rare and Manuscript Collections, Cornell University Library).

110. Grote, "On Genera of the Moths," *CE* 1, no. 6 (1875): 113–15. See also Grote, "Our Genera and the Law of Priority," 36–37; and "On Jacob Hübner and His Works on Butterflies and Moths," 131–35.

111. Grote, *Bulletin* 1 (1873–74): 1–2.

112. For a fine account of this reminiscence, see Ronald S. Wilkinson, "The Genesis of A. R. Grote's 'Collecting Noctuidae by Lake Erie,' " *Great Lakes Entomologist* 7, no.1 (1974): 16–18.

113. Grote, "Notes on Noctuidae," *CE* (October 1877): 196–99.

3. Beating Hearts

1. Alfred Russel Wallace, *The Malay Archipelago* (1880; repr., London: Periphus Editions, 1959), 257–58; and Wallace to Samuel Stevens, *Transactions of the Entomological Society of London* 5 (1858–61): 70. For Conrad's treatment of Wallace, see *Lord Jim* (1900; repr., New York: Oxford World's Classics, 2002), 151–52. For citations of the Wallace experience, see Adolph Portmann, *Tropical Butterflies* (New York: Iris Books, 1945), 7; Miriam Rothschild, *Butterfly Cooing Like Dove* (New York: Doubleday, 1991); and Sharman Apt Russell, *An Obsession with Butterflies* (New York: Basic Books, 2003), vi. In June 1949, the *Lepidopterists' News,* the most influential of all modern butterfly journals, quoted this passage in full "as the best example of enthusiasm in a butterfly hunter"; vol. 3, no. 7, p. 80.

2. Doherty was so aroused by it that he told his mother about it in two different letters. Doherty to his mother, February 20, 1883, and June 27, 1883, JMH. Eugene Pilate to Herman Strecker, June 2, 1875, HS-FM; Henry Edwards, "Notes on Noises Made by Lepidoptera," *Insect Life* 2, no. 1 (1889): 14. James Fletcher to Samuel Scudder, December 19, 1894, SS-BMS; and William Greenwood Wright, "Butterfly Hunting in the Desert," *American Naturalist* 17, no. 4 (1883): 363–69.

3. Arthur Shapiro, e-mail to author, October 28, 2009.

4. George Santayana, *The Sense of Beauty: Being the Outline of Aesthetic Theory* (1896; repr., New York: Dover 1961), 33, 51–54, 100–101. Freud wrote that "psychoanalysis has scarcely anything to say about beauty. . . . All that seems certain is its derivation from the field of sexual feeling. The love of beauty seems a perfect example of an impulse inhibited in its aim. 'Beauty' and 'attraction' are originally attributes of the sexual object." See Sigmund Freud, *Civilization and Its Discontents* (1930; repr., New York: Norton, 1962), 29.

5. Edward O. Wilson, "Paradise Beach," *Naturalist* (Washington DC: Island Press, 1994), 5–15; and Edward O. Wilson, *Biophilia* (Cambridge, MA: Harvard University Press, 1998). See also Roger Scruton, *Beauty: A Very Short Introduction* (New York: Oxford University Press, 2011), 49–66. Scruton's analysis is brilliant. Unfortunately, I read it too late to make adequate use of it in this book.

6. On color and wildness as the basis for attraction to natural beauty, see George

M. Trevelyan's 1931 lecture "The Call and Claims of Natural Beauty," in his *An Autobiography and Other Essays* (London: Longmans, Green, 1949), 92–106; and on early childhood encounters with natural color as the basis for later scientific work, see David Lee, *Nature's Palette: The Science of Plant Color* (Chicago: University of Chicago Press, 2007), ix–xi.

7. This argument again resembles Freud's in his *Civilization and Its Discontents* (see chapter 1).

8. James Tutt, "Address by the Vice-President of the City of London Entomological Society, and Natural History Society," *Entomologist's Record and Journal of Variation* 6 (February 15, 1895): 59–69; and Vladimir Nabokov, *Speak, Memory: An Autobiography Revisited* (New York: Putnam, 1966). For biographical material on Tutt, see Michael A. Salmon and Peter J. Edwards, *The Aurelian's Fireside Companion* (Hampshire, Eng.: Papia Publishing Ltd., 2005), 21–23, 229–41, and 370–74.

9. Augustus Grote, "Collecting Noctuidae by Lake Erie," *Great Lakes Entomologist* 7, no. 1 (1974).

10. On the BAAS, see Harriet Ritvo, *The Platypus and the Mermaid, and Other Figments of the Classifying Imagination* (Cambridge, MA: Harvard University Press, 1997), 51–68; on the AAAS, see Sally Gregory Kohlstedt, et al., *The Establishment of Science in America* (New Brunswick, NJ: Rutgers University Press, 1999), especially chapter 1, by Kohlstedt, "Creating a Forum for Science: AAAS in the Nineteenth Century," 7–49; and on the shaping impact of American entomologists on the AAAS, see W. Conner Sorensen, *Brethren of the Net: American Entomology, 1840–1880* (Tuscaloosa: University of Alabama Press, 1995), 242–52. On the advantages and drawbacks of establishing "agreed-upon norms," see Mary P. Winsor, "Practitioner of Science: Everyone Her Own Historian," *Journal of the History of Biology* 34 (2001): 229–45.

11. William Henry Edwards to Joseph Lintner, January 6, 1877, JL.

12. George Hulst, *Bulletin of the Brooklyn Entomological Society* (May 1880): 4.

13. E. Zimsen, *The Type Material of I. C. Fabricius* (Copenhagen: Munksgaard, 1964); Samuel Scudder, *BEUSC*, vol. 2, p. 1519; and Adalbert Seitz, *The Macrolepidoptera of the World* (Stuttgart: Seit'schen [Kernen], 1906–33), 352.

14. See Friedrich Schnack, *The Life of a Butterfly* (Boston: Houghton Mifflin, 1932), 30, 80; and R. P. Dow, "A Little Journey into Nomenclature," a talk given before the New York Entomological Society, March 3, 1908, in "Minutes of the New York Entomological Society," American Museum of Natural History, New York.

15. See William Holland to Scudder, October 30, 1886, SS-BMS. For this usage in its various forms, see Samuel Scudder, "The Names of Butterflies," in *BEUSC*, vol. 2, 785–88 and 1519; Zimsen, *The Type Material of I. C. Fabricius*; Seitz, *Macrolepidoptera of the World*, 352; Schnack, *Life of a Butterfly*, 30, 80; R. P. Dow, "A Little Journey into Nomenclature"; and Judith Wilson, *Describing Species: Practical Taxonomic Procedure for Biologists* (New York: Columbia University Press, 1999), 153. For a full discussion of current naming protocol, see Wilson, "Naming Species: Etymology," chapter 8 of *Describing Species*, 148–72. For the extensive application of Indian names to skippers, see William Howe, *The Butterflies of North America* (New York: Doubleday,

1975), 453–83, and Robert Pyle, *National Audubon Society Field Guide to North American Butterflies* (New York: Knopf, 1998), 796–826. "Linné [Linneaus]," observed Scudder, "applied the names of the Greek heroes in the Trojan war to a very large number of swallowtail butterflies, and his example has been followed by lepidopterists down to the present day."

16. Herman Strecker to George French, March 20, 1883, GF.

17. George Hulst to Strecker, October 31, 1878, HS-FM. On the "Delilah" moth, see Samuel Cassino, "A New Race of *Catocala delilah* Strecker," *Lepidopterist* (1918).

18. John Morris to Strecker, January 6, 1875, HS-FM. In a later letter, Morris insisted that "a name should express some quality of the object, and not be merely arbitrary"; Morris to Strecker, June 7, 1878, HS-FM.

19. Hulst to Strecker, November 10, 1875, HS-FM.

20. Charles Dury to Strecker, September 3, 1876, HS-FM.

21. Arthur Fuller to Strecker, June 27, 1875, HS-FM.

22. See Henry Edwards section on butterflies written for *The Standard Natural History*, ed. John Sterling Kingsley (Boston: S. E. Cassino, 1884), 494.

23. Hermann Burmeister, *A Manual of Entomology* (London, 1836), 624–32; originally in German.

24. The best discussion of the excesses of nineteenth-century naming is Ritvo's *The Platypus and the Mermaid*, 51–84.

25. On the Western imperialist side of naming, see ibid.

26. Emily Morton to Strecker, February 20, 1879, and William Holle to Strecker, November 2, 1876, HS-FM; and David Bruce to John Smith, December 10, 1888, Department of Entomology, Smithsonian Institution Archives, Record Unit 138, United States National Museum, Division of Insects, Correspondence, Washington, DC.

27. Strecker, *LRH*, 101–2.

28. See Edward Doubleday and John O. Westwood, *The Genera of Diurnal Lepidoptera; Comprising Their Characters, A Notice of Their Habits and Transformations, and a Catalogue of the Species of Each Genus* (London, 1846–52).

29. Edwards to Scudder, October 14, 1885, SS-BMS.

30. Edwards, *Catalogue of the Lepidoptera* (Philadelphia: American Entomological Society, 1877), 2; and, in another version of the catalog, "Catalogue of the Diurnal Lepidoptera of America North of Mexico," *Transactions of the American Entomological Society* 6 (1877): 1–67.

31. William H. Edwards, "Notes on Entomological Nomenclature," *CE* (May 1876): 81–94; *CE*, 113–19; and Fred Tepper to Strecker, May 16, 1876, HS-FM.

32. See Humboldt's introduction to *Cosmos* for diverse readings of "the face of Nature": volume 1 (1845; repr., Baltimore: Johns Hopkins University Press, 1997), 25–26.

33. *The Portable Darwin*, edited and with an introduction by Duncan M. Porter and Peter W. Graham (New York: Penguin, 1993), 151.

34. Scudder to Edwards, March 13, 1875, WHE-SA.

35. William Henry Edwards to Henry Edwards, June 22, 1876, HE; William Henry Edwards to Mead, February 4, 1877, TM; and William Henry Edwards to Lintner, February 6, 1873, JL.

36. Edwards, "Catalogue of the Diurnal Lepidoptera of America North of Mexico," 2–3. At the time of writing this letter, Edwards had just finished Haeckel's *History of Creation.* "Haeckel and Darwin," he wrote, "both expressly lay down the 'breeding line' as the origin of a 'good species.' " He loved the Haeckel book because it issued "the strongest affirmation of Darwin" and "bears awfully hard on Agassiz."

37. William Henry Edwards to Henry Edwards, March 8, 1876, HE. On Scudder's classification of lycaenids and swallowtails, see William Henry Edwards, "Notes on *Lycaena pseudodargiolus* and Its Larval History," *CE* 10 (January 1878): 1–14. For Edwards's attack on this argument, see especially the long footnote on p. 2.

38. Edwards to Lintner, February 6, 1873, JL. Here Edwards affirms the field over the laboratory.

39. Selim Peabody, *Cecil's Book of Insects* (Philadelphia: Claxton, Remsen, and Haffelfinger, 1872), 198; and S. H. Peabody to William Henry Edwards, December 15, 1875, and July 27, 1876, WHE-SA.

40. "Mr. Scudder's Butterflies," *CE* (December 1881): 246–50.

41. Kohlstedt et al., *The Establishment of Science,* 37; and Scudder, "Fossil Butterflies," in *Memoirs of the American Association for the Advancement of Science* (Salem, MA, 1875).

42. Edwards to Lintner, April 3, 1878, JL.

43. Edwards to Scudder, April 30, 1877, SS-BMS.

44. William Henry Edwards to Henry Edwards, November 16, 1876, HE.

45. See William Swainson, *Taxidermy with the Biography of Naturalists* (London, 1840), 8–12. Swainson mentions as standard "the fly net, elastic net, bag-net, hoop-net, landing-net, forceps, and digger" (8).

46. Theodore Mead to George Dodge, December 5, 1870, TM; and Mead to John Ridings, June 1, 1872, TM. On the invention of the net, see David Elliston Allen, *The Naturalist in Britain* (1976; repr., Princeton, NJ: Princeton University Press, 1994), 4.

47. Mead to Willie Edwards, March 18, 1870, Mead Letterbook, TM; and Mead to Arthur Whittemore, March 7, 1870, TM. On Schickel's and Akhurst's businesses, see Mead to Edward Nelson, August 22, 1871; Mead to Akhurst, May 28 and August 22, 1871; and Mead to Akhurst, March 22, 1874, TM. On "camping equipage," such as tents and hammocks, see *Psyche* 1, no. 18 (1876): 180–81.

48. George Crotch to Henry Edwards, March 5 and November 20, 1873, HE.

49. On the urban-based Romantics and their influence, see especially Keith Thomas's groundbreaking *Man and Nature* (New York: Oxford University Press, 1984), chapters 4–6; Anne Larsen Hollenbach, "Of Sanfroid and Sphinx Moths: Cruelty, Public Relations, and the Growth of Entomology in England, 1800–1840," *Osiris,* 2nd ser., 11 (1996): 201–20; and Kathryn Shevelow, *For the Love of Animals: The Rise of the Animal Protection Movement* (New York: Holt, 2008).

50. *New Checklist of North American Moths* (New York: D. Appleton, 1882).

51. Grote, "Moths and Moth-Catchers: Part II" *Popular Science Monthly* (July 1885): 377–89.

52. Grote, *The Hawk Moths of North America* (Bremen, 1886), 7.

53. Herman Strecker, *Butterflies and Moths of North America: Complete Synonymical Catalogue of Macrolepidoptera* (Reading, PA: B. F. Owen, 1878), 8–9.
54. Andrew Foulks to Herman Strecker, no date, but probably from the 1880s, HS-FM.
55. Grote, "Moths and Moth-Catchers," 377–89.
56. Eliza Fales to her sister, February 14, 1835, Fales Family Papers, Fales Library, New York University, New York, NY.
57. John Adams Comstock, "Presidential Address to the Twelfth Annual Meeting of the Lepidopterists' Society," *Journal of the Lepidopterists' Society* 16, no. 1 (1962): 248.
58. Grote, preface to *An Illustrated Essay on the Noctuidae of North America* (London, 1882), 22; and "Moths and Moth-Catchers," 377–89.
59. Alpheus Packard, letter to his father, c. 1862, in J. Walters Wilson's biography of Packard (Alpheus Packard Papers, Bowdoin College, Brunswick, ME). On Strecker's debt to Bridgham, see his portrait of *Catocala tristus* in *LRH*, 17, 37.
60. Alpheus Packard, *Guide to the Study of Insects* (New York, [1869] 1876), 239.
61. Neumoegen to Strecker, April 7, 1877, HS-FM.
62. Tepper to Strecker, April 25, 1876, HS-FM.
63. William Henry Edwards to Henry Edwards, June 5, 1872; March 8, 1874; and November 7, 1873, HE.
64. Grote to Strecker, March 17, 1873, HE; "Minutes of the Buffalo Society of Natural Sciences," Buffalo Society of Natural Sciences, May 2, 1873; and Strecker to Grote, March 27, 1873, Philipp Zeller Papers, NHM-LONDON. This letter was sent to Strecker, but a copy of it cannot be found in Strecker's papers in the Field Museum. Perhaps it was destroyed. Grote, however, retained it and mailed a copy to Zeller as part of his proof that Strecker was a thief.
65. *Bulletin,* vol. 1 (April 1873 to March 1874): 5–8, 23.
66. Strecker to Grote, March 27, 1873, Philipp Zeller Papers, NHM-LONDON.
67. Arthur Fuller to Strecker, December 9, 1873, HS-FM.
68. Roy Rosenzweig and Elizabeth Blackmar, *The Park and the People: A History of Central Park* (Ithaca, NY: Cornell University Press, 1992), 351–57.
69. Strecker to Grote, draft letter, April 15, 1873, HS-FM.
70. Grote to W. V. Andrews, November 13, 1873, HS-FM.
71. Grote to Zeller, March 16, 1874, Philipp Zeller Papers, NHM-LONDON.
72. William Hewitson to Strecker, August 28, 1875, HS-FM.
73. Grote, *An Illustrated Essay on the Noctuidae of North America* (London, 1882); "Moths and Moth-Catchers: Part II," 377–89; and *Checklist of the Noctuidae of America, North of Mexico* (Buffalo, 1876), 35.
74. Strecker to Mead, January 23, 1887, and May 1, 1881, TM.
75. "I doubt," Strecker argued, "if there can be much variation in the imago unless it existed in the earlier stages. Too much stress by far is laid on the circumstance of whether the larva differs or not from that of the ordinary form. If this were so conclusive, why is it then that the green and brown larvae of *Cer. Imperialis,* both bring precisely the same form of moth, or the tawny and green larvae of *Thyreus Abbotii,* produce the same results? No; if we have a varietal form or subspecies in the last stage of the insect we must just as reasonably expect to find it in the earlier stages." See his *Butterflies and Moths of North America,*

pp. 118, 155, and 158. Strecker immediately followed this observation with a very bad analogy: "Is the Albino off-spring of negro parents black when a child or with black or brown eyes? Certainly not; as an infant it has the same abnormal white cuticle to its body. . . . Again, would the child born with six toes and fingers on each foot or hand have but five to each extremity on attaining maturity?" Strecker's obvious mistake here is to equate childhood with the larval stage of butterflies. He pursued these anti-Edwards themes in his descriptions in *Proceedings for the Davenport Academy of Natural Sciences* in 1878. See his "Descriptions of Some Species and Varieties of North American Heteroceres, Mostly New," vol. 2 (1876–78): 270–78.

76. Strecker, *LRH,* 66–67, 84–85.

77. Ibid., 118–20, 53. Many other naturalists here agreed with Strecker, despite whatever else they thought of him.

78. William Henry Edwards to Joseph Lintner, November 7, 1878, JL.

79. Morris to Strecker, August 1878 (no day), HS-FM.

80. Edward Graef to Strecker, July 23, 1878; and Tepper to Strecker, August 19, 1878, HS-FM.

81. Lintner to Strecker, May 23, 1873, and July 16, 1877, HS-FM.

82. Hulst to Strecker, October 30, 1879, HS-FM.

83. Arthur Fuller to Strecker, May 4, 1874, HS-FM; on "cutting stone" as a dead end, Fuller to Strecker, May 29, 1874, HS-FM.

84. Mary Putnam to Strecker, January 12, January 17, and February 9, 1878, HS-FM.

85. Berthold Neumoegen to Strecker, April 2, 1877, HS-FM.

86. Heinrich Ribbe to Strecker, February 10, 1878, HS-FM.

87. Neumoegen to Strecker, April 28, 1879, HS-FM. On Doll, see George P. Engelhardt, "Chapters from the Life of a Butterfly Collector," *Brooklyn Museum Quarterly* 12, no. 4 (1925): 171–78.

88. Neumoegen to Strecker, April 28, 1879, and January 27, 1880, HS-FM.

89. When the Streckers failed to arrive for Christmas in 1876, the stockbroker wrote the stone carver, "I only wish you could be here, and I would really make it merry for you," by which he meant the wine and butterflies. See Neumoegen to Strecker, November 11 and December 22, 1876, HS-FM.

90. Neumoegen to Strecker, January 24, 1877, HS-FM.

91. In December 1877, he invited Strecker to his baby son's bris, or to "the circumcision of my boy. When may I expect you here?" Neumoegen wrote. See Neumoegen to Strecker, December 8, 1877, and January 29, 1878, HS-FM.

92. Strecker to O. H. Staudinger, January 27, 1874, HS-FM.

93. Strecker to Duncan Putnam, June 10, 1876; June 7, 1877; and January 14, 1878, DP.

94. Neumoegen to Strecker, February 8, 1877, HS-FM.

95. Neumoegen to Strecker, January 31, 1877, HS-FM.

96. See Charles Fernald's testimony to this effect, Fernald to Zeller, January 11, 1878, Zeller Papers, NHM-LONDON.

97. Titian Peale to Strecker, December 20, 1873; and Morris to Strecker, November 1879 (specific day unclear), HS-FM.

98. See letters from Strecker to George French, an Illinois naturalist, in which Strecker thanks French for his various payments: $8.50 in February 1878, $15

on July 1, 1878, $14 in June 1879, and so forth. Strecker to French, February 8 and July 20, 1878, and June 17, 1878, GF; and *The Naturalists' Directory*, ed. Samuel Cassino (Boston, 1886), 154.

99. William Henry Edwards to Henry Edwards, March 10, 1873, HE.

100. William Henry Edwards to Scudder, September 4 and 17, 1874, SS-BMS.

101. This history was laid out for me by John Willis, a direct descendant of Edwards, in an interview in the old Edwards house, May 16, 2002. See also Edwards to Scudder, November 12, 1874, SS-BMS.

102. Grote to Judge Clinton, May 20, 1880, George William Clinton Papers, Buffalo Museum of Natural Science.

103. Grote to Clinton, May 20 and 23, 1880, Clinton Papers.

104. In an 1875 letter to Strecker, the taxidermist John Akhurst echoed what both Edwards and Scudder believed. "In speaking of Grote, I hit him pretty hard on self conceit and vanity," John Akhurst (HS-FM).

105. Grote, *Checklist of the Noctuidae of America,* 35.

106. Grote to Strecker, November 6, 1875, and April 3, 1878, HS-FM.

107. W. V. Andrews to Strecker, October 12, 1876, HS-FM.

108. Preston Clark to William Schaus, June 22, 1916, Papers of B. Preston Clark, 1999-1, Carnegie Museum of Natural History Archives, Pittsburgh, PA.

109. For Barnes's quote, see Jeanne Remington's biography of Barnes, under "Brief Biographies," in the *Lepidopterist News* (April–May 1949): 54; and for the Clark citation, see Clark to William Schaus, June 22, 1916, B. Preston Clark Papers.

110. Strecker to Henry Skinner, June 7, 1900, HS-ANS.

111. Hulst to Henry Edwards, October 28, 1880, HE. The term "déclassé naturalist" is Harriet Ritvo's. See *The Platypus and the Mermaid*, 65.

112. See Levi Mengel to Henry Skinner, February 4, 1902, HS-ANS. In this letter, Mengel, Strecker's neighbor and also a butterfly enthusiast, explained how Strecker came to own his home: "You know Drexel gave him his old house, without security for the money advanced and was his friend until death." On Drexel's help with butterflies, see Drexel to Strecker, December 1, 1871, HS-FM. Drexel wrote: "I enclosed 18 pounds on our London House—this is payable to your order and I have endorsed it Pay Dr. Otto Staudinger on order. You must write your name under this when I have written it in pencil. By so arranging this will be a receipt and in case of any dispute the draft can be produced and will thus show the Doctor got the money from you."

113. Grote to Harrison Dyar, July 18 and September 6, 1898, HD.

114. Scudder, *Butterflies,* 242–43. I must say, however, that Scudder, during the time he spent under Agassiz, had already held something of this view. "My 'love for a collection,' " he told a European specialist in that year, "is only such that it may subserve the best interests of science and advance my own knowledge of its contents." See Scudder to Carl Robert Osten-Sacken, January 1, 1860, Samuel Hubbard Scudder Papers, EML.

115. Henry Edwards to Hermann Hagen, February 10 and March 18, 1880, Letters to Hermann Hagen, EML.

116. Henry Edwards to Hermann Hagen, February 8, 1881, EML.

117. Henry Edwards also made a point of discussing Charles Ishikawa at a meeting of the club, noting that he was "a most earnest student not only of Entomol-

ogy alone but of other branches of Natural History" ("Records of the New York Entomological Club, December 10, 1881, American Museum of Natural History). Edwards's singling out of Ishikawa is interesting. Was Ishikawa, in fact, Edwards's manservant who had been with him since Edwards had lived in San Francisco and whom Edwards had helped train as a butterfly expert? Was this man "Charlie"? The noted entomologist Herbert Osborne, in a sketch of Edwards in his history of entomology, argued that Charlie was definitely Japanese. As a young man, Osborne wrote an article for *Papilio,* so he must have known. But other evidence—James Behrens to Edwards, for instance—suggests he was Chinese (although, perhaps to Behrens, all Asians were Chinese or "Charlies"). The evidence points to Osborne.

118. William Henry Edwards, "On the American Form of *Papilio machaon,* Linn.," *Papilio* (May 1882): 74–77. Edwards tried to naturalize the original European form in Coalburgh. He released a hundred butterflies sent to him in chrysalid form by Theodore Mead from Europe. The experiment failed.

119. "Comments on Dr. Hagen's Paper in Nov.-Dec. No. of *Papilio,* on *P. machaon,* Etc.," *Papilio* 3, no. 3 (1883): 45–57.

120. On Henry Edwards's policy of giving out free copies, see Henry Skinner to Herman Strecker, June 9, 1891, HS-FM.

121. He had to "start all over from the lowest level of a broker," he told Strecker (Neumoegen to Strecker, April 20, 1882, HS-FM).

122. William Henry Edwards to William Greenwood Wright, May 16, 1883, WGW.

123. Strecker to George French, June 6, 1881, and July 18, 1883, GF.

124. One such effort was launched by Theodore Mead, probably with the urging of William Henry Edwards, in which he criticizes Strecker for his sloppy descriptions. See Mead, "*Limenitis eros* Versus Var. *floridensis,*" CE (April 1881): 79–80.

125. Neumoegen to Strecker, December 7, 1880, HS-FM.

126. Neumoegen to Strecker, March 9, 1882, HS-FM.

127. Quoted by Sigmund Freud in *Civilization and Its Discontents,* in the footnote on p. 57.

128. "New Moths from Arizona," *Papilio* 1, no. 9 (1881): 153–68. On Grote's attacks on Strecker as a cause for the collapse of *Papilio,* see also Harry B. Weiss, "The New York Entomological Club and 'Papilio,'" *New York Entomological Society* 56 (June 1948): 132–33.

129. Grote to A. Gunther, August 10, 1881; July 21, 1881; June 5, 1881; and April 2, 1881, NHM-LONDON; Grote to Arthur Butler, November 11, 1881, NHM-LONDON; and Grote to Hermann Hagen, April 29, 1880, Letters to Hermann Hagen, EML. On his proposals to Central Park and to the Museum of Comparative Zoology, see Grote to Hagen, April 29, 1880, EML, and Hagen to Henry Edwards, February 12, 1880, HE.

130. For a description of the collection, see Grote to Arthur Butler, February 25, 1881, NHM-LONDON.

131. Grote to Henry Edwards, Spring 1883; June 12, 1883; and June 15, 1883, HE.

132. *Papilio* 2, no. 5 (1882).

133. Samuel Aaron, "Notes and Queries," *Papilio* 4, no. 2 (1884): 42. For Strecker's article, see "*Citheronia infernalis* and *Catocala babayaga,* New Species," *Papilio* 4, no. 4 (1884): 73–74. Strecker described *C. infernalis* "from one female in my collection (!)" noting, "The male I was unable to examine, but was informed

it resembled the female." In an earlier February 1884 piece, Aaron had tried feebly to do William Henry Edwards some justice, by rehashing the question of *priority* that had appalled Edwards in the late 1870s. At that time, both Edwards and Strecker had received the same butterflies from the Texas collector Jacob Boll. Edwards published the description first in several journals, publishing being the *requirement* for taking credit for any described species. Strecker published them later in his own catalog, although to give the impression that he had published them earlier than Edwards, he inserted another date—September 1877—in his catalog. Aaron's article confirmed this maneuver by Strecker and must have much satisfied Edwards. Then Aaron negated what he had done by inviting his brother to attack Edwards.

134. To disprove this idea, Aaron quoted from one of Edwards's key theoretical sources, August Weismann, who argued in 1882 that the larval characters were too unstable for secure classification and that the adult imago should always remain the basis for specific identity. At the same time, Aaron's brother announced the publication of a new journal—*Entomologica Americana*—to be edited by John Smith and published by the Brooklyn Entomological Society (established in 1876), whose members—Hulst, John Smith, Edward Graef, Franz Schaupp, and others—had long felt estranged from the Manhattan naturalists.

135. William Henry Edwards to Henry Edwards, June 10, 1881, HE.

4. Word Power

1. William Henry Edwards to Hermann Hagen, May 16, 1871, Letters to Hermann Hagen, EML.
2. William Henry Edwards to Samuel Scudder, March 1, 1893, SS-BMS.
3. For a brief biography, see Arnold Mallis, *American Entomologists* (New Brunswick, NJ: Rutgers University Press, 1971), 52–54.
4. William Henry Edwards, *Autobiographical Notes* (privately printed, 1901), 220–21.
5. Joseph Lintner, "Report of the State Entomologist for the Year 1888," in New York State, *42nd Annual Report of the Trustees of the State Museum of Natural History* (Albany, 1888–89), 185–86.
6. According to Tim L. McCabe of the New York State Museum, Karner was the "site of intensive collecting by museum entomologists" in the late 1890s, but as Lintner observed, it had already had such a reputation in the 1860s. See Lintner, "On Lycaena Neglecta," *CE* (July 1875): 122; Tim L. McCabe, "The Karner Blue Butterfly," *Legacy: the Magazine of the New York State Museum* (June 2010): 16; and William Henry Edwards, under "Lycaena, II, III," *BNA*, vol. 2.
7. Lintner, *Entomological Contributions* (Albany, 1872), 46–47.
8. "List of Lepidoptera," in New York State, *Thirtieth Report of the State Museum of Natural History* (Albany, 1879), 141.
9. Eugene Pilate, "List of Lepidoptera Taken in and Around Dayton, O.," *Papilio* 2, no. 5 (1882): 65–71, and Charles Fernald, *Butterflies of Maine* (Augusta, 1884). On the moth catalogs and biological material on Fernald, see Mallis, *American Entomologists*, 141–50. Unillustrated and mostly for students and farmers, Fernald's lists lacked the color and life of the best natural history writ-

ing but contained much information about species never before presented. And Pilate's list was, well, a list.

10. "A List of the Butterflies of Philadelphia, PA," *CE* (August 1889): 145–53; and, on Skinner's rearing practices, see Henry Skinner to Herman Strecker, March 9, 1883, HS-FM.

11. On Hoy's catalogs, see William A. Field et al., *A Bibliography of the Catalogs, Lists, and Faunal and Other Papers on the Butterflies of North America and Mexico, Arranged by State and Province* (Washington, DC: Smithsonian Institution Press, 1974), 94. On Hoy's *CE* list, see October 17, 1884, pp. 199–200.

12. Arthur Shapiro, e-mail to author, November 17, 2010.

13. See Henry Edwards, "Moths and Butterflies," in *The Standard Natural History,* ed. John Sterling Kingsley, vol. 2, *Crustacea and Insects* (Boston: S. E. Cassino and Co., 1884), 435–502; Henry Edwards to Philipp Zeller, February 10, 1870, Philipp Zeller Papers, NHM-LONDON.

14. For William Henry Edwards on Cassino, see William Henry Edwards to Henry Edwards, May 22, 1886, HE; and Samuel Cassino to Henry Edwards, April 13, 1886, HE.

15. On Comstock, see Pamela Henson's essays "The Comstock Research School in Evolutionary Entomology," *Osiris* (1993): 159–77, and "The Comstocks at Cornell," in *Creative Couples in the Sciences,* ed. Helena M. Pycior et al. (New Brunswick, NJ: Rutgers University Press, 1996), 112–25. See also Henson's dissertation, "Evolution and Taxonomy: J. H. Comstock's Research School in Evolutionary Entomology at Cornell University, 1874–1930" (University of Maryland, College Park, 1990), pp. 124–25, 250–53.

16. On Linnaeus, see Lisbet Koerner, *Linnaeus: Nature and Nation* (Cambridge, MA: Harvard University Press, 1999), 40.

17. Ferdinand Ochsenheimer's 1804 bibliography of such books, "Sachsens entomologische Litteratur," gives a wonderful and authoritative glimpse into the extent of this literature. Ochsenheimer was one of the great early insect systematists. See his *Die Schmetterlinge Sachsens* (Dresden, 1805), 1–40. See also Jacob Hübner, *Beiträge zur Geschichte der Schmetterlinge* (Augsburg, 1786–89, 1790); Abel Ingpen (England), *Instructions for Collecting, Rearing, and Preserving British Insects* (London, 1827); and Achille Deyrolle (France), *Guide du jeune amateur de coléoptères et de lépidoptères* (Paris, 1847). For later little guides (which many Germans brought with them to America), see H. Rocksroh's *Anweisung, wie Schmetterlinge, gefangen, ausgebreitet, geordnet, bewohnt, und wie ihre Raupen und Puppen erkannt werden* (Leipzig, 1833) and *Hermanus Raupen und Schmetlerlingsjäger* (Leipzig, 1877). The British and Germans invented the leading practices and much of entomological gear. On the British contribution, made in the first half of the nineteenth century, see Michael A. Salmon, *The Aurelian Legacy: British Butterflies and Their Collectors* (Berkeley: University of California Press, 2001), 1–117, and David Allen, *The Naturalist in Britain* (Princeton, NJ: Princeton University Press, 1994), 83–107. The English imported what may have been the first butterfly net, a "bag net" from Holland (or Germany). See Salmon, *Aurelian Legacy,* 69.

18. Stainton's *Manual of British Butterflies and Moths* (London, 1857), aimed to get as much information as possible to as many people as possible, in the smallest size and at the cheapest price possible. He omitted all "hairsplittings" (as

he put it), Latin names, and synonymy, arguing that "those who collect insects, and who do not wish to be utterly 'isolated,' must learn to call [butterflies] by names by which other people will know them." See Stainton, vii. On Stainton, see Salmon, *The Aurelian Legacy,* 162–64.

19. Theodore Cocherell, "American Moths," *Dial* (January 16, 1904): 41–42.
20. William Walters to Strecker, August 27 and September 7, 1880, HS-FM.
21. Helen S. Conant, *The Butterfly Hunters* (New York, 1881), 139.
22. Julia Ballard, *Moths and Butterflies* (1880; repr., New York, 1889), 1.
23. Ibid., xxxiii–xxxiv.
24. George H. Hudson (of Plattsburgh, Pennsylvania) to John H. Smith (curator of entomology at the Smithsonian), September 24, 1886, Smithsonian Institution Archives, Record Unit 138, National Museum of Natural History, Division of Insects, Correspondence, Washington, DC.
25. William Henry Edwards to George French, February 20, 1886, GF. Edwards even helped advertise the book, after French sent him several circulars for that purpose. "I have distributed about all the circulars," he wrote French. "As I write letters to new parties I enclose one. I also sent out several to parties who were not likely to hear of the book at once. You may as well send me another score." For Edwards's praise to friends, see Edwards to Lintner, March 7, 1886, JL.
26. Scudder, "A Manual of North American Butterflies," *Science* 8, no. 194 (1886): 378.
27. Alpheus Packard, "Introductory," *American Naturalist* (March 1867): 4.
28. Augustus Grote, *The Hawk Moths of North America* (Bremen, 1886), 17; and on the subsidizing of *CE*, see Henry Skinner to Strecker, June 9, 1891, HS-FM.
29. Scudder, "Salutatory," *Science* (July 3, 1880): 6.
30. On the poor *Science* subscriptions as a cause of failure, see J. S. Kingsley's obituary of Scudder in *Psyche* 18 (1911): 175. *Science* "was ably edited," Kingsley wrote, "and rejoiced the hearts of scientific men of the day," but "there were not subscribers enough to pay the expenses and no thought of the later expedient of making it an organ to some large association."
31. Henry Edwards, "Notes on Noises Made by Lepidoptera," *Insect Life* 2, no. 1 (1889): 11–15 (Edwards mentions the Swinton piece at the outset); and Lord Walsingham, "Steps Toward a Revision of Chambers' Index, with Notes and Descriptions of New Species," *Insect Life* 2, no. 1 (1889): 51–54.
32. On the membership of the Brooklyn Entomological Society, see Edward Graef, "Some Early Brooklyn Entomologists," *Bulletin of the Brooklyn Entomological Society* 9, no. 3 (1914): 55–56.
33. William Henry Edwards to Henry Edwards, January 26, 1885, HE.
34. For a description of the farm, see William Schaus to Henry Skinner, September 11, 1916, HS-ANS.
35. Skinner to Strecker, March 31, 1896, HS-FM.
36. Skinner to Strecker, June 9, 1891, and January 30, 1897, HS-FM.
37. Skinner, *EN* (January 1897): 10; for a biographical sketch, see Mallis, *American Entomologists,* 322–33.
38. Flyer advertising the *Entomological News,* December 1, 1889, appended to letter to Strecker, HS-FM.
39. Grote, *The Hawk Moths of North America,* 58.

40. C. J. [Charles Johnson] Maynard, *The Butterflies of New England* (Newton-ville, MA: C. J. Maynard, 1886), iii–iv. Maynard published a revised edition in 1891, which added only an appendix with accompanying species described and "two hand-colored plates" of his own, pp. 67–78. On Maynard's reputation as a bird naturalist, see Mark V. Barrow Jr., *A Passion for Birds* (Princeton, NJ: Princeton University Press, 1998), 33–34. See also Maynard, "A Catalogue of the Birds of Loos, New Hampshire and Oxford Co., Maine," *Proceedings of the Boston Society of Natural History* 14 (1870–71): 365–85.

41. Maynard, "A Catalogue of the Birds of Loos," 1–2.

42. On Maynard's mounting of birds at the Boston Society of Natural History, see Barrow, *A Passion for Birds,* 33.

43. William Henry Edwards, *CE* (January 1887): 34–40.

44. Grote to Scudder, April 16, 1886, SS-BMS.

45. He started visiting the Bremen Museum of Natural History, doing what research he could there on American moths. He read philosophy and literature; he composed music, especially band music, with performances in Hamburg, Hannover, and Kassel. "That is doing pretty well in this musical country where competition is ruinous," he wrote a friend. His reputation as both a humanist and a naturalist flowered; in 1886, the Duke of Saxe-Coburg-Gotha awarded him a silver medal, the Princeps Musarum Sacerdos for Art and Science. See Grote to Charles Fernald, May 8, 1886, Charles Henry Fernald Papers (RG 40/11 C. H. Fernald), Special Collections and University Archives, University of Massachusetts, Amherst; on the medal, see *CE* (December 1886): 240.

46. Grote to William T. Davis, August 18, 1886, and April 2, 1887, William T. Davis Papers, Staten Island Museum History Archives and Library, Special Collections. In an otherwise scientific piece on "representative species," he invoked the "wanderings of Ulysses," an obvious reference to himself (*CE* [July 1888]: 177). "I am writing this in my European exile," he wrote in *The Hawk Moths of North America*, p. 13. He was critical, too, of German natural science for "its failure to generalize" and to show how "facts bear on each other, as Darwin had done." (But what of Weismann, Haeckel, and the others?) "The German mind is too apt to be satisfied with the mere accumulation of learning," he averred. American natural science was better, more advanced, or so he seemed to believe. But time and again, despite his efforts, he could not get a job in the United States; nor were there any offers. "Everywhere I hear there is no place for me, which is funny, the continent is so big," he wrote. "The crowning ambition of my life," he told Scudder in 1887, "was to be elected" to the National Academy of Sciences in Washington, and even though he made many overtures to friends (including to Scudder) to help him get elected, nothing ever came of it. See Grote to Davis, April 2, 1887, Davis Papers. There are reasons to believe that had he worked hard enough at it and courted enough people, he could have come back to a good job and acclaim, but the main obstacle had little or nothing to do with things in the United States: Grote's father-in-law feared that his daughter, Minna, and her children would suffer from the instability of living there; he hated the prospect of not seeing them and probably threatened to cut off Minna's inheritance, thereby impoverishing Grote.

47. Grote, *The Hawk Moths of North America*, 5, 60–61, and *An Illustrated Essay on the Nocturidae of North America* (London, 1882), 8.

48. *Popular Science Monthly* 22 (June 1885): 246–52, and 28 (July 1885): 377–89.
49. Grote, *CE* (January 1895): 1–6.
50. Grote, *CE* (June 1890): 108.
51. Grote, *The Hawk Moths of North America*, 6.
52. Strecker, *Butterflies and Moths of North America: Complete Synonymical Catalogue of Macrolepidoptera* (Reading, PA: B. F. Owen, 1878). The book was printed on the press of B. F. Owen in Reading and appears to have been published by Strecker himself. Although the print run is unknown, there were several copies in circulation.
53. Strecker, *LRH* (Reading, PA, 1872–78), 85.
54. Ibid., 78–79.
55. Ibid., 25–26.
56. The Whitman and Thoreau quotations can be found in Robert Finch and John Elder, eds., *The Norton Book of Nature Writing* (New York: Norton, 2002), 217, 229.
57. On this period, see William Henry Edwards to Scudder, January 8, 1889, SS-BMS; and on insects dying in transit, see Edwards's discussion "on transportation of eggs and young larvae" in his article "Notes on Certain Butterflies, Their Habits, Etc.," *CE* (February 1882): 24.
58. For a contemporary account of the impact of the railroad on natural science and collecting in the 1880s, see E. A. Schwarz, "Entomological Club of the American Association for the Advancement of Science—Annual Address of the President," *CE* (September 1892): 213–24.
59. William Henry Edwards, "Notes on Butterflies, with Directions for Breeding Them from the Egg," *CE* (May 1884): 81–89. By this time, Edwards had acquired great skill as a breeder. In the past, many of his caterpillars had escaped from his containers or died as a result of mold, shifts in temperatures, or the impact of smoke in the greenhouse. By 1880, however, he had learned to secure his larvae "within cylinders of fine wire set over" the food plants and "deep enough in the earth to prevent escapes." He had also discovered the necessity of "fresh air and moisture" for all his larvae. See "Argynnis VII," in *BNA,* vol. 2. "The life cycles of some of our common species are known" only from the eggs and larvae "described by William Henry Edwards a century ago," according to Thomas Emmel. See Thomas Emmel, Boyce A. Drummond, and Marc C. Minno, *Florissant Butterflies* (Stanford, CA: Stanford University Press, 1992), 13.
60. Edwards to Scudder, September 18, 1885, SS-BMS; Edwards to Bean, June 30, 1884, TB; and Edwards to Lintner, July 21, 1880, and July 23, 1889, JL.
61. It was not until 1914, with the publication of F. W. Frohawk's *Natural History of British Butterflies,* that the British would have their own catalog of butterflies, with a "complete series of drawings—of truly remarkable excellence of every phase of the life cycle of all our sixty-eight British butterflies," as Walter Rothschild put it in the preface to Frohawk's book. Frohawk, one of the leading English entomologists of the twentieth century, was wrong, however, to claim in his book that "before the year of 1898 it was the general belief that butterflies as a rule would not lay their eggs in captivity. Many entomologists also believed that in the case of these insects, each individual produced only a few eggs. . . . As a result knowledge of the earlier stages of many species was fragmentary and superficial." Apparently, he knew nothing about Edwards or

Scudder. See Frohawk, *Natural History of British Butterflies* (London, 1914). For biographical material on Frohawk, see Salmon, *The Aurelian Legacy*, 194.

62. Scudder, review of *BNA*, *Science* 11 (1888): 339a.

63. Fordyce Grinnell Jr., "The Work of W. H. Edwards," *Lepidopterist* 1, no. 12 (1917): 92.

64. Alexander B. Klots, *A Field Guide to the Butterflies* (Boston: Houghton Mifflin, 1951), 278.

65. Edwards, "Obituary," *CE* (1888): 140; *Autobiographical Notes*, 279; and Edwards to Scudder, October 14, 1885, SS-BMS.

66. Edwards to Henry Edwards, April 7, 1876, HE.

67. William Henry Edwards to Theodore Mead, April 11, 1884, TM.

68. William Henry Edwards to Scudder, October 14, 1885, SS-BMS.

69. William Henry Edwards to Scudder, December 20, 1892, SS-BMS.

70. William Greenwood Wright, "A Naturalist in the Desert," *Overland Monthly Magazine* (September 1884): 4.

71. Grace Wadleigh to J. Gerould (Dartmouth College), February 14, 1932, TB; and, on David Bruce, see Joseph and Nesta Dunn Ewan, *Biographical Dictionary of Rocky Mountain Naturalists* (Boston: W. Junk, 1981), 30.

72. William Barnes to Foster Benjamin, September 13, 1907, William Barnes Correspondence, Records of the Bureau of Entomology and Plant Quarantine, Record Group 7, National Archives, College Park, Maryland.

73. See Thomas Bean, quoted in Edwards, "Argynnis VI" and "Argynnis VII," *BNA*, VIII (unpaginated).

74. David Bruce to Henry Edwards, January 31, 1885, HE; on Bruce's work for Darwin and Doubleday, see Bruce to Henry Skinner, April 9, 1891, HS-ANS. In 1870, he sailed to Paris to see his youngest brother, then head gardener at the Château de Chantilly, the property of the Duc de Aumale. Together they got trapped in the cross fire of the Prussian invasion of that city. On "butterflying," see Bruce to Strecker, January 22, 1883, HS-FM; on the neighborhood "lads," see Bruce to Henry Edwards, December 23, 1884, HE.

75. Bruce to John Smith, December 6, 1888, Division of Insects Correspondence, Smithsonian Archives; William Henry Edwards to Bean, June 28, 1889, TB; and Thomas Bean to Edwards, March 17, 1890, TB.

76. William Henry Edwards to Bean, May 9, 1875, TB.

77. William Henry Edwards to Bean, February 12, 1889, May 28 and November 7, 1890, TB.

78. William Henry Edwards to Mead, September 6, 1891, TM.

79. William Howe, *The Butterflies of North America* (New York: Doubleday, 1975), 91, and James A. Scott, *The Butterflies of North America* (Stanford, CA: Stanford University Press, 1986), 243.

80. Bruce, from a long 1887 letter quoted by William Henry Edwards in *BNA*, vol. 3 (1898) in the section entitled "Erebia I."

81. William Henry Edwards to Scudder, April 1 and October 7, 1895, SS-BMS.

82. James Fletcher, "Annual Address of the President," in the *Twentieth Annual Report of the Entomological Society of Ontario* (Toronto, 1889), 2–8.

83. William Henry Edwards to Mead, September 6, 1891, TM; Edwards, "Papilio XII, XIII," *BNA*, vol. 2 (1884).

84. On eggs in quills, see Bean to William Henry Edwards, September 26, 1884,

WHE-SA; on eggs in cork, see William Henry Edwards to Bean, February 13, 1876, TB. For the morphine bottle reference, see Edwards, "Notes on Butterflies, with Directions for Breeding," 89.

85. William Henry Edwards to Henry Edwards, May 27, 1887, HE. He told Arthur Butler in 1886 that he planned to be finished in "four to five years," but he was, of course, dreaming. See Edwards to Butler, March 1, 1886, NHM-LONDON. After volume 2, Edwards planted more beans and potatoes in his garden with his wife, Catherine, than he had ever done in his life. He helped home-educate his first granddaughter and watched another, the only child of Edith and Theodore Mead, suffer and die from scarlet fever. Between 1884 and 1894 he served one term as president of the Board of Education for Kanawha County on the Republican line, establishing forty-nine schools and hiring fifty-six teachers, and another term as "road master," or Commissioner of Roads, during which time he "made a road that never was!" (as he wrote to Henry Edwards). "No pay and plenty of hard work," he informed William Wright. See Entomological Diary, January 18, 1889, WHE-SA; Edwards to Henry Edwards, April 18, 1889, HE; Edwards, *Autobiographical Notes;* Edwards to Wright, May 21, 1891, WGW; and Edwards to Henry Edwards, May 19, 1884, and May 27, 1885, HE.

86. William Henry Edwards to Henry Edwards, May 27, 1887, HE; and Edwards to Wright, March 12, 1891, WGW.

87. William Henry Edwards to Henry Edwards, October 10, 1883, HE. On Catherine Edwards's legacy, see Theodore Mead to his mother, February 8, 1900, TM. In 1900 Willie Edwards invested his mother's money in oil royalties, paying her $75 a month. Edwards had wealth in land, but he couldn't get to it because he hadn't paid taxes on it for years. See Willie Edwards to Theodore Mead, December 21, 1886, TM.

88. "Sometimes I have been $500 in their debt," he told Scudder in 1887. "They are remarkably kind people." And five years later: "That house has ever since my connection to it, nearly thirty years ago, paid in advance the bills of every description for *BNA*. Never once did they decline to send me a check at my request!" See William Henry Edwards to Scudder, August 27 and October 6, 1887, and January 6, 1892, SS-BMS.

89. Once Edwards promised William Wright that he would give him William Hewitson's books "if thereby I can get the caterpillar of *Rutulus*" (the western tiger swallowtail); Edwards to Wright, February 19, 1883, WGW. See also Edwards to Mead, February 4, 1890, TM. On the death of his mother, see Edwards to Lintner, November 18, 1887, JL.

90. William Henry Edwards to Henry Edwards, March 26, 1887, HE.

91. William Henry Edwards to Henry Edwards, May 5, 1886, HE.

92. William Henry Edwards to Holland, April 12, 1886, WH-CM.

93. William Henry Edwards to Scudder, August 28, 1888, SS-BMS.

94. William Henry Edwards to Scudder, October 6, 1887, and November 26, 1888, SS-BMS.

95. "By Jove that was grand you getting the $500 grant for Edwards," wrote Scudder's Canadian friend James Fletcher to Scudder. "It will give him grand help forward and he ought to be very grateful to you." See Fletcher to Scudder, January 9, 1892, SS-BMS.

96. See Scudder to Edward Nolan, December 31, 1888, and February 2, 1888, Director's Correspondence, Collection no. 567, Archives and Manuscripts Collection, Ewell Sale Steward Library, Academy of Natural Sciences, Philadelphia.

97. See Scudder, "Notes and News," *Science* 5, no. 11 (August 2, 1889): 339. James Fletcher, a Canadian butterfly man, also complained in the pages of the *CE* that Edwards had to "sacrifice" his insects in order "to continue his unselfish labors." See Fletcher, *CE* (April 1887): 72.

98. Fletcher to Scudder, March 1, 1893, SS-BMS.

99. On the "dreariness," see Scudder's preface to *Nomenclator Zoologicus* (Washington, DC: 1882–84), published in the *Bulletin of the United States National Museum*, no. 19. Scudder modeled his *Catalogue of Scientific Serials* after Agassiz's 1846 four-volume *Bibliographia Zoologicae et Geologicae* and on the single-volume update of Agassiz by the German naturalist Hermann Hagen, *Bibliotheca Entomologica* (Leipzig, 1862), which reached to 1862 and appeared in three languages (German, English, and French), with separate entries by Hagen, Carl Dohrn, and Henry Stainton. Agassiz's volumes offered a "comprehensive listing of books and articles relating to all phases of natural history published up to 1846, together with a guide to scientific periodicals and to the publications of societies and institutions of natural science"; see Edward Lurie, *Louis Agassiz: A Life in Science* (Baltimore: Johns Hopkins University Press, 1988), 106. Scudder, then, dealt fresh only with sixteen or seventeen years, but more had been published in that short span than ever in history. It demanded that he examine every American repository "book by book," as he put it, scrutinize every catalog and bibliography in whatever language on all scientific subjects, and solicit reliable scholars from other countries to double-check his lists for correctness and depth. Just in browsing it, the reader gets a feel for the wealth of the scientific achievement in the world, especially after 1860 and above all in Europe (German serials alone consume nearly one-third of the volume). Its full title is *Catalogue of Scientific Serials of All Countries Including the Transactions of Learned Societies in the Natural Physical and Mathematical Sciences, 1633–1876* (Cambridge, MA: Library of Harvard University, 1879). Scudder's second catalog, *Nomenclator Zoologicus,* combined older "lists" of genera by Agassiz and others with newer ones created by modern-day specialists in countries from Britain to Cuba, all contacted by Scudder. It was published in two parts; the first part contained nearly sixteen thousand entries, the second, eighty thousand.

100. See Scudder, "Fossil Butterflies," in *Memoirs of the American Association for the Advancement of Science,* vol. 1 (Salem, MA, 1875), 43, for an account of his European experience. Adolph Menge, a German archaeologist who exhumed many insects trapped in Prussian amber deposits, honored Scudder in 1880 by bequeathing to him all his amber fauna, one of the great collections in the world.

101. Scudder to Alexander Agassiz, April 17, 1880, Alexander Agassiz Letter Books, 1859–1910, EML. In 1886 he made summer trips to Florissant, a newly discovered fossil bed in Colorado, with thirty-five-million-year-old butterflies in it. Even earlier, in 1878, he had traveled to Florissant when it was a primitive, inhospitable place; he slept in a dusty post office near the rail station, treated the nearby Green River "as a toilet," and encountered "more drinking saloons" than he had ever wanted to see in his life. He went again in 1885, this time stay-

ing in a "nice hotel" and enjoying many "creature comforts." ("There was," however, he noted, "the same alignment of saloons.") In 1889 Scudder brought a tiny entourage with him—three boys, a cook, and several horses—but few amenities. He and the boys sat together in the dirt, hammering and chipping away at the shale in hopes that some fossil treasure might drop free from its adamantine obscurity. At one point, Scudder walked off alone, armed only with a net and a hammer, to quarry in what he called the "broiling, breathless valley." "I found just one bit of shade the entire distance by hugging a huge rock, and here I stripped to the skin to bathe in the air, the only element I had, and then pushed on, inspired by the thought of a coming plunge in the White River." "Hunting for Fossil Insects," *Thirty-Fourth Annual Report of the Entomological Society of Ontario, 1903* (Toronto, 1904), 101–3, an account of all his trips to Florissant; and, on his Canadian work, see Scott A. Elias, *Quaternary Insects and Their Environments* (Washington, DC: Smithsonian Institution Press, 1994), 3–8.

102. Emmel Drummond, and Minno, *Florissant Butterflies,* 3–5.
103. Scudder, "The Fossil Butterflies of Florissant," Part I (Washington, DC, 1889), 439–70.
104. Fletcher to Scudder, April 19, 1890, SS-BMS.
105. Lintner to Scudder, March 19, SS-BMS.
106. William Henry Edwards to Scudder, May 3, 1891, and May 5, 1894, SS-BMS.
107. In 1879 Scudder published *Butterflies: Their Structure, Changes, and Life-Histories,* a sort of trial run for the later volumes, delivered in 1878 or '79 as a lecture series at the Lowell Institute in Cambridge. Information on Scudder's Lowell Institute lectures comes from a review of *Butterflies* by William Henry Edwards in *Papilio,* January 1882, pp. 16–17.
108. Fletcher to Scudder, September 7, 1888, SS-BMS.
109. Ann Shelby Blum, *Picturing Nature: American Nineteenth-Century Zoological Illustration* (Princeton, NJ: Princeton University Press, 1993), 262.
110. Thaddeus Harris, *Treatise on Some of the Insects Injurious to Vegetation* (New York: Orange Judd, 1862), iii–iv.
111. Henry Bates to Scudder, September 11, 1869, SS-BMS.
112. Blum, *Picturing Nature,* 297–305. Blum situates Scudder's work within an evolving tradition of scientific representation in America.
113. For Grote, see Scudder, *BEUSC* (Cambridge, 1889), vol. 2, p. 835; for Higginson, *BEUSC,* vol. 2, p. 1627.
114. *BEUSC,* vol. 1, pp. 457, 519. See also, in vol. 1, pp. 8, 52, 127, 208, 280, 397, and 710.
115. Scudder's opening editorial for *Science* had noted that "scientific evolution, like the evolution of the species, requires complete conformity with the conditions of existence." See "The Future of American Science," *Science* (February 9, 1883): 3. Davis and Woodworth would acquire national reputations, Davis as creator of the science of geomorphology, and Woodworth as a leading economic entomologist in California (he also spent years in China, introducing the Chinese to pesticides, among other things).
116. Scudder, "The Names of Butterflies," excursus 25, *BEUSC,* vol. 2; and *BEUSC,* vol. 1, p. 111.
117. Earlier, in 1873, Scudder wrote: "Mr. Edwards and I have been helping each

other materially all through our connection with each other, though I always feel as if the gain were on my side, he is very liberal with his specimens. Recently, we have been over each other's drawings of earlier stages." Scudder to Henry Edwards, December 19,1873, HE.

118. Edwards to Lintner, November 27, 1885, JL. Edwards quoted from Scudder in this letter.

119. See *BEUSC,* vol. 1, excursus 20, "Three Pioneer Students of Butterflies in the Country," pp. 651–58; and *BEUSC,* vol. 1, excursus 24 on fossils, pp. 756–60.

120. Brian Boyd and Robert Pyle, eds., *Nabokov's Butterflies* (Boston: Beacon, 2000), 38, 66, 475, and 530.

121. William Henry Edward's footnote at the bottom of p. 70, *CE,* April 1889.

122. Scudder to Holland, February 11, 1894, WH-CM. In this letter, Scudder recalled his insomnia of "a year or two ago." When William Henry Edwards heard about Scudder's misfortune, he sympathized by blaming it on the Democrats, who had just gained control of the presidency. What could you expect from them who didn't care a fig about "science and knowledge," he told Scudder. But *they* weren't at fault, Scudder responded; the Senate Republicans, not the Democrats, had terminated the funding. Edwards, a passionate Republican in the tradition of Abraham Lincoln and Charles Sumner, couldn't believe it. "I see an investigation is in order," he wrote. "A Democrat would strike out the appropriations for pure cussedness." Edwards to Scudder, August 5 and July 27, 1892, SS-BMS.

123. On this adoption, see William Field's "Doctor Scudder's Work on Lepidoptera," *Psyche* 18 (December 1911): 180.

124. Scudder, *Every-day Butterflies: A Group of Biographies* (Boston: Houghton Mifflin, 1899), iii, 1–11, 24–28, 58–62, 67–72, 87–94, 106–13, 146–65, 248–53, 262–76, 311–16, 381–86; and *Frail Children of the Air: Excursions into the World of Butterflies* (1895; repr., New York, 1899).

125. Samuel Scudder, *Brief Guide to the Commoner Butterflies of the Northern United States and Canada* (New York, 1893), iv.

5. The Life and Death of Butterflies

1. Alexander Humboldt, *Cosmos,* vol. 1 (1845; repr., Baltimore: Johns Hopkins University Press, 1997), 39–40.

2. Herman Strecker, *Butterflies and Moths in Their Connection with Agriculture and Horticulture* (Harrisburg, PA, 1879), 17.

3. Augustus Grote, "The Nocturnidae of Europe and North America Compared," *CE* (August 1890): 145–50; and, on Darwin's governing perspective, see Peter J. Bowler, *Charles Darwin: The Man and His Influence* (Cambridge: Cambridge University Press, 1990), 75–85, 118–19.

4. See "Ancestry and Classification" in Samuel Scudder's *Butterflies: Their Structure, Changes, and Life-Histories* (New York: Holt, 1881), 226–48.

5. Charles Fernald to Scudder, December 29, 1879, SS-BMS.

6. Jakub Novak, "Alfred Russel Wallace's and August Weismann's Evolution: A Story Written on Butterfly Wings" (Ph.D. diss., Princeton University, 2008), 219.

7. On Riley, see Charles V. Riley, "Philosophy of the Pupation of Butterflies and Particularly of the Nymphalidae," *Proceedings of the American Association for the Advancement of Science* 28 (1880): 455–63, and Scudder, *BEUSC* (Cambridge, 1889), vol. 2, excursus 71, "The Act of Pupation," pp. 1693–95, and pp. 1554–58, in which he thanks and cites Riley. On Burgess, see Scudder, "The Services of Edward Burgess to Natural Science," *Proceedings of the Boston Society of Natural History* (December 16, 1891): 358–60, and Edward Burgess, "The Structure and Action of a Butterfly's Trunk," *American Naturalist* 14, no. 5 (1880): 312–19.

8. See especially, Scudder, *BEUSC*, vol. 1, "The Eggs of Butterflies," pp. 190–99; "The Modes of Suspension of Chrysalids," pp. 201–3; and vol. 2, "The Act of Pupation," pp. 1693–1711, and "How Butterflies Suck," pp. 1737–46. On Scudder's use of the microscope in the making of his wood engravings and lithographs, see Ann Shelby Blum, *Picturing Nature: American Nineteenth-Century Zoological Illustration* (Princeton, NJ: Princeton University Press, 1993), 292–97.

9. Scudder, *BEUSC*, vol. 1, p. 180.

10. William Doherty, "Notes on Assam Butterflies," *Journal of the Asia Society of Bengal* (1889): 126; and Doherty to Scudder, November 20, 1886, SS-BMS.

11. Scudder, *BEUSC*, vol. 2, p. 778.

12. For Edwards's culminating analysis, see William Henry Edwards, *BNA*, vol. 2 (Philadelphia: American Entomological Society, 1868–72).

13. Sir John Lubbock, *Fifty Years of Science: Being an Address Delivered at York to the British Association for the Advancement of Science in 1881* (London: Macmillan, 1882). Praise came from other quarters as well—for instance, from Raphael Meldola, another influential British evolutionary naturalist. In 1882 Meldola edited and translated August Weismann's *Studies in the Theory of Descent* (with an introduction by Charles Darwin), carefully incorporating into it a thirty-five-page appendix devoted entirely to an explication of Edwards's work on *Papilio ajax* (plates included), a great compliment. Joseph Lintner, too, in his 1878 address before the American Association for the Advancement of Science, singled out his friend for praise: "I commend to you the labors of William Henry Edwards in working out the histories of some of those butterflies which appear under different forms at different seasons of the year. The untiring zeal with which the work has been prosecuted and is being continued deserves the commendation which it has received from the most eminent European Entomologists." See Lintner, *CE* (August 1878): 173–74; and Weismann, *Studies in the Theory of Descent* (London, 1882), 126–60. See also, on the extent of Edwards's achievement, W. Conner Sorensen's breakthrough analysis in his *Brethren of the Net: American Entomology, 1840–1880* (Tuscaloosa: University of Alabama Press, 1995), 214–35.

14. E. O. Essig, *College Entomology* (New York: Macmillan, 1942), 503.

15. For a more extensive, fascinating analysis of the satyrids, see William Henry Edwards's *BNA*, vol. 2, "Satyrus II, III." See also Charles Remington, "Suture-Zones of Hybrid Interaction Between Recently Joined Biotas," *Evolutionary Biology* 1 (1968): 321–428.

16. Edwards never explained, however, why yellow females continued to multiply in the face of the supposed advantage of the black female. Nor has anyone

else, according to Michael Gochfeld and Joanna Burger, who write that there is no satisfactory explanation as to "why the polymorphism exists. We suspect that the predation pressure may never be high enough to eliminate the yellow females entirely." See their *Butterflies of New Jersey* (New Brunswick, NJ: Rutgers University Press, 1997), 125.

17. Edwards to Scudder, January 23, 1888; September 15, 1887; and January 2, 1888, SS-BMS.

18. Edwards, "On *Pieris bryoniae* Ochsenheimer, and Its Derivative Forms in Europe and America," *Papilio* 1, no. 6 (1881): 98; and Augustus Grote, "The Origin of Ornamentation in the Lepidoptera," *CE* (June 1888): 114–17, and "On Specific Names," *CE* (March 1889): 52.

19. Grote, "The Origin of Ornamentation in the Lepidoptera," 114–17.

20. Scudder, *Butterflies*, 234.

21. Scudder, excursus 50, "Variations in Habit and Life According to Locality and Season of the Year," *BEUSC*, vol. 2, pp. 1415–17.

22. Doherty, "A List of Butterflies Taken in Kumaon," *Journal of the Asiatic Society of Bengal* (1886): 103–6.

23. William Henry Edwards, "Libythea," in *BNA*, vol. 2. For lists of butterfly enemies, see Scudder, "Enemies," *BEUSC*, vol. 2, pp. 1612–13.

24. H. C. T. Godfray, *Parasitoids* (Princeton, NJ: Princeton University Press, 1994), 16.

25. Harold F. Greeney, "Emergence of Parasitic Flies from Adult *Actinote diceus* (Nymphalidae: Acraeinae) in Ecuador," *Journal of the Lepidopterists' Society* 55, no. 2 (2001): 79–80.

26. Archibald Weeks, "Method of Oviposition of Tachina," *Entomologica Americana* 3 (August 1887): 126.

27. On this history, Godfray, *Parasitoids*, 6–7; Richard R. Askew, *Parasitic Insects* (London: Heinemann, 1971), v, 117–21, 140–45; Nick Mills, "Parasitoids," in *Encyclopedia of Insects*, ed. Vincent H. Resh and Ring T. Carde (San Diego: Academic Press, 2003), 845–47; and David Grimaldi and Michael S. Engel, *Evolution of Insects* (Cambridge: Cambridge University Press, 2005), 427.

28. On Linnaeus and the Ichneumonidae, see James Duncan, *The Natural History of British Butterflies* (Edinburgh, 1835), 89–91.

29. Johann Meigen, William Kirby, Henry Spence, and James Duncan, among others, offered grisly portraits.

30. See Charles Darwin to Asa Gray, May 22, 1860, in *The Correspondence of Charles Darwin*, ed. Frederick Burkhardt et al., vol. 8 (Cambridge: Cambridge University Press, 1993), 224. I would like to thank Paul Farber for sharing with me this Darwin citation.

31. Alpheus Packard, *Guide to the Study of Insects* (New York, [1869] 1876), 196–97.

32. R. L. Duffus, *The Innocents at Cedro: A Memoir of Thorstein Veblen and Some Others* (New York: Macmillan, 1944), 52, 62–63. In the 1880s, Herbert Smith, a dedicated and respected young naturalist from upstate New York, studied Ichneumonidae in South America, where their numbers, he believed, reached many thousands of species (all unnamed). Smith's collection for the British Museum, assembled in this decade, was the largest of its kind ever donated there (see Smith to Ezra Cresson, December 6, 1892, Academy of

Natural Science Archives, Philadelphia). On the Smith and Edwards collections in the Natural History Museum in London, see Claude Morley, *A Revision of the Ichneumonidae* (London, 1912), vol. 1, preface and pp. 9–13; and vol. 2, pp. 16–17, 56–58, and 107.

33. Howard cites Edwards's work in his essay in Scudder, *BEUSC*, vol. 3, pp. 1880–81, 1883–84.

34. For Edwards's parasite collection, his sharing with Howard and Riley, and his accounts to Lintner, see Edwards to Lintner, July 22, December 24, and December 29, 1882, JL.

35. See Riley to Scudder, first page and date missing (but probably early 1870s and certainly before 1877, the year Riley left Missouri for Washington), SS-BMS.

36. *BEUSC*, vol. 3, pp. 1901–2, 1897–1911; Howard's contribution is on pp. 1869–96, and Williston's on pp. 1912–24.

37. William Field, the curator of entomology at the Smithsonian in the 1940s and a highly regarded authority, cited this work in his *Manual of Butterflies and Skippers of Kansas* as the *only* source of information on American parasites (Lawrence, KS: Bulletin of the Department of Entomology, no. 12, May 15, 1938), 50.

38. "The Enemies of Butterflies," *BEUSC*, vol. 2, p. 1613.

39. "*Aglais milberti*: The American tortoiseshell," *BEUSC*, vol. 1, pp. 428–29.

40. *BEUSC*, vol. 1, pp. 454–55.

41. Edwards called this butterfly *Thecla laeta*, when *Thecla* was a still a viable genus for most hairstreaks. Today, the Latin name is *Erora laeta*, designated by Scudder in 1872 and retained by modern catalogs. Scudder also chose "the spring beauty" as the vernacular name, which has *not* been kept. Reflecting the ongoing struggle over nomenclature, some contemporary authorities have called it the "turquoise beauty" (James Scott, *The Butterflies of North America* [Stanford, CA: Stanford University Press, 1986], 385), others the "early hairstreak" (Alexander D. Klots, *A Field Guide to the Butterflies* [Boston: Houghton Mifflin, 1951], 129, and Robert Pyle, *National Audubon Society Field Guide to North American Butterflies* [New York: Knopf, 1998], 467). Edwards would have opposed these names. For Scudder's names, see *BEUSC*, vol. 2, p. 819. See also Edwards, "Notes on the Collection of Butterflies Made by Mr. H. K. Morrison in Arizona, 1882," *Papilio* 3 (January 1883): 1.

42. "Limenitis I," *BNA*, vol. 2.

43. Grote, "Moths and Moth-Catchers: Part I," *Popular Science Monthly* (June 1885): 246–52.

44. *BEUSC*, vol. 2, p. 1603.

45. Excursus 51, "Southern Invaders," *BEUSC*, vol. 2, pp. 1332–34.

46. Excursus 46, "The Spread of a Butterfly in a New Region (with Map)," *BEUSC*, vol. 2, pp. 1175–90.

47. For a full rendering of the history of the work on monarch migrations, see Lincoln Brower, "Understanding and Misunderstanding the Migration of the Monarch Butterfly," *Journal of the Lepidopterists' Society* 49, no. 4 (1995): 304–85.

48. Charles Valentine Riley, "A Swarm of Butterflies," *American Entomologist* (September 1868): 28–29.

49. For Scudder's monarch life history, see *BEUSC*, vol. 1, pp. 720–48.

50. Scudder to Charles Fernald, September 15, 1888, Charles Henry Fernald

Papers (RG 40/11, C. H. Fernald), Special Collections and University Archives, University of Massachusetts, Amherst.

51. *BEUSC,* vol. 2, p. 742.

52. Grote, "Character of Protection and Defense in Insects," *CE* (July 1888): 155.

53. Grote, "On Insects Feigning Death," *CE* (June 1888): 120, and "Characters of Protection and Defense in Insects," *CE,* 155.

54. *BEUSC,* vol. 2, pp. 1036–37.

55. Doherty to his mother, April 18, 1883, JMH. Doherty wrote to Henry Elwes of his theory on "timidity as a source of protection" in 1889, which Elwes published in his article on Doherty's butterflies for *Transactions of the London Entomological Society* (1891): 256–57.

56. Grimaldi and Engel, *Evolution of Insects,* 603. For a recent account of all forms of defenses, see Graeme D. Ruxton, et al., *Avoiding Attack: The Evolutionary Ecology of Crypsis, Warning Signals, and Mimicry* (New York: Oxford University Press, 2004).

57. Henry Bates, "Contributions to an Insect Fauna. Lepidoptera: Heliconidae," *Transactions of the London Entomological Society* 23 (1862): 495–526. This discussion of Bates and mimicry owes a good deal to William C. Kimmler, "Mimicry: Views of Naturalists and Ecologists Before the Modern Synthesis," in *Dimensions of Darwinism,* ed. Marjorie Grene (Cambridge: Cambridge University Press, 1986), 98–127. I would like to thank Mary C. Winsor of the University of Toronto for directing me to the source.

58. On Bates, see Ruxton, et al., *Avoiding Attack,* 139–63. See also Thomas Belt, *The Naturalist in Nicaragua* (1874; repr., London, 1888); Major G. F. L. Marshall and Lionel de Nicéville, *Butterflies of India* (1881–82), vol. 1, p. 22; James J. Wood-Mason, "Description of Two New Species of *Papilio* from Northeastern India, with Preliminary Indication of an Apparently New and Remarkable Case of Mimicry Between the Two Distinct Groups Which They Represent," *Annals and Magazine of Natural History,* 5th ser., 9 (1882); and Roland Trimen, "On Some Remarkable Mimetic Analogies Among African Butterflies," *Transactions of the Linnean Society* 26 (1868): 497–522.

59. Excursus 62, "Color Relations of Chrysalids and Their Surroundings," *BEUSC,* vol. 2, pp. 1578–94. For the Wallace quote, see Andrew Berry, ed., *Infinite Tropics: An Alfred Russel Wallace Anthology* (London: Verso, 2002), 98–102, and David West, *Fritz Müller: A Naturalist in Brazil* (Blacksburg, VA: Pocahontas Press, 2003), 224–25. The Germans, too, had much to say, above all, Fritz Müller, an immigrant to Brazil living in Santa Catarina who, in the late 1870s, developed his own striking theory second only to Henry Bates's theory in historical significance; Müller discovered that many *unpalatable* species mimicked the color and patterns of *other unpalatable* and often unrelated species, with a natural advantage accruing to both *only* if the number of mimics remained smaller than the models. On Müller, see Ruxton, et al., *Avoiding Attack,* 115–31, and West, *Fritz Müller.*

60. Edward Poulton, *Colour of Animals* (London, 1889), viii.

61. Gerald H. Thayer, *Concealing-Coloration in the Animal Kingdom: An Exposition of the Laws of Disguise Through Color and Pattern—Being a Summary of Abbott H. Thayer's Disclosures,* with an introductory essay by A. H. Thayer (New York, 1909/1918), 3–12; and *BEUSC,* vol. 2, chap. 26, "Butterflies and

Moths," pp. 212–40. On Thayer, see Sharon Kingsland, "Abbott Thayer and the Protective Coloration Debate," *Journal of the History of Biology* 11, no. 2 (1978): 223–44; Hardy Blechman, *Disruptive Pattern Material: An Encyclopedia of Camouflage* (Tonowanda, NY: Firefly Books, 2004), 16–76; Nelson White, *Abbott H. Thayer: Painter and Naturalist* (Hartford: Connecticut Printers, 1951), 6–7; and Blum, *Picturing Nature*, 336–44. On Riley's discoveries, see his "Imitative Butterflies," *American Entomologist* 1, no. 10 (1869): 189–93.

62. Grote, *An Illustrated Essay on the Noctuidae of North America* (London, 1882), 35. Grote quoted, word for word, from an earlier article by D. S. Kellicott, "An Example of Protective Mimicry," which Grote had earlier published in the *North American Entomologist* 10 (1879): 1.

63. Grote, "Notes on Species of Lepidoptera," *CE* (March 1887): 52–54.

64. Excursus 57, "Nests and Other Structures Made by Caterpillars," *BEUSC*, vol. 2, pp. 1456–57; and excursus 74, "Odd Chrysalids," vol. 2, p. 1750.

65. Henry Smith to Samuel Henshaw, April 14, 1887, EML.

66. Doherty, "A List of Butterflies Taken in Kumaon," 106.

67. Henry Elwes, "List of Diurnal Lepidoptera Taken by Mr. W. Doherty (of Cincinnati) in Burma," *Transactions of the London Entomological Society* 21 (1891): 255.

68. Excursus 8, "The Means Employed by Butterflies of the Genus Basilarchia for Perpetuation of the Species," *BEUSC*, vol. 1, pp. 260–94.

69. William Henry Edwards rejected the idea that the two purples were different species, arguing, instead, that they were the same species, with the red-spotted a form of the white admiral. Naturalists today agree with Edwards. See Paul Opler, *Eastern Butterflies* (New York: Houghton Mifflin, 1992), 189. Opler also rejects Scudder's genus, *Basilarchia,* preferring instead the older European designation of *Limenitis fabricius,* which Edwards also embraced.

70. Frederick Clarkson, "Probable Origin of the Word Butterfly," *CE* 17 (March 1885): 45.

71. Excursus 23, "Mimicry and Protective Resemblance; or, Butterflies in Disguise," *BEUSC*, vol. 1, p. 719. This excursus was the longest Scudder wrote and the only one given a lengthy bibliography. See also his "Color Relations of Chrysalids and Their Surroundings," vol. 2, pp. 1578–89, in which he praises the work of Edward Poulton.

72. Excursus 26, "Hypermetamorphosis in Butterflies," *BEUSC*, vol. 1, pp. 804–8.

73. *BNA*, vol. 1, "Melitea I."

74. Edwards to Scudder, March 4, 1876, SS-BMS. The conflict over the *E. phaeton* webs is chronicled in the following Edwards letters to Scudder: February 3, February 14, February 17, February 25, and March 4, 1876, SS-BMS.

75. He had discovered the adult form of the spring azure (as it was also called, although *not* by Edwards). See Tom Allen, *The Butterflies of West Virginia and Their Caterpillars* (Pittsburgh: University of Pittsburgh Press, 1997), 104–5. For a modern analysis of the relationship between ants and Lycaenidae, see Naomi Pierce et al., "The Ecology and Evolution of Ant Association in the Lycaenidae (Lepidoptera)," *Annual Review of Entomology* 47 (2002): 733–71.

76. Edwards, "Notes on *Lycaena pseudargiolous* and Its Larval History," *CE* (January 1878): 1–15.

77. Edwards, Entomological Diary, June 10, 1878, WHE-SA.

78. William Henry Edwards to Henry Edwards, November 22, 1877, HE. In his introduction to volume 1 of *BEUSC,* in a subsection titled, "The Larva or Caterpillar" (pp. 15–16), Scudder attributed the discovery of the relationship between the ants and lycaenid larvae to two Germans, Christian Pezold and Eugen Esper, in the mid-eighteenth century.

79. Doherty, "A List of Butterflies Taken in Kumaon," 112–13.

80. "Lycaena, II, III," *BNA.*

81. Excursus 21, "Companionship and Commensalism Among Caterpillars," *BEUSC,* vol. 1, pp. 672–73; and excursus 35, vol. 2, pp. 962–64.

82. Grote, "The Origin of Ornamentation in the Lepidoptera," "The Friends and Associates of Caterpillars," *BEUSC,* 114–17.

83. See Karl Polyanyi, *The Great Transformation* (Boston: Beacon, 2001).

84. Karen Halttunen, *Confidence Men and Painted Women: A Study of Middle-Class Culture in America, 1830–1870* (New Haven, CT: Yale University Press, 1982), especially chapter 2, "Hypocrisy and Sincerity in a World of Strangers," 33–55.

85. Scudder, "Mimicry and Protective Resemblance; or Butterflies in Disguise," *BEUSC,* vol. 1, pp. 713–19.

86. As it turned out, however, Behr was wrong: the butterfly survived into the 1940s. See Robert Pyle, "Bring Back the Xerces Blue," *Whole Earth* (Spring 2001).

87. Grote, "Collecting Noctuidae by Lake Erie," *Entomologist's Record* 6, no. 5 (1895): 100.

88. Ibid., 98, and Grote's introduction to *Genesis I-II: An Essay on the Bible Narrative of Creation* (New York, 1880), 17. I discovered the Rousseau quote in Jacques Roger, *Buffon: A Life in Natural History,* trans. Sarah Lucille Bonnefoi (Ithaca, NY: Cornell University Press, 1997), 301–2.

89. Excursus 37, "Local Butterflies," *BEUSC,* vol. 2, pp. 884–85; excursus 46, "The Spread of a Butterfly in a New Region," *BEUSC,* vol. 2, p. 1175; and excursus 27, "The Best Localities for Collectors, *BEUSC,* vol. 2, pp. 817–19.

90. Emory L. Kemp, *The Great Kanawha Navigation* (Pittsburgh: University of Pittsburgh Press, 2000), 24–62.

91. See Scudder's account of the life history *Euphydryas phaeton* in *BEUSC,* vol. 1, p. 705; and Edwards, Entomological Diary, May 22 and June 1, 1884, WHE-SA.

92. Excursus 43, "Color Preferences of Butterflies: The Origins of Color in Butterflies," *BEUSC,* vol. 2, pp. 1101–4.

93. Excursus 15, "The Origins and Development of Ornamentation in Butterflies," *BEUSC,* vol. 1, p. 517. Scudder's preference for the beauty of butterflies compared to the aesthetic preferences of Charles Valentine Riley, also a gifted entomologist wide-ranging in his insect interests. To Scudder he wrote: "You know I see more beauty in the Heterocera and the other Orders than you do, so that my observations on the Rhopalocera are like stray shots—few and scattering." See Riley to Scudder, 187?, first page of letter missing, written from the "office of the State Entomologist" of Missouri, St. Louis, SS-BMS.

94. Excursus 70, "Sexual Diversity in the Form of the Scales," *BEUSC,* vol. 2, pp. 1681–82.

95. Ibid.; and Scudder, *Butterflies,* 235–38. In the mid-twentieth century, the great Swiss biologist Adolph Portmann would take a position very similar to

Scudder's, in defiance of the prevailing evolutionary wisdom. Portmann, too, accepted natural selection as causal but only for a special group of species that relied on protective coloration, warning signals, and deceptive resemblance, "a one-sided curio cabinet not to be compared to the wild profusion of form production in the garden of living creatures." He also noted, "The scales of butterflies give the appearance of metal, of gold and blue textures, appearances whose sense surpasses any merely adaptive ability." See Adolph Portmann, *Essays in Philosophical Zoology* (Lewiston, NY: Edwin Mellen, 1990), 27, and *Animal Forms and Patterns: A Study of the Appearances of Animals* (New York: Schocken, 1967), 122–23, 218–21. I would like to thank the noted lepidopterist Arthur Shapiro of the University of California, Davis, for introducing me to Portmann's work.

96. Excursus 43, "Color Preferences of Butterflies," *BEUSC,* vol. 2, pp. 1101–5, and excursus 70, "Sexual Diversity in the Form of the Scales," *BEUSC,* vol. 2, pp. 1681–83.

97. These words of Scudder's were quoted in a letter from William Henry Edwards to Theodore Mead, April 29, 1877, TM. The original Scudder letter is lost.

98. Scudder, *Butterflies,* 242–43; see Erica E. Hirshler, *A Studio of Her Own: Women Artists in Boston, 1870–1940* (Boston: MFA Publications, 2001), 242–43.

99. So, too, as he grew older, Wallace proposed yet another explanation for beauty: a higher power, not religious in any conventional sense but *spiritual,* serving mankind and working through natural selection, yet at the same time acting independently to produce all the splendor of the world. Alfred Russel Wallace, "The Colours of Animals and Plants," *Macmillan's Magazine* (September–October 1872): 641–42. See also Denis Diderot, quoted in Peter Gay, *The Enlightenment: An Interpretation,* vol. 2, *The Science of Freedom* (New York: Knopf, 1969), 162. I have been influenced here by Jakub Novak's discussion of Wallace in his "Alfred Russel Wallace and August Weismann's Evolution," 1–207.

100. Edwards to Scudder, January 13, 1888, SS-BMS.

101. Grote, "Moths and Moth-Catchers," 247. The late Thomas Eisner, a noted authority on chemical ecology at Cornell University, spoke the same passionate language of Scudder, Wallace, and Grote in regard to the "utter splendor that is embodied in the lepidopteran wing." Under "high magnification," he saw in the wings of many butterflies "a world of hidden dimensions . . . , a treasury of abstract art to be explored, pointillist in design, elegant in coloration, and infinitely pleasing. There is proof in these images that science and art, while dwelling separately in the confines of our consciousness, do merge in that vague domain of the subconscious that guides us in our passion." Thomas Eisner, "Hidden Splendor," *Wings* (Spring 2006): 5.

102. Grote, "Collecting Noctuidae by Lake Erie," 99.

6. In the Wake of Empire

1. Paul Maassen to Strecker, December 3, 1872, HS-FM.

2. Henry Stainton to Otto Staudinger, December 3, 1872 (Stainton's draft copy), Stainton Papers, NHM-LONDON.

3. Augustus Grote, *The Hawk Moths of North America* (Bremen, 1886), 61, and Theodore Mead to William Henry Edwards, December 17, 1872, WHE-SA.

4. On the entomology, see Roland Trimen, "On Some Remarkable Mimetic Analogies Among African Butterflies," *Transactions of the Linnean Society* 26 (1868): 503n; and Fa-Ti-Fan, *British Naturalists in Qing China* (Cambridge, MA: Harvard University Press, 2004), 11–56.

5. Alfred Russel Wallace, "The Colours of Animals and Plants," published in *Macmillan's Magazine* (September-October 1877) and, a month later, in *American Naturalist* 11, no. 11 (1877): 641–62.

6. Robert Rippon, *Icones Ornithopterorum* (London, 1898), 13. "It is safe to say," Rippon observed of Rothschild's collection, "that in no other part of the world is there such a vast collection of Lepidoptera, especially of Eastern Papilionidae: that among the Ornithoptera is to be found nearly every known form—and in such immense series and representing so many localities, that it may be considered that in this museum alone the materials for any account of research may be obtained."

7. See Hans Fruhstorfer's bills of sale to the museum, October 21 and November 12, 1898, MNK.

8. Rippon, *Icones Ornithopterorum*, 13.

9. William Holle (a German immigrant to Sheboygan, Wisconsin) to Herman Strecker, May 16, 1878, HS-FM.

10. See O. H. Staudinger, introduction, *Exotische Tagfalter* (Berlin, 1888–92); my translation.

11. George Louis Leclerc Buffon, "Of Insects of the Third Order," *Natural History of Birds, Fish, Insects, and Reptiles* (London, 1793), 5:274.

12. Figures for the mid-1700s appear in the *Encyclopedia Britannica* (London, 1775), vol. 3. For butterflies, see under "Papilio," 454. For Buffon's figures, see *Natural History of Birds, Fish, Insects and Reptiles*, 5:274. For the 1885 figure, see Samuel Scudder, *Butterflies: Their Structure, Changes, and Life-Histories* (New York: Holt, 1881), 226, and Scudder, *BEUSC* (Cambridge, 1889), vol. 1, p. 236. For recent figures, see Oakley Shields, "World Number of Butterflies," *Journal of the Lepidopterists' Society* 43, no. 3 (1989): 178.

13. On British practice as well as the general pattern, see Roland Oliver and G. N. Sanderson, eds., *The Cambridge History of Africa* (Cambridge: Cambridge University Press: 1985), 6:99; Ronald Robinson, John Gallagher, and Alice Denny, *Africa and the Victorians* (New York: St. Martin's, 1961), 7; and T. O. Lloyd, *The British Empire, 1558–1995* (London: Oxford University Press, 1995), 170.

14. P. J. Marshall, "The World Shaped by Empire," in *The Cambridge Illustrated History of the British Empire,* ed. P. J. Marshall (London: Cambridge University Press, 1996), 7–80.

15. Deepak Kumar, *Science and the Raj, 1857–1905* (Delhi: Oxford University Press, 1997), 110. On the imperialist significance of the railroad (and for the source of the quote), see Ronald E. Robinson, "Introduction: Railway Imperialism," in *Railway Imperialism,* ed. Clarence B. Davis and Kenneth E. Wilburn (New York: Greenwood, 1991), 3.

16. On this road, see Patrick Sweeter, *Streeter of Bond Street: A Victorian Jeweler* (London: Harlow, 1993), 121, 126–27.

17. Will Doherty to his mother, August 31, 1889, JMH; and Doherty to William

Holland, quoted in Holland's "Asiatic Lepidoptera, List of Diurnal Lepidoptera taken by Mr. William Doherty of Cincinnati in Celebes, June and July, 1887, with Descriptions of Some Apparently New Forms," *Proceedings of the Boston Society of Natural History* 25 (1890): 52–82. I thank Bernadette Callery, head of the research library of the Carnegie Museum, for directing me to this source.

18. The auctioning of Wallace's butterflies is mentioned in a letter from Richard Stretch to Henry Edwards, November 17, 1874, HE; see also Emily Grace Allingham, *A Romance of the Rostrum* (London: H. F. and G. Witherby, 1924), who dates the sale in 1872 (p. 131). Prefaced by Lord Walter Rothschild, who spent many hours at Stevens's, Allingham's book is the richest source for information on the place, tracing natural history sales from start to nearly finish, with a full chapter devoted to insect sales (pp. 128–52) and much information on related enthusiasms (birds and birds' eggs and orchids, for instance). On the power of the London auction as a means of making or establishing market prices, see Brian Learmont, *History of the Auction* (Iver: Barnard and Learmont, 1985), 24–65.

19. See William Kirby, "The Bird-Winged Butterflies of the East," *Nature* 51, no. 1315 (1895): 254–58. Although the British preference for all things English never disappeared, it did weaken, as Kirby realized more strongly with hindsight. "Since the death of Hewitson (1878)," he wrote, "new countries opened up, and wonderful butterflies reached Europe, never dreamed of by Hewitson, or which remained unobtainable objects of his desire, to the last. Chief among these may be mentioned the butterflies of Central Asia, a *terra incognita* (for the most part) in Hewitson's time; and the butterflies of the Eastern Archipelago, above all the brilliantly colored Ornithoptera (or bird wings) of these islands." According to Hewitson, Kirby's catalog was "a wonderful summary of all that has been done." Hewitson to Strecker, September 1873, HS-FM. See also Kirby, *A Synonymic Catalogue of Diurnal Lepidoptera* (London, 1871), with a supplement in 1877. Kirby, who briefly served as curator of insects in Dublin, hated the Irish and longed to leave Dublin "to start an insect business in London," as he told Henry Edwards, "for the express purpose of supplying foreign and provincial Entomologists with greater facilities for obtaining what they want. I expect to do more with foreign Lep. than anything else." Kirby to Edwards, September 22, 1874, HE. In the late 1870s, he got a job as a curator at the British Museum, thereby escaping his horrid Dublin, but he flopped in his effort to create a sideline business.

20. On Watkins & Doncaster, see Michael A. Salmon, *The Aurelian Legacy: British Butterflies and Their Collectors* (Berkeley: University of California Press, 2001), 71; on Janson's, see J. Harvey, Paula Gilbert, et al., "Janson Family Archives," in *A Catalogue of Manuscripts in the Entomological Library of the Natural History Museum, London* (New York: Mansell, 1996), 110. Winston Churchill, the future prime minister of England, the elder son of Randolph Churchill, and a butterfly lover, participated in this tropical collecting "mania." On a country vacation away from home, young "Winny," as he was then called, assured his mother that he was "never at a loss for anything to do, for I shall be occupied with 'Butterflying' all day (I was last year)." He specialized in native insects, of course, but in the mid-1890s, as a young officer stationed in Bangalore, India,

he caught numerous specimens, upwards of sixty-five kinds. His "bungalow" mates complained that the place "is degenerating into a taxidermist's shop." A nearby Bangalore garden thrilled him, "full of Purple Emperors, White Admirals, and Swallowtails and many other rare and beautiful insects." See Churchill to his brother, Jack, October 15, 1896, and Churchill to his mother, October 21, 1896, in Randolph Churchill, *Winston S. Churchill: Companion, Volume I, Part I, 1874–1896* (Boston: Houghton Mifflin, 1967), 690, 694–95.

21. William Weaver to Strecker, August 15, 1876, HS-FM.

22. Henry Elwes to Strecker, September 13, 1888, HS-FM.

23. Paul Mowis, a dealer in butterflies, lived in Darjeeling; he collected bravely in Tibet, harvesting rare insects, such as the male of the beautiful *Teinopalpus imperialis,* and selling them to museums throughout the world. Mowis sold male specimens to Berlin's Museum für Naturkunde for £25, "the largest sum he had ever received for an insect"; see "Notes and News," *EN* (April 1890): 57.

24. Henry Elwes, *Memoirs of Travel, Sport, and Natural History* (London: E. Benn, 1930), 77–79. On the size of Godman's collection, see Henry Stainton to Otto Staudinger, September 26, 1880, NHM-LONDON.

25. Alfred Wailly, *On Silk-Producing Bombyces and Other Lepidoptera* (London, 1880), 1.

26. Wailly to Strecker, January 21, 1880, and Henry Skinner to Strecker, April 18, 1883, HS-FM.

27. Wailly to Strecker, July 29, 1880, HS-FM.

28. Wailly to Strecker, December 15, 1879, and December 13, 1881, HS-FM.

29. The French played a lesser role, despite their imperialist claims on Indochina in the 1860s and sub-Saharan Africa in the 1890s; Marshall, "The World Shaped by Empire," 54–55. There was a lively exotic trade in Paris, managed partly by the Deyrolle family, which had operated a natural history business since the 1850s, selling abroad to Americans like Theodore Mead. Other major patrons were the Muséum Nationale d'Histoire Naturelle of Paris and the Oberthür brothers, Charles and René, amateur collectors living in reclusive splendor in the countryside just outside the city. A composer and poet, Charles Oberthür, published his own catalogs (*Études d'Entomologie*) in 1875 and beyond, on African, Asian, and South American butterflies, acquired by missionaries, resident Europeans, or experienced naturalists. See Charles Oberthür, *Études d'Entomologie* 1 (Paris, 1876–84), 1:13, 19; 3: ix, xi, 12; and vols. 6–10, preface. In the last, Oberthür credits especially the collecting in South America of Jean Stoltzmann, a "naturalist" of great "zeal" and "ardor" who had penetrated the "mysterious" ways of the tropics "with remarkable exactitude." Émile Deyrolle sold collecting equipment to Theodore Mead and Augustus Grote in the 1870s, and in 1879, Neumoegen purchased a rare green-and-gray iridescent African swallowtail, *Papilio zalmoxis,* one of the most beautiful butterflies in the world. See Émile Deyrolle to Neumoegen, c. June or July 1876, copied by Neumoegen for Strecker, in a letter to Strecker, August 8, 1876, HS-FM. See, for a list of other Paris dealers, Reinhard Gaedike, in Deutsches Entomologische Institute, *Nova Supplementa Entomologica* 6 (1995): 1–91. Gaedike supplies the dates of the Deyrolle merchants: Achille, 1813–1865; Émile, 1838–1917; Henri, 1827–1902; and Théophile, 1843–1923. He lists one insect merchant in Paris who may have preceded the Deyrolles—Henry Dupont, 1798–1873 (p. 102).

30. Doherty to his father, February 10, 1892; and to his mother, March 31, 1888, JMH.

31. German butterfly collectors, for instance, hunted for insects at the mouth of the Congo River during an expedition led by Paul Gussfeldt to West Africa between 1873 and 1876. Julius Falkenstein was among the collectors, his butterflies described and figured in catalogs by H. Dewitz. See Dewitz, "Afrikanische Tag-schmetterlinge," *Nova Acta,* vol. 41 (Halle, 1879). See also Paul Gussfeldt et al., *Die Loango-Expedition, 1873–1876* (Leipzig, 1879).

32. On Godeffroy, see Jacob Boll to Strecker, November 11, 1875, HS-FM; and, historically, see Lynn K. Nyhart, *Modern Nature: The Rise of the Biological Perspective in Germany* (Chicago: University of Chicago Press, 2009), 130–32.

33. Woldemar Geffcken to Strecker, no date, 1874; December 2, 1878; and September 9, 1882, HS-FM. On Bremen, see Bernhard Gerhard to Strecker, January 13, 1877. In his letters, Geffcken describes at some length the trade in butterflies in German towns. For an extensive—and very useful—1895 list of insect dealers (with a focus on lepidoptera) throughout Germany, see R. Friedlander and Sons, *Zoologisches Adressbuch: Namen und Adressen der lebenden Zoologen, Anatomen, Physiologen und Zoopalaeontologen* (Berlin: R. Friedlander, 1895), 1–76. This source also lists similar data for other countries, including France, England, and the United States. Hans Godeffroy, in Hamburg, had a transatlantic reputation (William Henry Edwards knew of him) and sold, among other things, Indian and Chinese butterflies.

34. Strecker to Holland, January 31, 1883, WH-CM. On the American use of Staudinger as a standard, see Berthold Neumoegen to Strecker, December 1, 1882, HS-FM: "I want to sell exotics . . . for cash only and my prices will be nearly as possible up to Staudinger's for perfect insects." See also Thomas Bean to William Henry Edwards, January 31, 1891, HE: "My prices are based chiefly on Staudinger's price list." And Will Doherty to his father, February 10, 1892, JMH: "Staudinger's price list is the basis of my contract with Doncaster, so that henceforth I can estimate the value of any of my catch with some closeness."

35. Staudinger to Isaac Martindale, July 9, 1892, Martindale Papers, Collection 533, Archives and Manuscripts, Academy of Natural Sciences, Philadelphia, and Andreas Bang-Haas to Skinner, June 26, 1891, HS-ANS, Collection 533.

36. Strecker to Staudinger (draft), July 16, 1894, HS-FM.

37. Strecker to Staudinger (draft), March 4, 1889, HS-FM.

38. Obituary, "Dr. Otto Staudinger," *Entomologische Zeitung* (Stettin: Entomolo-gischen Vereins zu Stettin, 1900), 389–95.

39. For Staudinger's trade with the museum, see, for example, Staudinger and Bang-Haas to the director of the Königlich Museum für Naturkunde, Berlin, March 24, 1890; October 14, 1893; November 1, 1893; November 10, 1893; October 6, 1894; March 8, 1895; January 12, 1897; February 26, 1897; March 20, 1897; March 23, 1897; October 28, 1897; March 24, 1898; and March 24, 1898, MNK. Most of these letters contain records of the money made, reaching into the thousands of marks. There are also several bills of sale in the museum's records. I would like to thank Hannalore Landsdorf, the chief archivist at this museum, for insight on Staudinger's favored relationship with the German king.

40. "Codizill to the last will and testament of Otto Karl Heinrich Richard

Staudinger," August 29, 1894, Staudinger Papers, MNK. Again, thanks to Hannalore Landsdorf for insight into the historical significance of this codicil.

41. "I agree with you," Fred Tepper, the American collector, wrote Strecker in 1876, "that Staudinger's list is rather poor in exotics." Tepper to Strecker, January 4, 1876, HS-FM.

42. Staudinger to Henry Stainton, January 12, 1880; and, on the butterfly auction, Staudinger to Stainton, October 1, 1880, Stainton Papers, NHM-LONDON.

43. See "Coleoptera—Liste VII (October 1888)"; "Coleoptera—Liste VIII (October 1889)"; "Conchylien—Preisliste Nr. II (November 1888)"; "Preisliste über Hymenopotera, Dipteren, Hemipteren, Orthopteren und Neuropteren (June 1890)"; and "Preisliste Frasstücke und biologische Objecte (December 1893)." I have seen these lists in the archives of Berlin's natural history museum (MNK).

44. Staudinger to Stainton, October 13, 1873, Stainton Papers, NHM-LONDON. Ribbe was an excellent naturalist, ambitious enough to leave the parent firm in 1877 to create his own business. His son, Carl, captured several new birdwing species on trips to the Malay Archipelago, mailing them to his father to put on the market. In the mountains of the Moluccas, just west of New Guinea, he netted seven thousand butterflies, later boasting that it took Alfred Russel Wallace eight years to get a mere thirteen thousand. See Ribbe to Strecker, July 17, 1878, and October 10, 1879, HS-FM (my translation). On Ribbe's departure from Staudinger's firm, see Ribbe to Strecker, July 8, 1877, and Neumoegen to Strecker, Feburary 5, 1877, HS-FM.

45. These figures are taken from a full run of Staudinger's catalogs, which cover the period from 1870 to 1901 and can be found in the Herman Strecker Papers, HS-FM, and in the William Holland Papers, WH-HSWP.

46. Stainton to Staudinger, September 26, 1880 (draft letter); and Stainton to Staudinger, January 12, 1880, Stainton Papers, NHM-LONDON.

47. See Henry Skinner to Strecker, April 14, 1893, HS-FM.

48. Staudinger to Stainton, September 22, 1880, Stainton Papers, NHM-LONDON.

49. On these Germans, see Richard H. Grove, *Green Imperialism: Colonial Expansion, Tropical Island Edens, and the Origins of Environmentalism, 1600–1860* (Cambridge: Cambridge University Press, 1995), especially 365–79, and Jorg Adelberger, "Eduard Vogel and Eduard Robert Flegel: The Experiences of Two Nineteenth-Century German Explorers in Africa," *History of Africa* 27 (2000): 1–29.

50. Albert Hyma, *The Dutch in the Far East: A History of Dutch Commercial and Colonial Empire* (Ann Arbor, MI: George Wahr, 1942), 185–86.

51. See Grove, *Green Imperialism*, 367–79.

52. The huge correspondence of William Doherty to his family gives a remarkable glimpse into the character of these places, JMH.

53. Doherty to his mother, August 20, 1886, JMH.

54. On Fruhstorfer's systematic contributions, see Phillip J. DeVries, *The Butterflies of Costa Rica, and Their Natural History,* vol. 2, *Riodinidae* (Princeton, NJ: Princeton University Press, 1997), 59.

55. Vladimir Nabokov to Dmitri Nabokov, May 30, 1970, in *Nabokov's Butterflies,* ed. Brian Boyd and Robert Pyle (Boston: Beacon, 2000), 673; and Fruhstorfer's section on the Brassolidae (a subdivision of the Nymphalidae, to which the morphos belonged), "The Macrolepidoptera of the American Faunistic

Region," in Adalbert Seitz's *The Macrolepidoptera of the World* (Stuttgart: Kernen, 1924), 5:353.

56. This experience obviously meant a good deal to Fruhstorfer, since he mentions it three times, with very little variation, in Seitz, *Macrolepidoptera,* 5:334, 340, 352.

57. L. Martin, "Hans Fruhstorfer," *Iris* (November 30, 1922): 1–8; and Hans Fruhstorfer (in Berlin) to Director Mobius, May 31, 1896, MNK; see, for a list of butterflies sold to the museum, c. May 1898, and, for bills for butterfly and beetle species, October 21 and November 12, 1898, MNK.

58. Fruhstorfer to Strecker, March 25, 1890, HS-FM.

59. Skinner to Strecker, December 2, 1889, HS-FM.

60. Henry Skinner, "The Effect of the War in Relation to Entomology," *EN* (September 1898): 173.

61. For an account of Thomas Horsfield's collecting, see his *Descriptive Catalogue of Lepidopterous Insects* (London, 1828).

62. Willis Weaver to Strecker, August 5, 1878, HS-FM.

63. Ibid.

64. Fred Knab to Strecker, February 23, 1885; September 11, 1885; and January 26, 1886, HS-FM. For Knab's ad, see "South American Lepidoptera," *Papilio* 4, no. 3 (1884): back page.

65. Knab to Strecker, January 26, 1886, HS-FM. Once he stayed overnight along the Amazon on a "terra calidas," an unstable land formation that slid into the river as he was sleeping, dragging him down and terrifying the life out of him.

66. Knab to his parents, May 13, 1885; and Knab Diary, 1884–86, Frederick Knab Papers, Smithsonian Archives, Washington, DC.

67. Oscar T. Baron to Strecker, February 12, 1883, HS-FM. See also Baron to Strecker, October 20 and 29, 1888, HS-FM; and Baron to Henry Edwards, October 23, 1881, HE.

68. Oscar T. Baron to Strecker, May 16, 1892, HS-FM.

69. Herbert Smith to Strecker, April 24, 1887, HS-FM.

70. Anna Weitzman and Christopher Lyal, "Integrating DNA Barcoding and Taxonomic Data" (Biodiversity Heritage Library, 2003).

71. Henry Elwes to Strecker, May 27, 1888, HS-FM; and *Biologia Centrali-Americana: Insecta-Lepidoptera-Rhopalocera,* vol. 1, *1879–1901* (London: R. H. Porter, 1915), pp. xxi, 262–357. The Smiths' species are sprinkled throughout.

72. Henry Elwes to Will Doherty, quoted in a letter from Doherty to his father, October 4, 1888, JMH.

73. Doherty to Ernst Hartert (from Senana, Sula Besi), October 8, 1897, Tring Correspondence, NHM-LONDON.

74. Wilhelm Petersen, *Lepidopteren-fauna von Estland* (Tallinn-Reval: Eesti, 1924).

75. Doherty to his mother, January 6, 1884, JMH. Even a missionary and his family in Calcutta, with whom Doherty was visiting, took up the pursuit, inviting Doherty to "joke about their attaching a butterfly net to their pastoral crooks."

76. Doherty to his mother, April 18, 1883, JMH.

77. Doherty to his father, August 31, 1883, JMH.

78. Doherty to his mother, February 17, 1884, JMH.

348 · NOTES TO PAGES 206-208

79. Doherty to his father, August 31, 1883, JMH.
80. Doherty to his mother, December 10, 1885, JMH.
81. Doherty, "A List of Butterflies Taken in Kumaon," *Journal of the Asiatic Society of Bengal* 2 (1886): 104, 121, 136. In the wide gorge of Sarju, below the Lesser Himalayas, through which the Sarju River passed on its way to the Ganges, he spent half a day trying to catch several specimens of a "magnificent species" of Morphidae that were soaring over the void, but they were "un-gettable," he wrote, "owing to the dangerous and precipitous nature of the place." He added, "They had true morphid flight, and always settled on the underside of leaves with folded wings" (121).
82. Ibid., 114.
83. Doherty to his mother, July 31, 1885, and September 14, 1885, JMH: "Thanks to the microscope, I am doing more new work in a month now than I used to do in three."
84. Michael Adas, *Machines as the Measure of Men: Science, Technology, and Ideologies of Western Dominance* (Ithaca, NY: Cornell University Press, 1989), 151. All the leading naturalists at the museum—James Wood-Mason, G. F. L. Marshall, and Lionel de Nicéville in particular—urged him to get something in print. They "fairly coaxed me into it," he wrote his mother. "It would make waves," they argued. Doherty to his mother, March 15, 1886, JMH.
85. He began with a discussion of the zones that divided up the Kumaon region, from the Great Indian Desert and the foothills and wet meadows at the base of the Himalayas to the forests of the Lesser and Greater Himalayas (the former three to seven thousand feet high, the latter up to ten thousand feet). The two Himalayan tracts, Doherty wrote, contained "all the typical Himalayan forms." Beyond ten thousand feet one found the alpine valleys, where the butterflies were "chiefly palearctic." Doherty examined the diversity of seasonal species of butterflies—the dry and wet forms—that distinguished the Kumaon. He speculated about the ocelli, or large eyespots, on the underside of the wings of wet-season genera (for protection against predatory birds). Before Edwards did (although both men were unaware of it), he observed the symbiotic relationship between small blue butterflies (Lycaenidae) and ants, which he had seen in nature. Finally, he presented a tentative new reclassification of butterfly genera based on eggs alone, something he had been thinking about for many months but only now had the courage to publish. Doherty knew that Scudder had planned to publish work on eggs as a basis for generic classification. But his system departed from Scudder's by resting on "ten genera instead of one (North America being very poor in genera)," and by superceding Scudder's "absolute Linnaean nomenclature, which correct or not, is certainly unintelligible to the majority of naturalists." Doherty to his father, March 15, 1886.
86. Doherty to his father, May 28, 1885, JMH.
87. W. Dönitz, "Ed. G. Honrath," in *Berliner Entomologische Zeitschrift* 39, no. 2 (1894): 319–20.
88. Doherty to his father, August 31, 1883, JMH.
89. Doherty to his mother, December 26, 1886, JMH; and Doherty to Arthur Butler, June 13, 1887, NHM-LONDON.
90. These words belong to Charles Dury, a Cincinnati naturalist and a close friend

of Doherty's. They appeared in an 1894 interview about Doherty for a Cincinnati newspaper; see typescript, Charles Dury, "On Butterfly Wings," JMH.

91. "The gist of it is that I'll be much richer than I thought." Doherty to his mother, November 24, 1888, JMH.

92. Doherty to William Holland, quoted in Holland's "Asiatic Lepidoptera."

93. "I thought I was gone," he told Henry Elwes in June 1890, "for there was no chance of my being found there, and it seemed quite impossible to do it. It took me eight hours hard work to do it." Quoted in Henry Elwes, "On Butterflies Collected by W. Doherty in the Naga and Karen Hills and in Perak: Part I," *Transactions of the London Entomological Society* (1891): 252.

94. On Doherty's unprecedented collecting in these places, see Doherty to Arthur Butler, June 13, 1887, NHM-LONDON; Doherty, "The Butterflies of Sumba and Sambawa: Part II," *Journal of the Asiatic Society of Bengal* 60, no. 2 (1891): 146; Elwes, "On Butterflies Collected by W. Doherty," 249–88; and William Holland, "The *Lepidoptera* of Buru," *Novitates Zoologicae* (1891–92): 54–55, 555–91.

95. Doherty to Charles Dury, no date, but, judging by the postmark, 1888; Charles Dury Papers, Cincinnati Museum of Science.

96. Elwes, quoted in Doherty to his father, May 9, 1891, JMH.

97. Doherty to his father, June 17, 1890, JMH.

98. Colonel G. B. Mainwaring, *Dictionary of the Lepcha-Language* (Berlin, 1898), 461–64, 489–90, 492, 512, and 516. The butterfly entry is on p. 468, demon on pp. 479–80.

99. Doherty to his mother, February 20, 1890, JMH.

100. Doherty to Julia Doherty, September 30, 1891; to his mother, December 31, 1891, JMH.

101. Ernst Hartert, *Aus den Wanderjahren eines Naturforschers* (Berlin: R. Friedlander, 1901–2), 209. On Doherty meeting Hartert at the Perak Museum, and on Wray, see pp. 214–15. On Hartert's career in England, see Paul Lovejoy's foreword, pp. xvii–xxviii, to Paul Staudinger's account of this expedition, *Im Herzen der Haussaländer* (Berlin, 1889), translated by Johanna Moody as *In the Heart of the Hausa State* (Athens: Ohio University Center for International Studies, 1990). On the men for whom Hartert collected, see Doherty to his mother, June 6, 1888, JMH.

102. See Lovejoy's foreword to Staudinger, *In the Heart of the Hausa State*. Staudinger did, in fact, collect a few insects, including some "very rare long-tailed Saturnia" and a new beautiful beetle species, *Simorrhina staudingeri*, caught in Zaria, Sudan. See Staudinger, *In the Heart of the Hausa State*, 8, 155.

103. Before the trip ended, they had toured the Hausaland, a tribal area in northern Nigeria and the Sudan. See Staudinger, *In the Heart of the Hausa State*, Lovejoy's foreword and p. 55.

104. Hartert, *Aus den Wanderjahren*, 3, 81; and Staudinger, *In the Heart of the Hausa State*, 8, 55, 126–27, 155, 198, 205, and Lovejoy's foreword.

105. "When he met Mrs. Kunstler in Perak," soon after her husband's death at sea in 1888, Doherty further explained to his mother, Hartert "outraged her feelings by asking her to tell him all the late K's tricks of the trade." See Doherty to his mother, June 6, 1888, JMH.

106. "It will be very good for me if he comes," Doherty wrote his mother, "because

I can learn practical bird stuffing. Also I can talk German with him which will be very good for me." Doherty to his mother, July 18, 1888, JMH.

107. Hartert, *Aus den Wanderjahren,* 228.

108. Doherty to his mother, August 18 and November 24, 1888; and Doherty to his sister, Tata, December 4, 1888, JMH.

109. Ernst Hartert, obituary of Doherty in *Novitates Zoologicae* 8 (1901): 497.

110. Hartert, *Aus den Wanderjahren,* 236 and 235.

111. Doherty to his mother, November 24, 1888, JMH.

112. Doherty to his father, December 27, 1891, JMH. On Doherty's thoughts on Daisy Smith, see Doherty to his sister Tata, October 4, 1888, JMH.

113. Doherty to his cousin May, July 20, 1889, JMH.

114. Doherty to Scudder, August 16, 1895, SS-BMS.

115. On the deafness, see Doherty to his mother, April 14, 1890, JMH; on the infections and diseases, see Doherty to his mother, May 18, 1888, and January 14, 1893, JMH; and Doherty to Hartert, December 1, 1899, Tring Correspondence, NHM-LONDON.

116. Doherty to his mother, April 18, 1883, JMH.

117. On these various figures, see Doherty to his mother, September 28, 1890; Doherty to Julia Doherty, May 15, 1892; and Doherty to his mother, January 19, 1893, JMH.

118. Doherty to his mother, November 11, 1889; and Doherty to his father, October 1, 1890, JMH.

119. Doherty to his mother, April 9, 1892, JMH. "Elwes and I correspond a great deal now," he wrote his mother in 1891 (Doherty to his mother, September 11, 1891, JMH).

120. He no longer wanted to live in the United States, he favored Trinidad, where he imagined bringing his Lepchas as well as the entire Doherty clan. Doherty to Harlan Doherty, March 30, 1891; and to his mother, September 28, 1890, JMH.

121. Doherty to Lili Doherty (from Macassar), June 20, 1896, JMH.

122. Doherty to Edward Janson, November 22, 1897 (from Ternate), and May 1898 (from Albion Place), Entomological Library, NHM-LONDON.

123. Doherty to his father (from Buru, Moluccas), March 27, 1897, JMH.

124. Doherty to his mother (from Dorey, New Guinea), June 12, 1897, JMH.

125. Doherty to Hartert (from Senana, Sula Besi), October 8, 1897, Tring Correspondence, NHM-LONDON.

126. Doherty to Lili Doherty, June 20, 1896, JMH.

127. Doherty to his mother (from Dorey, New Guinea), June 12, 1897; and Doherty to Lili Doherty, June 20, 1896, JMH.

128. Doherty to Harlan, March 30, 1891; to his mother, January 14, 1893; and to his father, November 1, 1893, JMH.

129. He requested especially the Franklin Square or Seaside Library editions of novels, early precursors of paperbacks that he could easily carry around with him. Doherty to his mother, August 11, 1892, JMH.

130. Doherty to his father, c. 1889, and from Buru, Moluccas, June 20, 1896, JMH.

131. Doherty to his father, November 24, 1895, JMH.

132. Elwes to Doherty, May 9, 1891; November 1892; and July 19, 1891, JMH. For Elwes's article on Doherty's butterflies, see Henry Elwes, "On Butterflies Collected by W. Doherty," 259.

133. Doherty to his mother, June 3, 1893, JMH.

134. Doherty to his father, November 24, 1895, JMH.

135. Doherty to his mother, June 3, 1893, JMH.

136. On Rothschild's naming, see Doherty to Rothschild, September 25, 1895, Tring Correspondence, NHM-LONDON. "Thank you for the Attacus dohertii," Doherty wrote. "It is very pleasant to have my name mentioned in connection with various species, as you have done."

137. Doherty to Rothschild, November 24, 1895, NHM-LONDON.

138. Doherty to his mother, June 1893, JMH.

139. "I can see the Highland House on the Neversink Mountain . . . from the window of my butterfly room," he wrote to Isaac Martindale, June 1890 (no day), Martindale Papers, Academy of Natural Sciences, Philadelphia, Pennsylvania.

140. Alexander Humboldt, introduction, *Cosmos*, vol. 1 (1845; repr., Baltimore: Johns Hopkins University Press, 1997), 40.

141. O. D. Foulks to Strecker, August 2, 1894, HS-FM.

142. Strecker to Holland, February 15, 1882, WH-CM.

143. On "three times the size," see Strecker to Hermann Hagen, October 6, 1881, Letters to Hermann Hapen, EML. The specimen figures come from many contemporary accounts, but see "Butterflies and Moths," *Rochester Morning Journal*, July 6, 1887; "Reading," in Morton Montgomery, *History of Berks County* (Philadelphia, 1886), 807–8; and *Cincinnati Communer Gazette*, November 26, 1892. These reports can be found in Box 60, folder 397, "Corporate Entities," HS-FM.

144. See Strecker to Holland, March 31, 1887, WH-CM. Hewitson paid $1,300 for the first two available specimens in 1875 (Hewitson to Strecker, April 4, 1875, HS-FM). Five years later, a dozen male specimens were scattered about in various European collections (Arthur Butler to Henry Edwards, February 5, 1883, HE). The Berlin collector Eduard Honrath paid 400 marks for his specimen, collected in Gabon by Fritz Krickledorff (Honrath to Strecker, September 21, 1881, HS-FM). In 1883 and '84, Watkins & Doncaster sold to Americans (see *Papilio* 3 [April 1883]: 85; and 4 [January 1884]. For a description of the insect, see Bernard d'Abrera, *Butterflies of the World* (London: Hill House, 2001), 147.

145. Elwes to Strecker, August 15, 1890, HS-FM.

146. On the terms "stock of monstrosities" and "my especial mania," and on the silver butterfly, see Strecker to Holland, November 29,1882, WH-HSWP.

147. Ibid.

148. Strecker to Samuel Henshaw, May 10, 1889, Museum of Comparative Zoology, Cambridge, Massachusetts.

149. See William Holland, "Maternal Ancestry," Box 4, folder 3, WH-HSWP; and Holland to Dale Pontius, December 5, 1923, WH-CM.

150. Holland, "Butterflies," *Mentor* (August 2, 1915): 1–2.

151. Holland to his parents, February 9, 1868, WH-HSWP; and Holland to Dale Pontius, December 5, 1923, WH-CM.

152. Holland, "Ecclesiastical Relations and Positions," in a typed sketch of his own life, "W. H. Holland," Box 16, folder 1, WH-HSWP; on the prestige of the Princeton seminary, see Barry Werth, *Banquet at Delmonico's: Great Minds, the Gilded Age, and the Triumph of Evolution in America* (New York: Random House, 2009), 76.

153. Holland to his parents, November 20, 1873, WH-HSWP.
154. Toward the end of his life, his parents long dead, Holland tried publicly to justify his decision in the face of criticism that he might have switched faiths for nonreligious reasons. "I was especially attracted to the Presbyterian Church," he explained in a journal on religion, "because it seemed to me more than any other communion in America to stand for liberty of conscience. It seemed to me to be of all denominations the most American in America. There appeared to be in it more of the spirit of the fathers, of the Republic, who had laid broad and deep the foundations of religious liberty throughout the land." See "Reminiscences and a Plea," undated draft but probably early 1920s, Box 16, folder 1, WH-HSWP.
155. Francis Holland to William Holland, November 25, 1873, and Holland to his parents, December 21 and 22, 1873, WH-HSWP.
156. Holland to his parents, November 7, 1873, WH-HSWP.
157. On Carrie's inheritances and Holland's executorships, see Holland to his parents, October 25, 1880, and December 7, 1889, WH-HSWP; on John Moorhead Sr. and Jr., see Holland to Scudder, December 25, 1889, SS-BMS, and John Ingham, *Making Iron and Steel: Independent Mills in Pittsburgh* (Cleveland: Ohio State University Press, 1981).
158. *Papilio* 3, no. 1 (1883): 27.
159. Holland to his parents, February 25, 1886; and William Henry Edwards to Holland, March 8, 1886, WH-HSWP.
160. Catalog 35, Staudinger, WH-HSWP.
161. Elizabeth Martindale to Strecker, March 15, 1892, HS-FM
162. Doherty to his father, January 14, 1890, JMH.
163. Doherty to his father, January 24, 1891, JMH.
164. Doherty to Holland, February 14, 1893, WH-CM.
165. Henry Skinner to Strecker, May 11, 1889, HS-FM.
166. See opening "notes," *Deutsche Entomologische Zeitschrift* 2 (1889); Neumoegen to Strecker, May 2, 1889; April 6, 1889; March 11, 1890; and April 28, 1893, HS-FM; and Staudinger to Scudder, April 6, 1889, SS-BMS. In 1884, when King Leopold of Belgium announced himself the sole "proprietor" of the African Congo, as if it had been "vacant land," and dispatched an expedition, Neumoegen purchased a piece of the enterprise, hoping to get "the best man" to find butterflies for him (*New York Times*, November 28, 1887). On Leopold, see Catherine Coquery-Vidrovitch, "Western Equatorial Africa," in *The Cambridge History of Africa*, ed. Oliver and Sanderson, 316–18.
167. Doherty to his father, November 1, 1887, JMH.
168. Doherty to his father, April 25, 1888; Doherty to his mother, November 11, 1889, JMH; and Neumoegen to Strecker, March 11, 1890, and April 28, 1893, HS-FM. For Doherty's fondness for Neumoegen, see Doherty (from Assam) to his father, November 11, 1889, and Doherty to his mother, November 11, 1889, JMH.
169. Doherty, "Green Butterflies," *Psyche* 6 (1891): 68.
170. "A Butterfly Without a Price," *Entomologist* 27, no. 368 (1894): 65–66; and Doherty, "The Butterflies of Sumba and Sambawa: Part II," 157.
171. Miller Evans of Pottstown, Pennsylvania, to Strecker, May 24, 1882, HS-FM.

172. *The Naturalists' Directory,* ed. Samuel Cassino (Boston, 1886), 154; and F. Cormack (dealer) to Strecker, September 13, 1897, HS-FM.

173. By 1895, many dealers in the United States were making their living from buying and selling insects. Among them were George Franck, old John Akhurst, and F. Cormack in New York City; Henry Engle in Pittsburgh (a seller of exotics); Charles Wiley in Manhattan, Kansas; L. E. Richsecker in Sonoma, California; F. W. Dieckman in San Francisco; H. H. Newcomb in Boston; and Ward's Natural Science Establishment in Rochester, New York, long a popular business, which in the 1890s decided to sell the world's insects. For citations of these men, see the *Zoologisches Adressbuch,* vols. 1 and 2. All the American dealers mentioned here can be found cited in these volumes, excellent sources for information on dealing, and on every other facet of naturalist activity in the world. For informal dealers, see Henry K. Burrison (who "devoted two evenings a week to my fad") and James Blake, Edward Owen, James Bailey, Oscar Baron, Adolph Conradi, Carl Braun, Edward Warren, and Herbert Loitloff. For the Burrison quote, see Burrison to Strecker, December 27, 1900, HS-FM.

174. Strecker to Charles MacGlashan, September 23, 1885, Charles F. MacGlashan Papers, c. 1878–1899, BANC MSS C-B, The Bancroft Library, University of California, Berkeley; and Augustus Mundt to Strecker, December 21, 1879, HS-FM.

175. Strecker to Holland, March 25, 1884; Edward Kremp to Strecker, October 12, 1884, HS-FM; and Strecker to George French, March 20, 1883, GF.

176. On the number of duplicates, see Strecker to Charles MacGlashan, September 23, 1885, HS-FM.

177. Charles Dury to Strecker, c. 1874, and August 10, 1876, HS-FM.

178. Adolph Eisen to Strecker, December 21, 1880, HS-FM. All the quotes in this paragraph come from this letter. See also Eisen to Strecker, December 1, 1880, HS-FM.

179. Simon Seib to Strecker, August 25, 1885, HS-FM.

180. O. D. Foulks to Strecker, August 8, 1891, HS-FM.

181. Mrs. M. E. Truman to Strecker, December 1, December 12, and December 25, 1892, HS-FM.

182. Emily Morton to Strecker, February 2, 1880; and Adrian Latimer to Strecker, October 17, 1879, and February 10, 1888, HS-FM.

183. Frank Snow to Strecker, September 24, 1879; and Elison Smyth to Strecker, December 19, 1883, HS-FM.

184. George Ehrman to Strecker, February 6 and March 11, 1886, HS-FM.

185. Ehrman to Strecker, May 7, 1891; July 9, 1894; June 8, 1888; and July 15, 1886, HS-FM.

186. Louis Glaser (of Allegheny City, Pennsylvania) to Strecker, January 6, 1885.

187. Ehrman to Strecker, February 26, 1893, HS-FM. On the damage inflicted by the gas wells, see William Holland, *The Moth Book: A Guide to the Moths of North America* (1903; repr., New York, 1968); on others who collected at the wells, see Louis Glaser of Beaver Falls, Pennsylvania, September 9 and 26, 1888.

188. George Santayana, *The Sense of Beauty: Being the Outline of Aesthetic Theory* (1896; repr., New York: Dover, 1961), 95.

7. Butterflies at the Fair

1. On Samuel Scudder at the fair, William Henry Edwards's family at the fair, and Edwards's inability to attend, see William Henry Edwards to Samuel Scudder, September 2, 1893, SS-BMS. Since their "introduction into our cities," Scudder observed in *BEUSC*, "entomologists have made use of electrical lights for the capture of insects, many nocturnal animals being attracted from the surrounding country by the brilliancy of the light" (vol. 1, p. 377).

2. Selim Peabody, *The White City (as It Was)* (Chicago, 1894), from his concluding statement, "The Story of the White City," unpaginated; *Official Guide to the World's Columbian Exposition*, handbook ed. (Chicago, 1893), 161; and Ralph Julian, *Chicago and the World's Fair* (New York, 1893), 114.

3. On Smith's display, see Herbert Smith to Ezra Cresson, December 2, 1892, Academy of Natural Sciences, Philadelphia. For David Bruce, see Bruce to Isaac Martindale, June 15, 1891, Isaac Martindale Collection 533, Academy of Natural Sciences, Philadelphia; and "Historical Department," in *Report of Board of World's Fair Managers of Colorado* (Denver, 1894), 47. On French's exhibit for Illinois, see the account covering "the geographical distribution of Illinois Butterflies shown by a set of specimens of the species: (a) Common to Illinois and the Atlantic Slope; (b) common to Illinois and the Pacific Slope; (c) common to Illinois and Europe; (d) found throughout Illinois; (e) found in northern Illinois only; (f) found in southern Illinois only." This exhibit "as well as the ornithological and ichthyological exhibition have been pronounced by scientists as superior to any exhibition of the kind heretofore attempted. The whole division will be found very interesting to the student of nature or even the casual observer." Anonymous, *The Illinois Building and Exhibits Therein at the World's Columbian Exposition, 1893* (Chicago: John Morris, 1893), 54–55. I have been unable to confirm that Smith actually made his display at the New York exhibit.

4. William Holland to Scudder, October 3, 1892, SS-BMS.

5. The Pennsylvania managers had written Strecker in July (see John Woodward to Herman Strecker, July 4, 1892, HS-FM). The stuffed bird and egg exhibition covering all the state birds at the Pennsylvania fair site was enormous, so one can only imagine what the butterfly display might have looked like had someone managed to fill it. "Probably no similar exhibit on the grounds elicited so much attention and commendation as this," the catalog asserted. See the text and photographs of the State of Pennsylvania, *Catalogue of Exhibits of the State of Pennsylvania and of Pennsylvanians at the World's Columbian Exposition* (Harrisburg: Clarence M. Busch, State Printer of Pennsylvania, 1893), 164–71. For the initial offer to Strecker and for its final withdrawal, see John Woodward to Strecker, December 25, 1892, and telegram dated January 21, 1893, delivered by the Philadelphia, Reading and Pottsville Telegraph Company (connected to Western Union), HS-FM.

6. Holland to Scudder, October 3, 1892, SS-BMS.

7. *Official Guide to the World's Columbian Exposition*, 61, 65, 96–97, 103; and Julian, *Chicago and the World's Fair*, 116, 140–41.

8. *Official Guide to the World's Columbian Exposition*, 46–47.

9. Katherine Peabody Girling, *Selim Hobart Peabody* (Urbana-Champaign: Uni-

versity of Illinois Press, 1923), 191; Henry Adams, *The Education of Henry Adams* (Boston, 1907; repr., 1918), 343–44. It would be a mistake to exaggerate the significance of the commercial at the cost of ignoring the purely cultural or educational (meaning noncommercial) character of the fair. According to the historian Jim Gilbert, in *Perfect Cities* (Chicago: University of Chicago Press, 1993), "the commercial impression generated by the White City, was not, perhaps, always successful as a strategy for the sale of goods. Indeed, several articles in the important advertising journal *Printer's Ink* suggest that attention to culture overwhelmed the regard for commerce" (102). The American people did not, then, automatically equate culture with commerce, which says something about them and this age. At the same time, the commercial and the cultural often became so entangled at so many points in the fair as to seem one and the same thing, a spectacular intermingling of old and new, nature and artifact, money and art. On the free darkrooms, see Julie K. Brown, *Contesting Images: Photography at the World's Columbian Exposition* (Tucson: University of Arizona Press, 1994), 94–97.

10. Peabody, *The White City*, in the unpaginated concluding section, "The Story of the White City"; *Official Guide to the World's Columbian Exposition*, 15, 49, 113; W. H. Edwards (American consul general in Berlin), "Europe at the World's Fair," *North American Review* 155, no. 432 (1892): 623–30; and Brown, *Contesting Images*, 27.

11. Emory L. Kemp, *The Great Kanawha Navigation* (Pittsburgh: University of Pittsburgh Press, 2000), 24–25, 71–72.

12. Julian, *Chicago and the World's Fair*, 97; and John C. Trautwine, "Two Great Railroad Exhibits at Chicago," *Cassirer's Magazine* (January-February 1894; repr., Lindsay Publications, 2004), 20.

13. Peabody, *The White City*, concluding statement, "The Story of the White City."

14. Harold Platt, *The Electric City: Energy and the Growth of the Chicago Area, 1880–1930* (Chicago: University of Chicago Press, 1991), 16–37, 59–92; David Nye, *Electrifying America: Social Meanings of a New Technology, 1880–1940* (Cambridge, MA: MIT Press, 1997), 29–38; *Official Guide*, 22–24, 52–54; and John Findling, *Chicago's Great World's Fairs* (Manchester: Manchester University Press, 1994), 82–83.

15. Brown, *Contesting Images*, 25, 30.

16. *Guide Through the Exhibition of the German Chemical Industry: Columbian Exposition in Chicago, 1893* (Berlin: Julius Sittenfeld, 1893), 58, 63–65. German chemical companies led all other German businesses in support of exhibiting at the fair (1).

17. I want to note here that previous "industrial expositions" in the United States on the regional and local levels were often mixed events, displaying items from "natural history" as well as manufactured goods and machines. See, for instance, the program of the "Fifth Cincinnati Industrial Exposition, 1874" with a special "department" on "natural history," including "best collection of insects and cocoons" (in the Charles Dury Correspondence with Strecker, HS-FM).

18. George Santayana, "The Genteel Tradition at Bay," written in 1931 and republished in *The Genteel Tradition* (Cambridge, MA: Harvard University Press, 1967), 185; and *The Sense of Beauty: Being an Outline of Aesthetic Theory* (1896; repr., New York: Dover, 1961), 64, 108.

19. Lester Ward, "Art Is the Antithesis of Nature," an 1884 lecture republished in Henry Steele Commager, ed., *Lester Ward and the Welfare State* (Indianapolis: Bobbs-Merrill, 1967), 11, 77; and Comte de Buffon, "Initial Discourse," in *From Natural History to the History of Nature,* ed. John Lyon and Phillip Sloan (Notre Dame, IN: University of Notre Dame Press, 1981), 111.

20. This argument has been partly influenced by George Basalla's *The Evolution of Technology* (New York: Cambridge University Press, 1988), especially chapter 1, in which he compares natural with "artifactual diversity," 1–15.

21. See Wolfgang Schivelbusch, *The Railway Journey: The Industrialization and Perception of Time and Space* (Berkeley: University of California Press, 1987), 89; and Robert Kohler, *All Creatures: Naturalists, Collectors, and Biodiversity, 1850–1950* (Princeton, NJ: Princeton University Press, 2006), especially pp. 1–46.

22. On the creation of "twilight zones" caused by railroads, see Robert Kohler's fine discussion in *All Creatures,* 1–46. Kohler also uses "inner frontiers" as a substitute for "twilight zones."

23. Ibid., 18. On the nature literature produced during this period, see the books of Liberty Hyde Bailey.

24. Samuel Klages to Henry Skinner, June 19, 1899, HS-PAS.

25. William Henry Edwards to Theodore Mead, October 19, 1888, and October 26, 1872, TM.

26. Edwards to Joseph Lintner, December 2, 1878, JL.

27. William Henry Edwards to Henry Edwards, May 20, 1873, and January 30, 1875; and Edwards to William Greenwood Wright, October 23, 1893, WHE-SA.

28. William Edwards to Theodore Mead, December 2, 1886, TM.

29. William Henry Edwards describes this activity in his Entomological Diary, July 11, 1884, WHE-SA. In 1886, Willie filed his own patent for a "novel use of natural gas," or gas so powerfully magnified and channeled through a heated blowpipe as to convert slag and melted metal into Bessemer steel; see Willie Edwards to Theodore Mead, December 21, 1886, TM.

30. Edwards to Wright, October 23, 1893, WGW.

31. On clientele, see Edwards to Scudder, May 1, 1894; and Edwards to Wright, October 13, 1896, WGW.

32. Entomological Diary, June 23 and 25, August 15, and August 10, 1894.

33. Edwards to Wright, June 29, 1895, WGW. Three years later, Edwards commemorated the date again for Wright: "Three years ago, this AM I waked up at Glenwood Spr. and took my first walk up Grand River, and saw *Dionysius* in the flesh. What a good two months I had!" (June 29, 1897).

34. For Kirby and Spence, see *Introduction to Entomology,* vol. I (1843; repr., Elibron Classics, 2005), 532; on Gustav Belfrage, see Samuel Geiser, *Naturalists of the Frontier* (Dallas: Southern University Press, 1952), 236; on light trap development, see "Chronological History of the Development of Insecticides and Control Equipment from 1854 to 1954" (Washington, DC: USDA Agricultural Research Service, 1954); and on Dury, see Dury to Strecker, no date but c. 1873, HS-FM. The same thing happened in other countries at the same time. See Von F. Wesely, "Das elektrische Licht und Die Schmetterlinge," *Ento-*

mologische Zeitschrift (July 1888): 37; and Henry S. Saunders, "Collecting at the Electric Light, 1886," *CE* (February 1887): 1–2.

35. *BEUSC,* vol. 1, p. 377. "Nowadays," echoed Charles Valentine Riley in 1892, "the electric lights in all large cities furnish the best collecting places, and hundreds of species may be taken in almost any desired quantity." C. V. Riley, *Directions for Collecting and Preserving Insects* (Washington, DC, 1892), 51. David Bruce used many "lights" to collect moths. "I am trying to arrange a portable Lime Light (or Calcium light)," he wrote Henry Edwards in 1884, "so I can have something more effective than the Kerosene lamp when I go to the mountains next year—last year I spent two weeks at Buffalo Creek, forty miles from Denver and about 1000 feet higher, and had good luck with a reflector, Kerosene Lamp" (Bruce to Edwards, December 23, 1884, HE). Bruce also relied on the new electrical globes in Denver and elsewhere.

36. George Franck to Strecker, July 16, 1884, HS-FM; George L. Hudson to John Smith, August 9 and November 8, 1886, Smithsonian Archives, Record Unit 138, Division of Insects, Incoming Correspondence, 1878–1906; and Edward Warren to Strecker, June 23, 1884, HS-FM.

37. John Morris to Strecker, December 20, 1884, HS-FM.

38. Max Albright to Strecker, November 5, 1894. The Albright material comes from these other letters to Strecker: October 30, 1893; April 15, 1894; August 5, 1894; and June 14, 1896, HS-FM.

39. See editorial, *EN* 2, no. 8 (1891): 8–9. On the historical significance of the halftone process, see Neil Harris, "Iconography and Intellectual History: The Halftone Effect," first published in 1979 and republished in Harris, *Cultural Excursions* (Chicago: University of Chicago Press, 1990), 304–17.

40. F. N. Doubleday to Holland, April 13, 1900, WH-HSWP; and Neltje Blanchan, *Bird Neighbors* (New York: Garden City Publishers, 1922; from the 1904 Doubleday edition), x.

41. Holland to his parents, November 6–9, 1891, Box 24, folder 7, WH-HSWP.

42. *Pittsburgh Dispatch,* October 29, 1899, in Holland Scrapbook, p. A718; and Holland to his parents, December 3, 1892, WH-HSWP. By 1899, students from England, Germany, Mexico, and Canada were attending the university.

43. Carnegie began calling "at our cottage," and then "we called on his." Quoted in *The Life of Andrew Carnegie* by Burton Hendrick and Daniel Henderson (Garden City, NY: Doubleday, Doran, 1932), 1:227–28. On the summer resort, see 1:114. Holland performed the role of family naturalist, or "the butterfly man," to Margaret, Carnegie's young daughter. See Holland to Scudder, May 8, 1890, SS-BMS.

44. Holland to Scudder, July 11, 1890, SS-BMS.

45. Holland to parents, February 10, 1890, WH-HSWP.

46. Even before the museum had left the drawing boards, he had bought outright for Holland a very costly collection of butterflies caught near Sikkim, one of the first of its kind ever made. "He does this at *my* suggestion," Holland reported to his parents. "He is a very generous man." Holland to Isaac Martindale, October 3, 1892, Martindale correspondence, Academy of Natural Sciences, Philadelphia; and Holland to his parents, November 23, 1892, WH-HSWP.

47. Edward Klages to Henry Skinner, December 29, 1903, HS-ANS.

48. Holland to W. Baxter, May 9, 1902, Holland Letterbook, WH-CM.

49. On Carnegie's bequest, see William Schaus to Henry Skinner, October 20, 1919, HS-ANS: "I didn't know that Carnegie had left Holland 5000 a year, and as he has a rich wife, I think you or I could have made better use of the money, but to him who hath much shall be given."

50. See his handwritten remembrance beginning "In the year 1879," c. 1920, Box 24, Holland Papers, WH-HSWP.

51. Holland to his parents, February 10, 1890, WH-HSWP.

52. Edwards to Scudder, April 27, 1899, SS-BMS.

53. Holland to Scudder, June 21, 1898, WH-HSWP. On Doubleday's distribution and advertising, see Doubleday to Holland, October 6, 1898, WH-HSWP.

54. George Iles, *Flame, Electricity, and the Camera* (New York: Doubleday and McClure, 1900).

55. Holland to William Wesley and Son, January 20, 1899, Director's Correspondence, WH-CM.

56. Holland to Professor John K. Richardson, Wellesley Hills, Massachusetts, January 25, 1904, Director's Correspondence, WH-CM.

57. Vladimir Nabokov to Edmund Wilson, January 18, 1944, in Brian Boyd and Robert Pyle, eds., *Nabokov's Butterflies* (Boston: Beacon Press, 2000), 299 (see also p. 620). Holland, *The Butterfly Guide: A Pocket Manual for the Ready Identification of the Commoner Species Found in the United States and Canada* (New York: Doubleday, 1915, 1925). Holland dedicated this guide "To the Boy Scouts of America."

58. Clench's review of *The Butterfly Book* in *Lepidopterist's News* 1 (May 1947).

59. See Mark V. Barrow Jr., *A Passion for Birds* (Princeton, NJ: Princeton University Press, 1998), 127–53; and Jennifer Price, *Flight Maps: Adventures with Nature in Modern America* (New York: Basic Books, 1999), 57–109.

60. I discovered this shift while reading a very useful German sourcebook, *Zoologisches Adressbuch,* published in two editions (1895 and 1900), in German, English, and French by Friedlander and Sons in Berlin, comparisons of which reveal for the United States a remarkable decline in bird collecting with a consequent rise in butterfly collecting, all over only five years, and due without doubt to laws passed in those years against bird collecting. An unparalleled source, these volumes give the names, addresses, and nature specialty of men and women, traders and collectors, throughout Europe, England, and the United States.

61. Lewis to Holland, April 24, 1899, WH-HSWP.

62. Schlarbaum to Holland, January 14, 1901, WH-HSWP.

63. Cary to Holland, July 23, 1899, WH-HSWP.

64. "The disappearance of the Rosy Maple Moth," Holland wrote in his *Moth Book,* his 1903 sequel to *The Butterfly Book,* was "due no doubt to the gas wells and furnaces, which licked up in their constantly burning flames other millions of insects." Holland, *The Moth Book: A Guide to the Moths of North America* (1903; repr., New York, 1968), 95.

65. Kenneth D. Frank, "Effects of Artificial Night Lighting on Moths," in *Ecological Consequences of Artificial Night Lighting,* ed. Catherine Rich and Travis Longcore (Washington, DC: Island Press, 2006), 306. This essay appears in part 5 of the volume, which deals exclusively with invertebrates.

66. James Sinclair, *Instructions for Collecting and Preserving Valuable Lepidoptera* (Ocean Park, CA, 1917), 23.

67. Kenneth D. Frank, "Effects of Artificial Lighting on Moths," and Gerhard Eisenheis, "Artificial Night Lighting and Insects: Attraction of Insects to Streetlamps in a Rural Setting in Germany," in Rich and Longcore, eds., *Ecological Consequences of Artificial Lighting*, 281–344.

68. The term "suppressing" comes from Patrick Maynard, *The Engine of Visualization: Thinking Through Photography* (Ithaca, NY: Cornell University Press, 2000), 77.

69. Anna Botsford Comstock, *Handbook of Nature Study* (Ithaca, NY: Comstock Publishing, 1911), 16.

70. Herbert Smith to Olive Thorne Miller, March 29, 1897, and Smith to L. Hays (sp.?), April 22, 1897, William Holland Letterbook, vol. 2; for Comstock's views, see her *Handbook of Nature Study*, p. 8.

71. In a letter to Scudder in 1888, Riley wrote that "it is perhaps gratifying to our 19th century vanity, but I feel that, with all our modern processes of illustration, when it comes to naturalness, accuracy, and clearness, the old illustrations of Sepp and of Rosel Von Rosenoff beat anything that we do today, and when it comes to adolescent stages I cannot even except Edwards' figures or your own." See Riley to Scudder, September 12, 1888, SS-BMS.

72. Frederic Clements to Holland, February 7, 1901, Holland Papers, WH-HSWP. On Clements, see Donald Worster, *Nature's Economy* (New York: Cambridge University Press, 1994), 209–20.

73. Gene Stratton-Porter, *Moths of the Limberlost* (New York: Doubleday, Page, 1912), 13–14, 111; and *A Girl of the Limberlost* (Garden City, NY, 1909), 108, 149–51.

74. Philip Ball, *Bright Earth: Art and the Invention of Color* (Chicago: University of Chicago Press, 2003), 220.

75. Strecker to Skinner, May 21, 1899, HS-ANS.

76. F. C. Schaupp, "Insect Life on Coney Island," *Bulletin of the Brooklyn Entomological Society* 2, nos. 10–11 (1880): 79–81.

77. Ernest Oslar (of Denver) to William Barnes, May 21, 1919, William Barnes Correspondence, Records of the Bureau of Entomology and Plant Quarantine, 1863–1956, Record Group 7, National Archives at College Park, MD.

78. Preston Clark to William Barnes, May 23, 1916, Barnes Papers, Smithsonian Institution, Washington, DC; and Buchholz obituary, "Otto Buchholz," *Journal of the Lepidopterists' Society* 13, no. 1 (1959): 27–29.

79. On the new industrial order in West Virginia, see John Alexander Williams, *West Virginia* (New York: Norton, 1976), especially chapter 5, "Paint Creek," pp. 130–58.

80. Ronald L. Lewis, *Transforming the Appalachian Countryside: Railroads, Deforestation, and Social Change in West Virginia, 1880–1920* (Chapel Hill: University of North Carolina Press, 1998), 35–37, 72, and 132. Lewis also observes that "although the band saw was first patented in Britain in 1808 and in the U.S. in the 1830s, problems securing the endless steel band kept it out of general production until the late nineteenth century."

81. Michael Williams, *Deforesting the Earth: From Prehistory to Global Crisis* (Chicago: University of Chicago Press, 2006).

82. David Lowenthal, *George Perkins Marsh* (Seattle: University of Washington Press, 2000), 272–74.

83. Wallace, quoted in *Infinite Tropics: An Alfred Russel Wallace Anthology,* ed. Andrew Berry (London: Verso, 2002), 150–51. August Weismann, "On the Mechanical Conception of Nature" (1877), reprinted, in *Studies in the Theory of Descent,* trans. Raphael Meldola (London, 1882), 651.

84. William Seymour Edwards, *Coals and Cokes in West Virginia* (Cincinnati, 1892), 7.

85. See Mead,"Theodore L. Mead . . . An Autobiography," published in *The Yearbook of the Amaryllis Society* (1935), 2: 3–14.

86. Barrow, *A Passion for Birds,* 133; and Brian Czech and Paul R. Krausman, *The Endangered Species Act: History, Conservation Biology, and Public Policy* (Baltimore: Johns Hopkins University Press, 2001), 16–17.

87. For Donald Worster, see chapter 8 "Private, Public, Personal: Americans and the Land," in his *The Wealth of Nature: Environmental History and the Ecological Imagination* (New York: Oxford University Press, 1993), 101; and for Peter Huber, see *Hard Green: Saving the Environment from the Environmentalists* (New York: Basic Books, 1999), xiv–xv, xxi, xxx.

88. On tourism in the parks, see Richard West Sellars, *Preserving Nature in the National Parks* (New Haven, CT: Yale University Press, 1997).

89. Lewis, *Tranforming the Appalachian Countryside,* 278.

90. See Williams, *Deforesting the Earth,* 284; James E. McWilliams, *American Pests: The Losing War on Insects from Colonial Times to DDT* (New York: Columbia University Press, 2008), 10–25.

91. On secondary succession, see Scott L. Ellis, "Biogeography," chapter 1 in *Butterflies of the Rocky Mountain States,* ed. Clifford D. Ferris and F. Martin Brown (Norman: University of Oklahoma Press, 1980); and David R. Foster, *Thoreau's Country: Journey Through a Transformed Landscape* (Cambridge, MA: Harvard University Press, 1999), 73–74.

92. This paragraph draws especially on Steven Stoll, *The Fruits of Natural Advantage: Making the Industrial Countryside in California* (Berkeley: University of California Press, 1998). See also David Vaught, *Cultivating California: Growers, Specialty Crops, and Labor, 1875–1920* (Baltimore: Johns Hopkins University Press, 1999). See also Norman Boltin and Christine Lang, *The World's Columbian Exposition* (Urbana: University of Illinois Press, 2002), 74–75, 85.

93. J. F. M. Clark, *Bugs and the Victorians* (New Haven, CT: Yale University Press, 2009), 134.

94. On Buffon, see, for example, *Histoire Naturelle,* vols. 1–5 (London, 1792), and, on Linnaeus, see Lisbet Koerner, "Carl Linnaeus in His Time and Place," in *Cultures of Natural History,* ed. N. Jardine (Cambridge: Cambridge University Press, 1996), 150. On the United States, see Clark A. Elliott, *Thaddeus William Harris (1795–1856)* (Bethlehem, PA: Lehigh University Press, 2008), 75–96.

95. See "French Exhibition of Economic Entomology," in the column "General Information," in the *Entomological Journal* (July 1868): 50–51; and on French and British applied agriculture, see Clark, *Bugs and the Victorians,* 132–215.

96. Herbert Osborn, *Fragments of Entomological History* (privately printed, 1937), 7–8; L. O. Howard, *A History of Applied Entomology* (Washington,

DC: Smithsonian Institution, 1931), 53–57; W. Connor Sorensen, *Brethren of the Net: American Entomology, 1840–1880* (Tuscaloosa: University of Alabama Press, 1995), 92–126.

97. James Fletcher to Scudder, December 19, 1894, SS-BMS.

98. "Annual Address of James Fletcher, President of the Entomological Club of the A.S.S.A., 1889," reprinted in *Entomologica Americana* 4 (January 1890): 1–9.

99. Howard, *A History of Applied Entomology*, 74–75, 101. On Howard's role, see Hae-Gyung Geong, "Exerting Control: Biology and Bureaucracy in the Development of American Entomology, 1870 to 1930" (PhD diss., University of Wisconsin, 1999), 33–52.

100. CRS Report for Congress, "Salaries of Members of Congress: A List of Payable Rates and Effective Dates, 1789–2008," Order Code 97-1011 GOV, compiled by Ida A. Burdick, Analyst on the Congress, Government, and Finance Division; and Howard, *A History of Applied Entomology*, 177–80.

101. Osborn, *Fragments of Entomological History*, 7–8.

102. Augustus Grote, *An Illustrated Essay on the Noctuidae of North America* (London, 1882), 21. Grote tried to get a job on the Entomological Commission, run by Riley, but Riley detested him as an egomaniac (others viewed Riley as one, too!) and blackballed his chances. Both men remained hostile to each other for the rest of their lives.

103. Strecker, "Butterflies and Moths of North America, in Their Relation to Horticulture and Floriculture" (paper, Pennsylvania Fruit Growers' Association, Reading, 1879), 17.

104. Joseph Lintner, "On the Importance of Entomological Studies," *Papilio* 1, no. 1 (1881): 1–2.

105. *EN,* January 1897.

106. Fletcher to Scudder, January 18, 1893, SS-BMS.

107. See Jakub Novak, "Alfred Russel Wallace's and August Weismann's Evolution: A Story Written on Butterfly Wings" (PhD diss., Princeton University, 2008), 244; and Soraya De Chadarevian, "Laboratory Science Versus Country-House Experiments: The Controversy Between Julius Sachs and Charles Darwin," *British Journal for the History of Science* 29 (1996): 17–41.

108. Fordyce Grinnell, "The Spirit of the Naturalist and of Natural History Work," *Lepidopterist* 1, no. 1 (1916): 1–2.

109. Henry Bird to Frank Jones, April 29, 1911, Frank Jones Papers, Academy of Natural Sciences, Philadelphia.

110. On the separation of entomology from ecology and Riley's role in it, see Paolo Palladino, "Entomology and Ecology: The Ecology of Entomology" (PhD diss., University of Minnesota, 1989), 26–31. On Riley's new biological approach to insect control, see Geong, "Exerting Control," 8, 22–33. On the importance of *Insect Life,* see Howard, *A History of Applied Entomology*, 68–69. On the history of economic entomology, see Howard, *A History of Applied Entomology;* Arnold Mallis, *American Entomologists* (New Brunswick, NJ: Rutgers University Press, 1971); Sorensen, *Brethren of the Net;* Edmund Russell, *War and Nature: Fighting Humans and Insects with Chemicals from World War I to "Silent Spring"* (Cambridge: Cambridge University Press, 2001); and, most recently and comprehensively, McWilliams, *American Pests.*

111. Henry Bird to Frank Jones, October 7, 1948, Frank Morton Jones Papers, Bio-

logical Correspondence, record number 565, Archives and Manuscripts Collection, Academy of Natural Sciences, Philadelphia.

112. George Hulst to William Henry Edwards, August 2, 1878, WHE-SA.

113. For the end to farming in Brooklyn and the spread of suburban real estate, see Marc Linder and Lawrence S. Zacharias, *Of Cabbages and Kings County: Agriculture and the Formation of Modern Brooklyn* (Iowa City: University of Iowa Press, 1999), especially chapters 2, 3, and 11. Of the Long Island prairie, one butterfly man observed in May 1912, that it could "be reached one hour from Brooklyn or New York, combining as it does prairie and pine barren, and it ought to receive special attention for collectors during the present season, for the original prairie is fast yielding to cultivation." See George Engelhardt, in the minutes of the "Annual Meeting of the New York Entomological Society," May 7, 1912, American Museum of Natural History, New York.

114. Anna Comstock, *Nature-Study Review* 1, no. 4 (1905): 143–46.

115. Theodore Roethke, "The Small" and "All the Earth, All the Air," in *The Collected Poems of Theodore Roethke* (1958; repr., New York: Anchor Books, 1991), 117, 142.

8. Death of the Butterfly People

1. John S. Garth and J. W. Tilden, *California Butterflies* (Berkeley: University of California Press, 1986), 45; and William Henry Edwards to William Greenwood Wright, January 6, 1904, WGW.

2. Thomas Bean to Herman Strecker, April 23, 1893; December 14, 1894; June 2, 1894; and February 25, 1899, HS-FM.

3. Bean to William Barnes, March 15, 1910, William Barnes Correspondence, Records of the Bureau of Entomology and Plant Quarantine, 1863–1956, Record Group 7, National Archives at College Park, MD.

4. Edwards to Samuel Scudder, April 18, 1901, SS-BMS, and Edwards to William Holland, 1904, WH-CM.

5. Edwards to Scudder, December 7, 1887, SS-BMS.

6. Edwards to Scudder, December 6, 1892, SS-BMS. On Mrs. Bowen's death, see Edwards to Theodore Mead, August 11, 1888, TM.

7. Preface to volume 3, *BNA* (Philadelphia: American Entomological Society, 1868–72).

8. Edwards to Scudder, November 23, December 6, and December 20, 1892, SS-BMS.

9. Karl Jordan and Walter Rothschild, "A Revision of the American Papilios," *Novitates Zoologicae* 13 (August 1906): 425.

10. Holland to William Schaus, September 15, 1930, William Schaus Papers, Record Group 7100, Smithsonian Institution Archives, Washington, DC.

11. Neumoegen to Bean, July 3, 1893, Bean Letterbook, American Museum of Natural History, New York; Doll to Strecker, May 28, 1891, HS-FM; and Neumoegen to Edwards, February 26, 1894, WHE-SA.

12. "Resolutions on the Death of Mr. Neumoegen," minutes, March 19, 1895, Annual Meeting of the New York Entomological Society, p. 59, American Museum of Natural History.

13. *EN* (September 1899): 208–9.

14. Edwards to Scudder, April 13, 1891, SS-BMS.

15. William Henry Edwards to Henry Edwards, February 9, 1891, HE.

16. Henry Edwards, *A Mingled Yarn: Sketches on Various Subjects* (New York: G. P. Putnam's Sons, 1883), 106.

17. See clipping, "Harry Edwards's Funeral," in the Entomological Diary of William Henry Edwards, April–May 1891, WHE-SA. The details in this paragraph also come from the following letters: Edwards to Wright, May 10, 1891; May 21, 1891; and June 15, 1891, WGW.

18. Henry Edwards to Holland, August 6, 1891, WH-CM. On his advertisements, see Edwards to Henry Skinner, August 1, 1891, HS-ANC.

19. Henry Edwards to Holland, August 6, 1891, WH-CM; and Edwards to Skinner, August 1, 1891, HS-ANC.

20. Henry Edwards to Arthur Butler, October 14, 1885, NHM-LONDON.

21. Harry Weiss, "*The Journal of the New York Entomological Society,* 1893–1942," *Journal of the New York Entomological Society* 6 (December 1943): 285–94.

22. Holland to Scudder, January 2 and 8, 1892, SS-BMS.

23. Henry Edwards to Scudder, December 2, 1890, SS-BMS.

24. Holland to Schaus, November 30, 1923, Schaus Papers, Smithsonian Institution, Washington, DC; and Frank E. Lutz, "Amateur Entomologists and the Museum," *Natural History* 24, no. 3 (1924): 339; and, on Slosson and the founding of the New York Entomological Society, see Charles Leng, "History of the New York Entomological Society, 1893–1918," typescript, American Museum of Natural History.

25. On the donations, donors, and the fund, see William Beutenmuller to Annie Trumbull Slosson, November 24, 1891; William Beutenmuller to Mr. James, December 24, 1891; Beutenmuller to A. M. Palmer, December 28, 1891; and Beutenmuller to Morris Jessup, November 5, 1891, and May 26, 1892, Beutenmuller Scrapbook, American Museum of Natural History, New York.

26. Edwards to Wright, June 11, 1891, WGW.

27. "Old Buzzfuzz" mentioned in a letter from Skinner to Strecker, around 1890–91, HS-FM.

28. The quote relating to Neumoegen, that he could "keep friends with everybody," appears in Strecker to Charles Valentine Riley, January 9, 1882, draft letter, HS-FM.

29. Strecker to F. Cormack, no date but c. 1897, HS-FM; and O. W. Barrett, "Cheap Tropical American Butterflies," *EN* (October 1902): 238.

30. Russell Robinson to Strecker, March 24, 1900, and April 8, 1899, HS-FM. Strecker's letters to Robinson don't exist; I have inferred his responses from Robinson's letters. There were other letters, but these have disappeared as well. See, on Wirt Robinson, Austin Clark, "Notes on the Butterflies of Margarita Island, Caracas, and Carupano, Venezuela," *Psyche* (February 1905): 1–11; and his obituary, in "Notes and News," *Auk* 58 (1941), 131–32.

31. John Morris to Strecker, August 4, 1895, HS-FM.

32. Fifty years later, Bird still delighted in telling this story to his friends; see Henry Bird to Frank Jones, November 15, 1950, and October 2, 1952, Frank Jones Papers, Academy of Natural Sciences, Philadelphia.

33. William Henry Edwards to Henry Skinner, August 1, 1891, HS-ANS; Smith to Strecker, May 5, 1900, HS-FM.
34. Smith to Strecker, May 5, 1900, HS-FM.
35. Preface to "Supplement No. I," of *LRH* (Reading, PA, 1872–78).
36. Skinner to Strecker, January 7 and 16, 1899, HS-FM.
37. Strecker to Skinner, June 7, 1900, and September 16 and January 17, 1899, HS-ANS.
38. Strecker to Skinner, June 7, 1900, HS-ANS.
39. Strecker to Andrew Weeks (draft), March 18, 1895, HS-FM.
40. Strecker to Harrison Dyar, December 16, 1900, HD.
41. Strecker to Skinner, May 3, 1900, and November 27, 1900, HS-ANC.
42. William Kearfott to Strecker, January 17 and 21, 1901, HS-FM.
43. Levi Mengel to Henry Skinner, July 18, 1901, HS-ANS; on the bout of influenza, see Strecker to Henry Skinner, September 19, 1901, HS-ANS.
44. Mengel to Skinner, November 30 and December 17, 1901, HS-ANS.
45. Holland to Sir George Hampson, June 24, 1902, WH-CM.
46. Mengel to Skinner, January 21, 1906, HS-ANS, collection no. 210; William Gerhard, *EN* (July 1909); and David Walston, "The Sculptor Who Collected," *Field Museum of Natural History Bulletin* 46 (January 1995): 1.
47. D. F. Hardwick, "An Analysis of the Heliothidine Types (Noctuidae) of Herman Strecker with Lectotype Designations," *Journal of the Lepidopterists' Society* 32, no. 1 (1978): 49.
48. Will Doherty to his father, March 27, 1897, JMH.
49. Doherty to Ernst Hartert, January 31, 1900 (from the Norfolk Hotel), Tring Correspondence, NHM-LONDON.
50. Theodore Roosevelt, *African Game Trails* (New York: Scribner's, 1909), 2–3.
51. Doherty to Mr. Durrant, February 5, 1900, NHM-LONDON.
52. Doherty to Hartert, January 31, 1900, NHM-LONDON.
53. José Steinbach to Holland, February 10, 1909; and on Steinbach's "first sendings" to Holland, see Steinbach to the "Director of the Carnegie Museum," March 19, 1900, WH-CM. On the total number of birds sent by Steinbach to the museum over thirty years, see W. E. C. Todd (curator of ornithology) to Edwin G. Conklin, May 3, 1939, WH-CM. By 1914 he had collected for the Carnegie museum "25,000 moths in 3000 species." See Steinbach to William Schaus (care of Douglas Stewart), April 24, 1914, Director's Correspondence, WH-CM. On Holland's fear of being cheated, see Steinbach to Holland, July 1, 1915, WH-CM. It is not clear that the 1900 Lacey Act banning transport of wild birds (Steinbach's specialty for the museum) and animals across American state lines had any legal impact on the international transport of dead or living birds. Probably it did, therefore making the Carnegie museum, and other similar museums, a clear violator of if not the letter of the law, its "spirit."
54. Tyler Townshend to Strecker, February 7, 1900, HS-FM. Townshend also wrote that Holland insisted "that the collection contained several hundred specimens short of my count" but that, "out of the generosity of his heart," he would "give me $100 for the lot. The price I agreed to collect for was $10 per 100 [he caught 1,800 altogether], Holland to get the entire catch for the season. I accepted the $100 offer of course—couldn't do otherwise, since I needed the

money at once. But I wrote him that I knew that my count was approximately correct, and that I should never again send him a single specimen of anything. He wrote me several abusive letters, and said that it was fortunate for me that I was out here where I am or he would make me smart for my insinuations. Wouldn't that crinkle your hair!"

55. Holland to Hans Fruhstorfer, January 18, 1902, WH-CM.

56. Herbert Smith to Holland, January 25, 1893; August 7, 1894; and December 8, 1894, WH-CM.

57. Smith to Holland, Director's Letter book, October 11, 1892, WH-CM.

58. The title of Smith's expedition appears in a letter from C. F. Baker to George Clapp, October 10, 1898, Smith Papers, Geological Museum of Alabama, Tuscaloosa. I have seen the title nowhere else, however. Baker was a field botanist from Auburn, Alabama, who may or may not have accompanied the Smiths on their journey. This letter indicated that he did go, but probably not for long.

59. Agreement between Holland and Smith beginning "It is mutually agreed," February 22, 1898, Box 27, folder 11, WH-HSWP. Smith also signed contracts with the rich British naturalists, Herbert Druce and Lord Walsingham, who promised to pay well for rare moths and butterflies. He signed another with George Clapp, a Carnegie trustee and naturalist, who arranged with Smith to act as his partner and agent, dealing with money matters, receiving payments for lots, and establishing research agendas. Smith to Managing Committee of the Carnegie Museum, December 8, 1897, and Smith to George Clapp, March 7, 1898, Carnegie Museum Archives, Pittsburgh.

60. Holland to Strecker, September 5, 1888, HS-FM. A year earlier, his article on the butterflies of West Africa—a first stab at such fame—appeared in the *Transactions of the American Entomological Society,* the journal that launched the career of William Henry Edwards and others. On the basis of only one or two specimens, he identified several "new" species collected by Christian ministers. In an accompanying piece on Japanese butterflies, he took the liberty to name one insect "after my faithful Japanese assistant, Tora-san," and another after Mrs. Mabel Louise Todd, "who did so much to enliven the stay at the old castle of Shirakawa." See "Contributions to a Knowledge of the Lepidoptera of West Africa," *Transactions* (1887): 73, 76.

61. Holland to his parents, March 21, 1894, WH-HSWP.

62. Holland to Skinner, October 19 and November 9, 1892, HS-ANS; and Holland to Scudder, February 2 and 3, 1891, SS-BMS.

63. Holland to Doherty, February 10, 1900, Director's Letter book, WH-CM.

64. Doherty to Hartert, March 13, 1900, NHM-LONDON.

65. Doherty to Hartert, March 15, 1900; March 1900 (undated), and March 13, 1900, NHM-LONDON.

66. Doherty to Hartert (Norfolk Hotel), no date but c. January-March 1900, Tring Correspondence, NHM-LONDON.

67. Holland to Doherty, January 11 and 31, 1901, Director's Letter book, WH-CM.

68. Holland to Doherty, January 11 and 31, 1901, and November 12, 1900, Director's Letter book, WH-CM. In the last letter, Holland reported visiting London and seeing "our mutual friends, Mr. Rothschild and Dr. Hartert." "Mr. Rothschild," he wrote, "expressed great solicitude in reference to you. And the first

question that he asked me when I met him at Tring was, 'Dear Doctor, have you heard from Doherty.' He blurted out the inquiry as soon as he had ended the customary salutations, and seemed to be greatly anxious to know what had become of you."

69. Holland to Doherty, January 11, 1901, Director's Letter book, WH-CM.
70. Ibid.
71. George Clapp, "Herbert Huntington Smith," *Nautilus* (July 1899): 137.
72. Smith to Holland, July 20, 1899, and January 6, 1900, WH-CM.
73. Holland to Robert Berryhill of Iowa City, December 19, 1903; and Holland to Charles T. Scott, April 23, 1903, director's letter book, WH-CM.
74. Holland referred to himself as Carnegie's "obedient servant" in a letter to E. Ray Lankester, director of the British Natural History Museum, July 14, 1904, NHM-LONDON. In 1900 Carnegie and Holland went to England together, for a celebration of the British Museum of Natural History, bearing with them a replica of the beast, which the British had dearly wanted. Other European capitals got their replicas, too. Holland even traveled to Buenos Aires, to present a replica to the leading museum in Argentina. See Holland to Lankester, July 14, 1900, WH-CM, and Holland, *To the River Plate and Back: The Narrative of a Scientific Mission to South America* (New York: G. P. Putnam's Sons, 1913), 1–11. Holland made the trip in 1911.
75. Holland to Andrew Carnegie, March 27, 1899, Director's Letter book, WH-CM.
76. Holland to Stephen Downey of Laramie, Wyoming, May 10, 1899; and Holland to Carnegie, March 27, 1899, Director's Letter book, WH-CM.
77. Holland to Dr. J. L. Wortman, August 15, 1899, Director's Letter book, WH-CM.
78. Holland to Professor J. B. Hatcher, August 13, 1902, WH-CM.
79. Ibid.
80. "As I look back at what he accomplished here," Daisy wrote a friend, "I am astonished. I *must* stay until the work is ended—for there is no one else who could do it." No husband-wife bond could have been closer; they had labored and traveled together throughout Latin America, and now canoed the rivers of Alabama in search of freshwater shells together, camping in a full, satisfying isolation together, and without the tropical nightmare. See Daisy Smith to George Clapp, undated, c. 1919, and Smith to Clapp, December 25, 1905, Smith Papers, Alabama Museum of Natural History, Tuscaloosa. See also Eugene A. Smith, "Explanatory Note" to an article by Calvin Goodrich, "The Anculosae of the Alabama River Drainage," Museum Paper No. 6, Alabama Museum of Natural History, Tuscaloosa (July 1, 1922), 1.
81. Doherty to his mother, October 21, 1900, JMH.
82. Doherty to Hartert, November 4, 1900, NHM-LONDON.
83. Doherty to Edward Janson, February 10, 1901; Doherty to Oldfield Thomas, November 4, 1900, "Letters on Mammalia, 1900, A-M," NHM-LONDON.
84. Doherty to Hartert, November 4, 1900, NHM-LONDON. See also Doherty to his mother, October 21, 1900, JMH; and, for a description of *Papilio rex*, see Bernard d'Abrera, *Butterflies of the World* (London: Hill House, 2001), 147, 151.

85. Doherty to his mother, October 21, 1900, JMH.
86. J. M. Doherty to Hartert, no date, c. 1902, and January 26, 1902, NHM-LONDON; Doherty to his mother, October 21, 1900, JMH; and Doherty to Hartert, November 4, 1900; December 26, 1900; February 9, 1901; February 14, 1901; January 26 and May 12, 1902, NHM-LONDON. From these letters I have also extracted a narrative of Doherty's activities until his death.
87. Augustus Grote to Dyar, August 28, 1895, HD.
88. On Smith replacing Strecker, see George Horn to Isaac Hays, March 1, 1897, Isaac and I. Minis Hays Papers, American Philosophical Society, Philadelphia.
89. J. F. Gates Clarke, the chief lepidopterist at the Smithsonian in the mid-twentieth century, wrote to F. Martin Brown, March 13, 1972: "I am a firm believer in and advocate of the use of genital characters in classification. They present, usually, the best set of definitive characters I know of. Admittedly they fail in some cases beyond presenting generic differences but when they do other characters can be used for specific separation." See Record Group 427, Box 2, Division of Lepidoptera Records, 1963–1990, Smithsonian Institution, Washington, DC. In an attack on William Henry Edwards, John Smith told Henry Skinner, "I should never base a classification on larvae, unless there was something in the imago to bear them out. My test in every case rests only with the genitalia." See Smith to Skinner, December 15, 1896, HS-ANS, collection 150. In a letter to Spencer Baird, Edwards wrote of Smith in 1886, when Smith was the curator of insects at the Smithsonian and the editor of *Entomologica Americana,* "Your curator knows next to nothing about butterflies." In another letter to Baird, he wrote that "any determination by [Smith] would carry little weight. I don't care to subject myself to his impertinence in the magazine he edits, and therefore say what I do to you in confidence." See Edwards to Baird, October 12, 1886, Spencer Baird Papers, Smithsonian Archives, Washington, DC.
90. Grote to Dyar, May 13, 1895, and September 27, 1896, HD.
91. See his unpublished paper "The Epic of Papaipema" (1940), especially pp. 1–5. The full draft is in the Henry Bird Papers, American Museum of Natural History, New York.
92. Grote to Bird, circa 1900; the letters can be found in the back pages of Bird's "The Epic of Papaipema."
93. On Harrison Dyar, see Marc Epstein and Pamela Henson, "Digging for Dyar: The Man Behind the Myth," *American Entomologist* 38 (Fall 1992): 148–69. On John Comstock, see Pamela Henson's "The Comstock Research School in Evolutionary Entomology," *Osiris* 8 (1993): 159–77, and "The Comstocks at Cornell," in *Creative Couples in the Sciences,* ed. Helena M. Pycior et al. (New Brunswick, NJ: Rutgers University Press, 1996), 112–25, as well as her dissertation, "Evolution and Taxonomy: J. H. Comstock's Research School in Evolutionary Entomology at Cornell University, 1874–1930" (PhD diss., University of Maryland, College Park, 1990).
94. Grote to Dyar, November 6 and 17, 1896; December 16, 1895; January 28, 1896; and April 17, 1896, HD.
95. John Henry Comstock and Anna Botsford Comstock, *How to Know the Butterflies: A Manual of the Butterflies of the Eastern United States* (1904; repr.,

New York: D. Appleton, 1915), "with forty-five full-page plates from life reproducing the insects in natural colors."

96. Ibid., 93–94, 113–14, 194–95, 245–46, and 50.

97. Ibid., 41–42, and Alexander B. Klots, *A Field Guide to Butterflies* (Boston: Houghton Mifflin, 1951), 51.

98. Grote to Dyar, June 17, 1895, HD.

99. Grote to Dyar, January 17, 1896, HD.

100. These generalizations appeared in the *CE*, as "Diphyletism in the Diurnal Lepidoptera" (December 1899, pp. 290–91), and "The Principle Which Underlies the Changes in Neuration" (October 1900, pp. 289–92). They synthesized for Americans much of what he had published in the following articles: "Die Saturniiden," *Mittheilungen aus dem Roemer Museum, Hildesheim* (June 1896): 1–32; "Die Schmetterlingsfauna von Hildesheim" *Mittheilungen aus dem Roemer Museum, Hildesheim* (February 1897): 1–45; "Specializations of the Lepidopterous Wing," *Proceedings of the American Philosophical Society* (May 17, 1898); "The Descent of the Pierids" *Proceedings of the American Philosophical Society* (January 1900): 3–67; and "Fossile Schmetterlinge und der Schmetterlingsflügel," *Verhandlungen K.K. Zool.-bot. Gesellschaft Wien* (1901).

101. Grote to Dyar, December 12, 1898, HD.

102. Grote, "A Reply to the Critic of *Psyche*," *Psyche* (August 1897): 106.

103. Scudder, "The Butterflies of Hildesheim," *Psyche* (June 1897): 83; and Grote to George Horn, October 1897, Hays Papers, American Philosophical Society, Philadelphia. Modern research supports Scudder's contention that the swallowtails are, in fact, lower in evolutionary development than the skippers. Author's interview with Charles Remington, September 13, 2003.

104. Staudinger, quoted in *CE* (July 1902): 185.

105. Dyar, preface, *Insecutor Inscitiae Mensruus* 1 (January 1913).

106. Grote to Dyar, June 13, 1898, HD.

107. Grote to Dyar, February 11, 1896, HD; all of the italics are mine.

108. Grote to William Henry Edwards, July 4, 1895, WHE-SA; Grote to Charles Fernald, August 28, 1896, Charles Henry Fernald Papers (RG 40/11 C. H. Fernald), Special Collections and University Archives, University of Massachusetts, Amherst; and Grote to Dyar, January 17, 1896, HD.

109. Grote to Dyar, May 15, 1898, HD.

110. Scudder to close friend Charles Fernald, November 12, 1896, Fernald Papers; Edwards to Scudder, November 21, 1896, SS-BMS; and Scudder to young colleague Lawrence Bruner, September 18, 1896, LB.

111. Scudder to Bruner, January 5, 1897, LB. This letter to Bruner contains the only comment by Scudder I know of on his son's death. There is no doubt he destroyed most of his letters bearing on the illness and demise. All of Fletcher's letters written between 1896 and 1898, for instance, were removed by Scudder.

112. Scudder to Bruner, March 27, 1897, and November 23, 1901, LB.

113. Scudder to Holland, February 17, 1902, WH-CM.

114. Scudder to Bruner, October 21, 1895, LB.

115. Albert P. Morse, "The Orthoterlogical Work of Mr. S. H. Scudder, with Personal Reminiscences," *Psyche* (December 1911): 18; and Alfred Goldsborough Mayor, "Biographical Memoir of Samuel Hubbard Scudder, 1837–1911,"

Memoirs of the National Academy of Sciences, vol. 17 (Washington, DC, 1924).

116. U.S. Geological Survey, Monograph 40 (1900). On prayers, see *Psyche* 8 (1897): 142–43; on cave crickets, *Psyche* 9 (1902): 312; and on pink grasshoppers, see *EN* 12 (1901): 129–31. On the guide to grasshoppers, see *Guide to the Genera and Classification of the North American Orthoptera Found North of Mexico* (Cambridge, 1897); and on the later catalog, see "Catalogue on the Described Orthoptera of the United States and Canada," *Proceedings of the Davenport Academy of Natural Science* 8 (1900): 1–101.

117. Scudder to Samuel Henshaw, August 22, 1900, EML.

118. Scudder to Henshaw, April 13, 1907, EML.

119. Theodore Cockerell, *Science* 34 (September 13, 1911): 338–42; and Mayor, "Biographical Memoir of Samuel Hubbard Scudder," 81–86.

120. Entomological Diary, January 1895, WHE-SA.

121. Edwards to Wright, October 13, 1896, WGW.

122. Edwards to Wright, March 20, 1897, WGW.

123. Edwards to Wright, March 20, 1897, WGW; see also Edwards to Wright, October 13, 1896, WGW.

124. Edwards to Wright, June 14, 1905, WGW.

125. Edwards *Timothy and Rhoda Ogden Edwards and Their Descendents: A Genealogy* (Cincinnati: The Robert Clark Company, 1903).

126. Edwards, *Shakesper Not Shakespeare* (Cincinnati, 1900), 178. On current debates about the authorship of the plays, see Doug Stewart, "To Be or Not Be Shakespeare," *Smithsonian Magazine* (September 2006): 62–64.

127. Edwards to Scudder, April 18, 1901, SS-BMS. In his 1909 obituary of Edwards, Charles Bethune, the editor of the *Canadian Entomologist,* wrote incorrectly that "Mr. Edwards was seventy-five years old when he gave up his studies of butterflies, feeling, no doubt, that his advanced age precluded him from carrying on further investigations with the ability and success that he had so remarkably displayed." Why Bethune made this claim about Edwards is not clear, since Edwards had never felt healthier or more able to do work on butterflies. He was perfectly prepared to begin a fourth volume. See Bethune, "William Henry Edwards," *CE* (August 1909): 217.

128. Edwards to Wright, November 20, 1898, and January 6 and 29, 1904, WGW.

Index

Page numbers in *italics* refer to illustrations.

Aaron, Eugene M., 114, 126
Aaron, Samuel, 114
Abbot, John, 6–7, 40, 140
Academy of Natural Sciences, 55, 61,
    100, 109, 126, 127, 141, 177, 250,
    264
Acraea excelsior, 272
Adams, Henry, 228
adaptation, 166–9, 174–5, 177
admirals, 173, 174, 238
adult, *see* imago
aesthetic entomology, 51–2, 67, 77
Africa, xvi, 55, 61, 64, 170, 195, 196,
    200, 205, 210, 265–9, 272–4
Agassiz, Louis, 33–40, 42, 43, 44, 52,
    78, 91–2, 143, 155, 173, 184
Agriculture Department, U.S., 130, 163,
    205, 241
    Entomology Division, 125, 249–50
Akhurst, John, 13, 52, 65, 95, 110
Albright, Max, 234–5
alpines, 87, 137, 139, 226, 255
amateurs, collectors aided by, xxiv–xxv,
    24–6, 42, 64–5, 118–19, 132,
    135–9, 255–61
Amazon, 11–12, 170, 233
American Association for the
    Advancement of Science (AAAS),
    86, 88, 89, 90, 93, 125, 140, 249,
    251–2, 260
American Entomological Society, 18,
    117, 126, 141, 263
American Entomologist, 32, 168, 171
American Museum of Natural History,
    26, 99, 108, 110, 113, 236, 259,
    262
American Naturalist, 123

"Ancestry and Butterflies" (Scudder),
    187
"Ancestry and Classification"
    (Scudder), 186–7
androconia, 184, *185,* 187–8
angle-wings, 23–4, 49
Angus, James, 65
Antimachus swallowtail, 64
anti-Semitism, 65–6, 197, 213, 215,
    261
ants, 176–8
Apanteles atalantae, 164–5
Apanteles edwardsii, 164
Apatura, 212
aposematic, 171
Appalachia, 48
Appalachian Mountain Club, 48
Appias nero, 87
arctics, 40, 44, 49, 78, 256
Aristotle, xix, 162
arsenic, 13
art
    of butterflies, *see* illustration,
        illustrators
    at Chicago Fair, 187, 227–8
    and nature, 6, 240
    and science, xxii–xxiii, 51–2, 67,
        76, 110–11, 116, 143, 186–7,
        275
arthemis, 173
Art Nouveau, *187*
Asia, xvi, 193, 195, 201, 205
Atlantic Monthly, 74, 123
Atlas moth (*Attacus dohertyi*), 216
Audubon, John, 18, 52
Audubon Society, 238
Aurivillius, Christopher, 265

Australia, 29–30, 61, 83
automobiles, 243–4, 247

Baird, Spencer, 14, 104
Ball, Philip, 242
Ballard, Julia, 122–3
baltimore checkerspot (*Euphydryas phaeton*), 25, 134, 175–6, 183, 237, 245
banded purples, 34, 40
Bang-Haas, Andreas, 200–1, 213, 222, 264
Bang-Haas, Carmen Staudinger, 201
Bang-Haas, Otto, 213
Barnes, William, 109, 243
Baron, Oscar T., 30, 203
Barthélemy-Saint-Hilaire, Jules, 281
Basilarchia (*Limenitis*), 173
Bates, Henry, 12, 21, 143, 157, 196, 252, 277
  mimicry theory of, 170–1, 179
Bean, Thomas, 136–9, 156, 232, 255
beauty
  as motivation of collectors, xxv, 64, 66, 98, 253
  natural vs. artificial, 230, 238, 242, 283
  nature and, xv, xxv, xxvi, 8–9, 41, 51–2, 67, 79, 114–15, 152–3, 183–9, 253
  passionate reaction to, 82–5, 276, 283
  purpose of, 183–8, *185*
"Beauty of the World (J. Edwards), 8, 282
Behr, Hans Hermann, 30, 181
Behrens, James, 30
Belfrage, Gustave, 234
Bell, Alexander Graham, 124–5
Belt, Thomas, 171
"Best Localities for Collectors, The" (Scudder), 182
Bethune, Charles, 124
Beutenmuller, William, 258–60, 262
Bhutanitis lidderdalii, 217
Bickmore, Albert, 26
Biologia Centrali-Americana, 204
Bird, Henry, 252–3, 262, 274–5

Bird Neighbors (Blanchan), 235, 237
birds
  of Australia, 29–30, 172, 235
  collecting of, 10–12, 14, 54, 127, 128, 204, 214, 216, 218, 266
  killing of, 11, 26, 27, 96, 235, 266
  protection of, 238–9, 246, 266
Birds of North America (Audubon), 10
birdwings, 55, 62, 64, 82, 103, 120, 199, 214, 223, 225
"Black Marias" (notebooks), 274
black swallowtails (*Papilio polyxenes*), xvii
Blanchan, Neltje, 235
blues, 15, 19, 24, 46, 65, 120, 130, 158, 277
Bohemian Club, 30, 258
Boisduval, Jean Baptiste, 3–4, 6, 18, 32, 44, 46, 61, 63, 73, 248
Bonpland, Aimé, xxi
"boss masons" (collectors), xxiv–xxv, 270
Boston Society of Natural History, 37, 40, 128, 154
Bowen, Lydia, 18–19, 130, 135, 158, 178, 188, 255, 283
Bowles, Wesley, 25
Braconidae, 164
Brazil, 6, 58, 86, 172, 201–2
breeding, 13, 19–22, 25, 66, 98, 119, 121, 130, 133, 156, 168, 197, 276
Bridgham, Eliza Fales, 97–8
Bridgham, Joseph, 97–8
Bridgham, Samuel, 97
Brief Guide to the Commoner Butterflies of the Northern U.S. and Canada (Scudder), 149
British Museum, 5, 12, 32, 114, 216, 258, 264, 265, 266
Brooke, Gustavus Vaughan, 29
Brooklyn, N.Y., 50, 62, 65, 70–1, 160, 242–3, 254, 277
Brooklyn Entomological Society, 66, 125–6, 242, 254
Brooklyn Institute of Arts and Sciences (Brooklyn Museum), 257
brown elfins (*Incisalia augustinus*), 31
Browning, Robert, 146

Bruce, David, 26, 89, 136–9, 226, 232, 233, 243, 255, 282, 283
Bruner, Lawrence, 279
Buchholz, Otto, 244
Buffalo Museum of the Natural Sciences, 74, 76
Buffalo Society of Natural Sciences, 73, 99, 108
Buffon, George-Louis Leclerc, Comte de, xix–xx, xxii–xxiii, xxiv, 5, 7, 17, 19, 36, 37, 42, 45, 133, 159, 195, 230, 248
Burgess, Edward, 44, 154, 155
Burke, Edmund, 152
Burke, Jenny, 24
Burke, John, 24
Burmeister, Hermann, 88
Butler, Arthur, 32, 208, 258
butterflies (Lepidoptera)
    beauty of, xv, 64, 66, 77, 183–9, 283
    bonding and community in, 175–9
    at Chicago Fair, 226–7
    displaying of, 13–14, 34, 70, 94, 119, 208, 216
    distribution of, 5, 48–9, 78, 147, 207
    hierarchies of, 5, 44, 46, 90, 92, 123, 127, 147–8, 277
    identification of, 5, 6, 7, 19–20, 36, 90–1, 106, 159, 172, 179, 276
    known species of, 58, 195
    life and death of, 152–90
    life cycles of, see life cycles, life histories
    mutations of, 218, 242
    pricing of, 59–60, 195, 198–9, 220, 222, 236, 261, 267, 268, 272
    sexual selection in, 158, 171
    survival mechanisms of, 165–81
    threats to, 160–83
    worldwide distribution of, 193–225
    see also butterfly collecting, collectors; specific families and species
Butterflies and Moths of North America (Strecker), xviii, 101, 130
Butterflies of India, Burma, and Ceylon (Nicéville), 206

Butterflies of New England, The (Maynard), 127
Butterflies of North America (W. H. Edwards), 18–19, 24, 32, 175, 240–1
    volume 1, 22, 25, 28, 31–2, 42, 47, 117, 134, 283
    volume 2, 25, 114, 134, 139, 141, 178
    volume 3, 139–41, 233, 240, 244, 256, 281
Butterflies of the Eastern United States (French), 123
Butterflies of the Eastern United States and Canada, The (Scudder), 41–2, 107, 142–51, 144, 154–7, 155, 161, 163, 168, 174, 176, 185, 186, 226, 234, 237–8, 251
Butterflies of the West Coast (Wright), 255
Butterfly Book, The (Holland), 235, 237–41, 257, 263
butterfly collecting, collectors
    amateur contributions in, xxiv–xxv, 24–6, 42, 64–5, 118–19, 132, 135–9, 255–61
    apparatus of, xix, 39, 69, 70, 79, 94, 95–8, 115, 121, 208–9, 233–4
    as commercial endeavor, see commercial collecting
    dangers of, 28, 205, 214, 265–6, 272
    death of, 255–83
    debate over goal of, 86, 97–8, 110, 138
    disparagement of, 52, 90, 114, 251–2
    ethical concerns in, 59–60, 97, 181–3, 222, 225
    European vs. American, 46–7, 52
    fieldwork vs. indoor study in, 79, 92, 129, 131
    German-American, 50–81
    hierarchy debate among, 90–5
    nomenclature debate among, xxv, 43–5, 48, 85–95, 102, 113, 127–9, 133, 147–8
    passion and zeal of, 58, 61, 62, 65–6, 82–115, 239

butterfly collecting, collectors
    (continued)
    as pathway to nature, xix, xxvi, 10,
        115, 240–1
    physical challenges of, 4, 48, 136–9,
        209, 272–3
    rivalry and conflict within, 85–115,
        166; see also specific rivalries
    sharing among, 12, 28–9, 32, 42, 58,
        65, 97–8, 106, 163, 217, 251
    split in community of, 99–106,
        110–11
    technological advancements in,
        231–9
    worldwide, 193–225
    Yankee, 3–49
Butterfly Guide (Holland), 238
Butterfly Hunters, The (Conant), 122

cabbage butterflies (Pieris rapae), 167
Cabinet of Curiosities (Seba), xxii
Calephelis borealis, 80
California, 3–4, 17, 30, 42, 82, 83, 119
California dogface, 4
Calvinism, 8–9, 21, 33, 35
Canadian Entomologist, 25, 76, 80, 93,
        99, 119, 124, 126, 128, 129, 133,
        176
Carnegie, Andrew, 190, 231, 236,
        258–9, 264, 266, 270
Carnegie Museum of Natural History,
        xxvi, 190, 231, 236, 238, 258, 267
Cassino, Samuel, 120–1
catalogs, 116, 127–30, 141, 196, 200,
        205, 262, see also specific books
Catalogue of North American
    Butterflies (Weidemeyer), 13, 51
Catalogue of Scientific Serials (Scudder),
        141
Catalogue of the Described
    Transformations of North
    American Lepidoptera (W. H.
    Edwards), 120
"Catalogue of the Lepidoptera of
    America North of Mexico" (W. H.
    Edwards), 90
caterpillars, see larva
Catocala agrippina, 130

Catocala amestris, 130
Catocala relicta, 65
Catocala sappho, 130
cats, 26, 27, 28, 96, 136
Cecil's Book of Insects, 93
Cecil's Books of Natural History
    series, 93
Center (Karner), N.Y., 118
Central America, 13, 51, 56, 167–8,
        204, 267
"Central Park Affair," 100–1, 108–10
Cethosia bonpland, xxi
Chadbourne, Paul, 34–5, 39
Charon's nymphalid, 142
Chedi (cook), 210, 213
Chesapeake and Ohio Railroad, 107,
        133, 140
Chicago Academy of Sciences, 93, 226
Chicago Columbian Exposition
    (Chicago World's Fair; 1893), 93,
        226–30, 228
children
    in collecting, xxvi, 238, 240–1
    literature for, see literature, butterfly,
        for children
China, 54, 193, 198
Chionobas alberta peartiae, 256
Chionobas ivallda, 31, 44, 206
Chionobas varuna, 256
chloroform, 96
Churchill, Randolph, 196
Civilization and Its Discontents
    (Freud), 84
Clark, Preston, 109, 244
Clarkson, Frederick, 173
classification, xix, xx, 32, 36, 78–9,
        126, 133, 147, 153, 277
    debate over, xxv, 5, 43–8, 90–5,
        164
Clements, Frederic, 241
Clench, Harry, 238
Clinton, George, 74
Coalburgh, W.Va., 14, 27, 46, 47, 117,
        175, 183, 232–3, 251
coal mining, xxv, 12–13, 15, 49, 133,
        196
    environmental damage from, 16–18,
        183, 244, 245

Coals and Cokes in West Virginia (Willie Edwards), 245
Cockerell, Theodore, 121–2, 281
Colias alticola, xxi
Colias meadii, 28
Colorado, 137, 142, 233, 243–4
"Color Preferences of Butterflies" (Scudder), 184
"Colours of Animals and Plants, The," (Wallace), 193
Columbian emerald butterflies, 196
comma (*Polygonia comma*), 28
commercial collecting, xvi, 50, 58–60, 104–5, 138, 190
   worldwide market in, 193–225
common sulphur (*Colias philodice*), 87, 134
Comstock, Anna Botsford, 121, 150, 240, 254, 275
Comstock, John, 121, 274–6
Conant, Helen, 122–3
Confidence Man, The (Melville), 180
Conrad, Joseph, 82
coppers, 42, 130
Corps of Engineers, U.S., 183, 229
Cosmos, The (Humboldt), xx, 55, 152
countershading (obliterative shading), 171
Cramer, Pieter, xxii, 3, 55, 131
Cresson, Ezra Townshend, 100, 109, 177
Crèvecoeur, Hector St. John de, xviii
Crotch, George, 95
Cuvier, Georges, 36, 37
Cynthia cardui, 30

dams, 183, 229
Darwin, Charles, xix, xxi, xxii, xxiv, 10, 20, 21, 38, 43, 67, 78, 91, 117, 137, 157, 159, 162, 169, 170, 183, 188, 245, 252, 282
   see also Darwinism; evolution, theory of; natural selection
Darwinism, xxi, 8–9, 19–21, 24, 45–8, 83, 90, 91, 121, 125, 134, 147, 206, 275
   debate over, 33, 35, 37–9, 47, 64, 90–1

evolving perceptions of, 76–8, 152–90
Davenport Museum of Natural Sciences, 104
Davis, William Morris, 147
deforestation, 16, 244–5, 247
Diana fritillary (*Speyeria diana*), 3, 15, 25, 238, 244, 245
Dickinson, Anna, 26
Dickinson, Emily, 145–6
Dictionary of the Lepcha-Language, 209
dimorphism, 4, 22–3, 135, 157–8, 159, 174, 207, 283
"Dipterous Parasites, The" (Williston), 163
Doherty, Harlan, 214
Doherty, James Monroe, 204
Doherty, Will, xvi, xxv, 83, 156, 160, 169, 172, 178, 189, 194, 196, 201, 205, 215, 217, 232, 250
   as commercial collector, 203, 207–16, 211, 261, 265–6, 268–9, 271–2
   death of, 265, 273–4, 273
   early years of, 204–6
   financial setbacks of, 221, 268
   and Hartert, 210–16
   and Holland, 218, 220–1, 266, 268–9, 271
   ill health of, 204, 206, 212, 213, 221, 265, 268, 273
   mental decline of, 213–15
Doll, Jacob, 105, 257
Doncaster, Arthur, 215, 216
Donovan, Edward, 55
Doubleday, Edward, xviii, 4–6, 12, 29, 40, 47, 63, 80, 90, 137
Doubleday, F. N., 235, 236–7
Dramatis Personae (Browning), 146
Drexel, Joseph, 54, 61, 110
Drexel, Lucy, 54
Drury, Dru, 18, 55, 131, 140, 193, 223, 234
Dryocampa, 72
Duffus, R. L., 162–3
Dury, Charles, 88

Dyar, Harrison, 110, 262, 263–4, 274–6, 277–8

East Africa Protectorate, The (C. Eliot), 273
eastern tiger swallowtail (*Papilio glaucus*), 3, 21–2
Ecological Consequences of Artificial Night Lighting (Frank), 239
ecology, xix, 245
  human threat to, 16, 181–3, 196, 213, 225, 229, 230–1, 239–54
  protective measures in, 31, 238–9, 247–8
economic entomology, 21, 41, 76, 118, 247–53
Edison, Thomas, 124
Education (Adams), 228
Edwards, Catherine Tappan, 13, 25, 139, 281
Edwards, Edith, 25
Edwards, Henry, 29–31, 29, 42, 83, 86, 119–20, 125, 193, 234
  collection of, 258–9
  death of, 258, 278
  kind and generous nature of, 26, 28–9, 257, 260
  and *Papilio*, 111–14, 251, 259, 260
  stage career of, 17, 28, 29–30, 111, 257–8
  and Strecker, 61–2, 63, 101, 110, 260–1
  and W. H. Edwards, 14, 20, 25–6, 31–2, 92, 107, 120, 135, 136, 140, 157, 233, 260, 282
Edwards, Jonas, 11, 12
Edwards, Jonathan, xvi, 8–10, 21, 24, 33, 35, 244, 282, 283
Edwards, Polly Brooke, 29, 31, 258–9, 279
Edwards, William (father), 10
Edwards, William (grandfather), 9
Edwards, William Henry, xv, 8–28, 8, 31–3, 51, 52, 54, 57, 64, 67, 77, 78, 86, 96, 98, 111, 117, 120–2, 125–6, 136, 156, 174, 204, 207, 218, 226, 231, 238, 240, 248, 250, 251, 257, 274, 276, 278, 282

  academic contributions of, 8, 12–13, 15, 19–20, 63–5, 112, 116, 129, 147, 168–9
  acclaim for, 32, 134–5, 141, 157
  aids and helpers of, 24–6, 135–9, 255–61
  on beauty, 188–9
  on bonding and community, 175–9
  coal mining business of, xxv, 6, 12–13, 15, 24, 49, 107, 133
  collection sold by, 107, 140, 220, 237–8
  Darwinism of, 8–9, 19–20, 24, 37, 45–8, 90–1, 121, 134, 153, 156–9, 188, 245, 275
  early years of, 8–11, 24, 25
  ecological contradictions of, 181–3, 229, 243–6
  financial setbacks of, 107, 139–41, 233, 281
  and H. Edwards, 14, 20, 25–6, 31–2, 92, 107, 114, 120, 135, 136, 140, 157, 233, 260, 282
  as inspiration to amateurs, 24, 138–9
  life history approach of, 7, 132, 139, 148, 156–7, 175
  literature of, 76, 131–41, 143, 146; *see also specific works*
  in nomenclature debate, 87, 90–5, 127–9
  and railroads, 232–3, 243–6
  Scudder's friendship with, 10, 13, 24, 26, 43, 46, 47, 140–2, 279, 283
  Scudder's rifts with, 8, 43–8, 80–1, 90–5, 99, 107, 123, 147–9, 175–6, 186
  Strecker vs., 60, 61, 99–103, 110
Edwards, Willie, 139, 233, 244–6, 281
  and Mead, 26–7, 112, 245–6
"Effect of the Glacial Epoch upon the Distribution of Insects in North America, The" (Grote), 78
eggs, xv, 6, 19, 39, 83, 90, 135, 138–9, 143, 147, 149, 155, 160, 162, 164, 173, 174, 197, 200, 206, 233, 249, 281
Ehrman, George, 224–5
Eisen, Adolph, 223

Eisenheis, Gerhard, 240
electric lights, 226, 228, 229, 230,
    250
  and moths, 233–4, 239–40
Eliot, Charles, 273
Eliot, George, 145
Eliot, Ida Mitchell, 42, 156
Eliot, Samuel, 156
Elwes, Henry, 197, 204, 209, 213, 215,
    217, 221, 266
Emerson, Ralph Waldo, 17–18
Emerson, William, 71
Emmons, Jeremiah, 10
emperor moth, 98
"Enemies of Butterflies, The" (Scudder),
    160, 164
England, xvi, 4, 11–12, 207
  butterfly market in, 195–8
Entomological Americana, 124–6, 160,
    251
Entomological News, 124, 126, 202,
    235, 251, 267
Epitome of the Natural History of
    Insects of India (Donovan), 55
Erie, Lake, 79–80
Essay on Classification (Agassiz), 36
Eudaemonia jehovah, 86–9, 101
Europe
  American collectors disparaged in,
    52, 90, 114
  natural science tradition of, 3–4, 36,
    50, 63, 85, 121
Everyday Butterflies (Scudder), 150
Evolution, 75
evolution, theory of, xxi, xxv, 21, 37–9,
    76, 91, 107, 152, 156, 159–60,
    173, 179, 188, 276
"Evolution and Taxonomy"
    (Comstock), 121
excursuses, 148–50, 166, 174, 182, 183
exotics, 50, 58, 60–2, 66, 104, 106,
    127
  worldwide market in, 193–225
Exotische Tagfalter (Exotic Butterflies;
    Staudinger), 195, 200, 221

Fabricius, Johann, 3, 87
falcate orangetip, 172

farming
  as butterfly habitat, xvi–xviii, 14,
    150, 254
  industrial, 247–8, 253–4
Felder, Rudolf and Cajetan, 73
Fernald, Charles, 119, 155, 168, 278
Field Museum, 51, 109, 265
Fifty Years of Science (Lubbock), 157
fish, 36–7, 38–9, 42
Fletcher, James, 82–3, 142, 143, 249
food plants, 5, 6, 14, 15, 23, 25, 28,
    40, 72, 166, 175, 176–7, 203,
    249
"Fossil Butterflies of Florissant, The"
    (Scudder), 142
fossils, 38, 270
  of butterflies, 39, 43, 93, 141–2
Foulks, Andrew, 97
Frail Children of the Air (Scudder), 150
Franck, George, 234
Frank, Kenneth, 239–40
Fraser's swamp, 24–5, 175, 183, 229,
    245
Frederic R. Grote and Sons, 74
French, George, 123, 226
Freud, Sigmund, vii, 84
fritillaries, 49, 68, 101, 120, 136, 137,
    238
Fruhstorfer, Hans, 201–2, 213–14, 266,
    268
Fuller, Arthur, 63, 88, 103–4, 106

gas lamps, 234
gas wells, 224, 239
genera, as term, 4
Genera of Diurnal Lepidoptera, The
    (Doubleday and Westwood), 63–4
Genesis I–II (Grote), 182
genitalia, 44, 46, 143, 147, 154, 155,
    159, 274, 276
Geology of the Wisconsin Survey, 119
Gerhard, William, 265
German-American butterfly collectors,
    13, 50–81, 239
Germans
  collecting by, xvi, 193, 195, 201–2,
    207, 209, 210
  technological inventiveness of, 228–9

Girl of the Limberlost, A (Stratton-
   Porter), 241
Glenwood Springs, Colo., 233, 243–4
Glover, Townsend, 130
Godeffroy, Hans, 198
Godman, Frederick, 197, 204, 215, 267
Goethe, Johann Wolfgang von, xxii, 51
Gortyna, 274–5
Gould, John, 11–12, 14
Graef, Edward, 70–1, 103, 125, 277–8
grasshoppers, 279–80
Gray, Asa, 37
great spangled fritillaries (Cybele), 14
"Green Butterflies" (Doherty), 221
Grinnell, Fordyce, Jr., 134–5, 252
Gross-Smith, Henley, 216, 266
Grote, Anna Radcliffe, 70, 73
Grote, Augustus Radcliffe, xv, xvii,
   xxiv–xxv, 32, 51–2, 67–81, 67,
   85, 86, 98, 111–15, 116, 117, 124,
   125, 130, 133, 143, 159, 165, 166,
   182, 183, 204, 248, 250, 251, 252,
   261
   academic contributions of, 76–81
   acclaim for, 70
   on beauty, 188–9
   classification system of, 277
   collection of, 79, 113–14
   C. Robinson and, 73–4, 76, 77
   daughters of, 74, 107, 114
   death of, 274, 278
   early years of, 70–4
   in Europe, 274–5, 277–8
   and evolving perception of
      Darwinism, 76–8, 107, 153
   in feud with Strecker, 67, 99–106,
      108–10, 112, 113, 263, 274
   on killing of specimens, 96
   on mimicry, 169, 172, 174, 180
   moths as preference of, 71–3, 76,
      79–80, 108, 129, 166
   as multifaceted, 67, 74–5, 129
   and *Papilio*, 111–14
   personal misfortunes of, 67, 70, 73,
      74, 107–12, 113–14
   poetry of, 145–6
   publications of, 76, 128–30
   resentful nature of, 277–8

Grote, Frederic, 70, 73, 74, 107
Grote, Julia Blair, 74, 77, 277
Guenée, Achille, 73
guidebooks, 70, 116, 119, 121–4, 149,
   237
   see also specific books
Guide to the Study of Insects
   (Packard), 98
Guild, Clarissa, 42

Haeckel, Ernst, 245
Hagen, Hermann, xxiv, 35, 90,
   111–12, 147, 177
hairstreaks, 34, 165
halftone, 235, 263
Hampson, George, 264, 266
Handbook of Nature Study
   (Comstock), 240
Handbuch für Schmetterlingsliebhaber
   (Meigen), 68–9, 70–1
Hard Green (Huber), 246
Harris, Moses, 276
Harris, Thaddeus, 40–2, 143, 145
Hartert, Claudia, 216
Hartert, Ernst, 210–16, 211, 266, 268,
   272, 273
Harvard, 35–6, 40, 83, 141, 145,
   149
harvester butterfly (*Feniseca
   tarquinius*), 71
Hatch Act, 249
hawk moths (sphinxes), 72, 99, 122,
   131, 172, 234
Hawk Moths of North America, The
   (Grote), 129
Heine, Heinrich, 113
heliconians, 171, 172
Heliconius humboldt, xxi
Henshaw, Samuel, 172, 218, 280
hermaphrodites, 98
Herrich-Schaeffer, Gottlieb, 73, 131,
   276
Hesperia catullus, 87
Hesperia massasoit, 62
Hewitson, William, 63–4, 100, 130,
   131, 188, 193, 197, 217
hibernacula (nests), 175
hibernation, 167, 169

Higginson, Thomas Wentworth, xviii, 16, 145–6
Hill, W. W., 119
Histoire naturelle (Buffon), xx
"hod carriers" (amateur aids), xxiv–xxv, 24–6, 65, 132
Holland, Carrie Moorhead, 219, 235
Holland, Daniel, 219
Holland, William, 219, 255, 263, 275
  and Carnegie, 190, 231, 236
  collection of, 226, 232, 235, 236, 258
  as commercial collector, 190, 218–21, 266–71
  death of, 257
  disparaging and critical approach of, 266–71
  opportunistic, self-serving nature of, xxv–xxvi, 218–19, 227, 231, 236, 267–71
  parsimonious nature of, 140, 220–1, 259, 264, 266, 268
  religious background of, 218–19
  wealth and power of, 235–8, 266
Holle, William, 89
Honrath, Eduard, 208, 221–2
Hopkins, Mark, 33, 35, 39
Horsfield, Thomas, 202–3, 218
Houghton, Mifflin and Company, 140, 143, 281
Howard, Leland O., 163–4, 249
How to Know the Butterflies (A. Comstock), 275
Hoy, Philo Romayne, 119
Huber, Peter, 246–7
Hübner, Jacob, 32, 44–5, 55, 87, 147
Hudson, George L., 234
Hulst, George, xxv, 59, 65, 86, 88, 103, 110, 125, 254
Humboldt, Alexander, xix, xxi–xxiii, xxiv, 11, 34, 36, 39, 40, 41, 52, 55, 80, 124, 133, 159
  views on nature of, 16, 17, 51, 131, 152, 181, 188, 217
hummingbirds, 11–12, 13
Huxley, Thomas, 117
"Hypermetamorphosis in Butterflies" (Scudder), 174–5

hyperparasites (secondary parasites), 163–4

ichneumon wasps, 42, 70–1, *161*, 162–5, 173, 177–8, 241, 250
Icones Ornithopterorum (Rippon), 194, 225
Illustrated Essay on the Noctuidae of North America, An (Grote), 129
illustration, illustrators, 24, 32, 97, 121, 143, 251, 255
  black and white, xxii, 263
  color, xxii, 18–19, 120, 127, 129, 194, 217
  importance of, 55–6, 117, 131
  photography vs., xxiii, 235, 237–8, 240–1, 263, 275
  plates in, xxii, 18–19, 63, 69–70, 130, 143, *144*, 147, *155*, *161*, 178, *185*, 216, 283
  of Strecker, 53–4, 63–4, 104, 109, 130, 145, 240, 251, 263
  woodcuts, 121, 143, 150
  see also photographs, photography; *specific artists*
Illustrations of Exotic Entomology (Drury), 55
Illustrations of Natural History (Drury), 18
imago (adult form), xv, 3, 6, 58, 90, 147
imperialism, xvi, 225
  evils of, 197, 201, 202, 210–11, 213
  showcased at Chicago Fair, 228
India, 35, 83, 170, 172, 178, 195–7, 201, 205–7, 210, 211–13, 217, 267
Insect Life, 124–5, 249
"Insect Life on Coney Island" (Schaupp), 242
Introduction to Entomology (Comstock), 121
Introduction to Entomology (Kirby and Spence), 11, 170, 233
"Iron and Its Relation to Civilization" (H. Edwards), 17
Ishikawa, Charles, 112

James, William, 76, 83, 188
Janson, Edward, 214

Janson's (store), 95, 196, 214, 215
Jefferson, Thomas, xvi, 17
Jesup, Morris, 259–60
John Muir's hairstreak (*Callophrys muiri*), 31
Jordan, Karl, 149, 256
Judaism, Jews, 65–6, 75, 197, 213, 215, 261

Kanawha River, 13, 14, 15, 169, 183, 229, 244
Kanchu (Lepcha), 210
Karner blue, (*Lycaeides melissa samuelis*), 118
Kearfott, William, 264
Kellicott, D. S., 172
Kern, Benjamin, 53
Kern, Edward, 53
Kern, Richard, 53
Ketterer, Edward, 256
killing, in collecting, 11, 26, 27, 28, 96–7, 122–3, 182, 204, 240–1
Kirby, William F., 11, 170, 196, 216, 233
Klages, Edward, 236
Klapperthal Glen, 56
Knab, Fred, 203
Kohler, Robert E., 231
Kramer, John, 50
Kumaon, 206–7

"Laborer in Politics, The" (Grote), 75
Lacey Act (1900), 246, 266
LaFarge, John, 187
Lamarck, Jean-Baptiste, 37
Landis, H., 59
Langtry, Lillie, 30
lappet moths, 98
larva (caterpillar form) xv, 3, 22–3, 40, 42, 60, 72, 90, 92, 131, 135, 138, 147, 149, 154, 155, 157, 172, 174, 200, 207, 233, 237, 249, 255, 276, 283
    collection of, 15, 62–3
    in identification, 6, 19, 114, 159, 274
    parasites of, 71, 160, 162–3, 165, 173, 177
Latimer, Adrian, vii, 65, 224
Latrielle, Pierre, xxi, 3, 87

Le Conte, John, 3
Leidy, Joseph, 55
Lepchas, 209–10
Lepidoptera of Ceylon (Moore), 178
"Lepidoptera of the Adirondack Region, The" (Lintner), 119
Lepidoptera: Rhopaloceres et Heteroceres (Strecker), 63, 64, 87, 101–2, 130–1
Leslie, Mrs., 18, 255–6
Lewis, Ronald L., 244, 247
libraries, 116–17
Lichenee blue, 82
life cycles, life histories, xv, xix, xxv, 17, 28, 40, 76, 119, 132, 139, 147, 148, 152–90
    in identification, 5, 7, 19–20, 36, 90–1, 106, 159, 172, 179
Life of a Butterfly, The (Scudder), 150
limelight, 233–4
limited transmutation, xix
Linnaeus, Carl, xxiv, 5, 17, 89, 121, 130, 162, 195, 248
    binomial nomenclature system of, xix–xx, xxv, 3, 32, 36, 45, 48, 87, 132–3, 147, 209
Linnean Society of New England, 37, 205
Lintner, Joseph, 86, 92, 93, 102, 103, 117–19, 133, 142, 148, 163, 172, 177, 232, 249, 251
"List of Butterflies Taken in Kumaon, A" (Doherty), 178, 207
literature, butterfly, xxiii, 116–51
    affordability of, 151, 238
    for children, 93, 119, 121–3, 149–51, 241
    instructional, 119–23
    lists in, 117–19
    magazines and journals, 116, 123–4
    publication costs in, 19, 120, 143, 149, 238, 263
    see also catalogs; guidebooks; illustration, illustrators; *specific works*
lithography, xxii, 63, 127, 143, 256, 263
"Local Butterflies" (Scudder), 182

Longfellow, Henry Wadsworth, 283
Lord Jim (Conrad), 82
Lorquin, Pierre, 3–4, 17, 61
Lorquin's admiral (*Limenitis lorquini*), 4
Lorquin Society, 252
Lubbock, John, 157
lumber industry, 244–5
"lumpers," 43, 45, 133
luna moth, 4, 197–8, 278
Lycaenidae, 31, 65, 144, 178, 212, 237, 267
Lyman, Henry, 281

Malaysia, 82, 194, 196, 201, 205, 208–10, 211, 212–14, 216, 221, 265
Man and Nature (Marsh), 245
Mann, Helen, 10
Manual of British Butterflies and Moths (Stainton), 121
Manual of Entomology, A (Burmeister), 88
Marsh, George Perkins, 16, 245
Marshall, G. F. L., 170, 267
Maynard, Charles, 127–8
McGlashan, Charles, 28–9
Mead, Sam, 26, 28, 58
Mead, Theodore "Ted," 25, 95, 96–7, 125, 176, 208, 220, 226, 232, 233
death of, 257
and *Papilio,* 111–13
and Strecker, 99–102, 110
and W. H. Edwards, 25–9, 32, 91, 93, 136, 281
and Willie Edwards, 26–7, 245–6
meadow fritillary (*Boloria bellona*), xvii
Meigen, Johann, 68, 70–1
Melville, Herman, 180
Merian, Maria Sibylla, xxii, 76
Meske, Otto, 118
metamorphosis, xv, 19, 154, 174, 203
see also life cycles, life histories
Meyer, Julius, 66
microlepidoptera, 215, 272
microscope, 36, 39, 44, 45, 66, 76, 92, 136, 143, 184, 206, 207, 256
migration, 72, 77, 129, 141, 165–9

Milbert's tortoiseshell, 164–5
milkweed butterflies, 168
mimicry, 155, 169–75, 179–81, 207, 248, 212
"Mimicry and Protective Resemblance" (Scudder), 181
mimics (*Hypolimnas misippus*), 135
monarchs (*Danaus plexippus; Anosia plexippus*), xvii, 3, 25, 48, 49, 93, 141, 150, 154, 167–9, 173, 248, 280
Monongahela National Forest, 247
Montúfar, Carlos, xxi
Moore, Frederic, 178
Moorhead, John, 219–20
morphos, 58, 65, 201–3, 223–4
M. achilles, 55
M. cypris, 223, 224
M. menelaus, 64, 87, 195, 223
M. sulkowski, 223
Morris, John, 50–1, 52, 66, 101, 106, 216, 234, 262
in nomenclature debate, 87–8
Morrison, Herbert, 26
Morton, Emily, 89, 156, 224
Mother Jones, 244
moths, 4, 51–2, 54, 58, 62–3, 65, 86–7, 100, 118, 129, 166, 172, 209, 212, 215, 241, 250, 252, 262, 264, 274
butterflies compared to, 71–2
Grote's preference for, 71–3, 76, 79–80, 108, 129, 166
snaring of, 68, 79, 225, 233, 239
swallowtailed, 6–7
Moths and Butterflies (Ballard), 122
"Moths and Moth-Catchers" (Grote), 129
Moths of the Limberlost (Stratton-Porter), 241
mountaineering, 48, 141
mourning cloak (*trauermantel*), 68–9, 122, 147
Muir, John, 31, 97
multibrooding, 167, 169
Museum of Comparative Zoology, xxiv, 36, 37, 39, 95, 114, 147, 172, 218, 279, 280
Mycalesis mineus, 207

Nabokov, Vladimir, 43, 118, 149, 201–2, 238
National Park Service, 247
natural history, xv–xxvi, 67, 82
  expanding interest in, 179–80, 189
  as motivation for collecting, 85
Natural History Museum (Indian Museum; Calcutta), 178, 206
Natural History Museum (London), 198, 208
Natural History of Selbourne (White), 10
Natural History of the Rarer Lepidopterous Insects of Georgia (Abbot and J. Smith), 6
Naturalist (Wilson), 84
Naturalist's Directory, The, 106
natural selection, xxi, 9, 37–8, 77, 152, 162, 170–1, 174
  beauty and, 183–6, 188
nature
  accessibility of, xvii–xviii
  and art, 6, 240
  beauty as elemental to, xv, xxv, xxvi, 8–9, 41, 51–2, 67, 79, 98, 114–15, 152–3, 183–9, 253
  for benefit of human beings, xx, 248
  economic impact on, xvi, xxvi, 6, 7, 16
  environmental damage to, 9–10, 16–18
  as evolving and unstable, xxi, 45
  featured at Chicago Fair, 227
  opposition to human abuse of, 31
  passionate response to, 82–5, 254
  and religion, 34–5
  Romantic tradition in, 50–81
  spirituality of, 36, 184–5, 254
  as unified and interdependent, xx, xxi, xxiii, 18, 36, 51–2, 78, 131, 153
  utilitarian uses of, 17–18, 152, 248, 250–1
  see also ecology
Naturphilosophie, 37
nest-building, 175–6
nets, xix, 70, 94–6, 115, 137, 208

Neumoegen, Berthold, 98, 101, 125, 197, 200, 221–2, 232, 258, 283
  death of, 199, 257
  financial setbacks of, 113, 199, 257, 261
  and Papilio, 111–13
  and Strecker, 65–6, 104–6, 110, 113, 222, 260–1
Neumoegen, Rebecca, 257
Neversink Mountain, 56, 216
New Infidelity, The (Grote), 75, 108, 129
Newman, George, 22
Newton, Isaac, 8, 21
New York Entomological Society, 111–12, 257, 259, 264
New York State Museum of Natural History, 118
Nicéville, Lionel de, 170, 178, 206, 267
Noctuidae, 72, 80, 212, 274
Nokomis fritillary (Argynnis nokomis), 283
Nomenclator Zoologicus (Scudder), 141
nomenclature, 141, 216, 217, 222, 248, 256, 265
  binomial, xix–xx, 3, 71, 130
  debate over, xxv, 43–5, 48, 85–95, 102, 113, 127–9, 133, 147–8
  distribution in, 78
  importance of, 88–9, 122
  native, 209
North American Entomologist, 76
Novak, Jakub, 155
Nymphalidae, 144, 147, 175

Oberthür, Charles, 214, 215, 221, 268, 272
obliterative coloration, 171
observation, xxii, 21, 30, 34, 38–9, 43, 152, 167, 252
ocelli (eye spots), 98, 172
Oeneis semidea, 40, 44, 49, 78
Olmsted, Frederick Law, 227
On the Origin of Species (Darwin), 20, 35, 37, 91
"Origin of Ornamentation in the Lepidoptera, The" (Grote), 180

Orinoco River, xxi, 11, 52
Ornithoptera
  O. brookiana (*Troides brookiana*),
    64
  O. croesis, 64, 82, 120, 132, 214
  O. dohertyi (*Troides dohertyi*), 194
  O. paradisea, 217
  O. priamus, 55, 62, 132
  see also birdwings
orthochromatic process, 235
Oslar, Ernest, 243–4

Packard, Alpheus, 98, 123–4, 162,
    175
Paint Creek, 15, 18, 19, 24, 57, 140,
    238, 244
painted ladies (*Vanessa cardui*), xviii,
    42, 207
palearctics, 200, 272
pale swallowtail, 4
palpi, 143, 144
Pambu (Lepcha), 210, 213, 214
Papaipema, 252–3, 262
Papilio
  P. antenor, 64
  P. antimachus, 55, 217, 225
  P. hospiton, 60
  P. jacksoni E. Sharpe, 265
  P. machaon, 112, 207
  P. marchandii, 130, 132
  P. neumoegeni, 221–2
  P. rex, 272
Papilio (journal), 111–14, 119, 124–6,
    207, 220, 259, 260, 261
Papillons Exotiques (Cramer), 55
parasites, 42, 70–1, 160, 163–5, 177
parasitoids, 160–5, *161*, 173, 176–7,
    187
Paris Exposition (1878), 205
parnassians, 87, 277
pasture species, xvii
Peabody, Cecil, 93
Peabody, Selim, 93, 226, 228
Peale, Charles Willson, 6
Peale, Titian, 6–7, 13, 40, 52, 66, 106,
    202–3
pearl crescent (*Phyciodes tharos*), 28,
    134, 158

Peart, John, 135
Peart, Mary, 18–19, 130, 135–6, 145,
    158, 178, 188, 241, 255–6, 283
Perak Museum, 210, 211
"Perils of the Egg, The" (Scudder),
    165
Personal Narrative of Travels
    (Humboldt), 11, 16, 52
pests, 41, 74, 118, 247–50
  see also economic entomology
Petersen, Wilhelm, 205
Phalaenae, 122
Philadelphia and Reading Railroad, 56
Philadelphia Centennial Exposition
    (1876), 119, 229
photographs, photography, 26, 55,
    143, 145, 250
  as alternative to killing of specimens,
    241
  color, 236–7, 263, 275
  estrangement from nature through,
    240–1
  exchange among collectors of,
    xxiii–xxiv, 46, 66, 105
  of living specimens, 277
  and mimicry, 179–80
  showcased at Chicago Fair, 229–30
  vs. illustration, xxiii, 235, 240–1
"phylogony of species," 159
Pilate, Eugene, 82, 118–19
pipevine swallowtail, 40, 150, 158
Pittsburgh, Pa., 224–5
poetry, 116, 143, 145–7, 237–8, 251
poisons, 96–7, 123, 241
polymorphism (polyphemism), 22–3,
    44, 127, 156–7, 159, 174, 207,
    233
Popular Science Monthly, 124, 129
Poritia hartertii, 212
post office, collecting abetted by,
    xxiii–xxiv, 57, 66, 133, 232
Poulton, Edward, 171
Practical Entomologist, 21, 77
predators, 160–5, 169–71, 176
printing presses, xxiii
proboscis, 72, 155
Psyche, 48, 118, 124, 126, 221,
    267

pupa (chrysalis), xv, 3, 92, 122, 135,
136, 147, 154, 155–6, 160, 172,
174, 197, 200, 207, 237, 241, 252
in identification, 6, 19, 90
purples, 34, 40, 49, 158, 173, 174
Putnam, Duncan, 104, 106
Putnam, Mary, 104

question mark (butterfly), 23

railroads, xvi, xvii, xxiii, 73, 113, 180,
247, 265–6, 271
collecting abetted by, 28, 117, 133–4,
231–3, 242
as engine of imperialism, 195–6
environmental damage by, 16, 56,
242–6, 250, 254
and mining, 107
showcased at Chicago Fair, 229
rattleweed (Crotolaria retusa), 176
Reakirt, Tryon, 60–1
red admiral (Atalanta), 163, 164, 165
red-spotted purple, 173, 174, 179
Reed, W. H., 270, 271
regal fritillary (Speyeria idalia), 71, 150,
185
religion
eschewed by butterfly collectors, 26,
33, 53, 155
Grote on, 75–6
nature and, 34–5
vs. science, 35, 37–9, 155
Reliquia, 83
Rhodophora florida, 172
Ribbe, Heinrich, 104, 200
Ridings, John, 22
Riley, Charles Valentine, 32, 77, 125,
155, 163–4, 168, 171–2, 241,
248–9, 253
Ring and the Book, The (Browning),
146
Rippon, Robert, 193, 225
River Styx nymphalid, 142
Robinson, Coleman, 73, 74, 76, 77
Robinson, Ronald E., 195
Robinson, Russell, 56, 261–2
Robinson, Wirt, 56, 262
rockslide alpine, 139

Rocky Mountain parnassian
(Parnassius smintheus), 256
Rocky Mountains, 28, 137–8, 168, 243
Roethke, Theodore, 254
Rogers, William Barton, 37–8
Romantic Enlightenment, xx
Roosevelt, Theodore, xvi, 243, 246, 266
Rosenhof, August Rösel von, xxii
Rothschild, Walter, 149, 193, 215–16,
236, 256, 266, 268
Royal Museum of Natural History,
194, 200, 202, 264
Rural New Yorker, 63, 104

Saint-Hilaire, Geoffroy, 37
saloons, 50
Salvin, Osbert, 204, 267
Sammlung exotischer Schmetterlinge
(Hübner), 55
Santayana, George, 83–4, 152, 188–9,
225, 230, 253
Saturniidae, 98
satyrs, 31, 44, 136, 206, 256, 283
Saunders, William, 124
Schaupp, Franz, 125, 242–3
Schaus, William, 257
Schonborn, Henry, 65
science
and art, xxii–xxiii, 51–2, 67, 76,
110–11, 116, 143, 186–7, 275
vs. religion, 35, 37–9, 75–6, 155
Science Magazine, 123, 124, 148
Scopelosoma, 80
Scribner's, 283
Scudder, Charles, 33, 35, 39
Scudder, David, 35
Scudder, Ethelinda Blatchford, 40,
42–3, 44, 46, 47, 56, 142
Scudder, Gardiner, 42, 46, 141, 149
death of, 278–9, 280
Scudder, Samuel, xvii, xxv, 33, 51, 54,
57, 62, 67, 74, 77, 78, 83, 86, 96,
111, 116–19, 128, 136, 139, 152,
154–7, 162, 166–9, 175, 182, 185,
187, 195, 204, 206, 207, 213, 221,
226–7, 232, 234, 236, 237, 238,
241, 249, 251, 252, 255, 256, 259,
267, 274, 282

acclaim for, 142, 143, 149
and Agassiz, 33–9
all-inclusive approach of, 7, 41–2
anti-Darwinism of, 33, 47
on beauty, 183–8
Canada trip of, 39
classification and nomenclature of,
43–8, 90, 127–9, 173, 277
Darwinism of, 153, 155, 159–60,
173
data gathering of, 41–2, 47–8
death of, 278, 281
disability of, 279–80
ecological concerns of, 182, 183
as editor, 124–5
Europe and North Africa trip of,
42–7
and fossil butterflies, 141–2
on grasshoppers, 279–80
and Harris, 40–1
literature of, 131–4, 141–51, 160
on mimicry, 173–5, 179, 181
in nomenclature debate, 87, 90–5,
147–8, 277
on parasites, 163–5
personal misfortunes of, 42–3, 56–7,
149, 278–81
religious nature of, 33, 46, 279
Strecker vs., 60, 99–102, 110
W. H. Edwards's friendship with, 10,
13, 24, 26–7, 43, 46, 47, 140–2,
279, 283
W. H. Edwards's rifts with, 8, 43–8,
80–1, 90–5, 99, 107, 123, 147–9,
175–6, 186
seasonal variation, 5
Seba, Albertus, xxii, 131
Sehnsucht (heart yearning), 52, 58, 61,
104, 106, 109, 216
Sense of Beauty, The (Santayana), 83–4,
230, 253
sex, passion for nature compared to,
83–4
"Sexual Diversity in the Form of the
Scales" (Scudder), 184
Shakespeare, William, 29, 30, 34, 282
Shaksper not Shakespeare (W. H.
Edwards), 282

Shapiro, Arthur, 83, 119
Sharpe, Emily, 265, 272
silk moths, 130, 197
silvery blues, 276
Sinclair, Thomas, 143
Skinner, Henry, 119, 126–7, 200, 202,
251, 262–4, 267
skippers (Hesperiidae), 35, 49, 62, 87,
120, 204, 212, 237, 263, 277
in hierarchy dispute, 5, 90–1, 123,
147–8, 277
see also specific species
Slossen, Annie Trumbull, 259
Smerinthinae, 72
Smerinthus myops, 62
Smith, Amelia Woolworth "Daisy,"
203–4, 212, 267, 269–71
Smith, Herbert, 172, 203–4, 212, 226,
240–1, 267, 269–71
Smith, John B., 6, 125, 262–3, 274, 277
Smithsonian Institution, 14, 47, 104,
120, 131, 141, 257
Smythe, Elison, 224
snouts, 142
Snow, Frank, 224
social reformers, 74–5
Song of Hiawatha, The (Longfellow),
283
Soule, Caroline G., 42, 156
South America, xxi, 61, 70, 200,
202–3, 266–7, 270
Spanish Gypsy, The (Eliot), 145
spanner moths, 172
Specimen Days and Collect (Whitman),
133
Spence, William, 11–12, 170, 233
Sphingidae (hawk moths; sphinx
moths), 72, 99, 122, 131, 172, 234
spicebush swallowtail, 3
spiders, 8–9, 97
"splitters," 43, 45, 133
"Spread of the Butterfly in a New
Region" (Scudder), 182
spring azure (Lycaena pseudargiolus),
176–7
Sproesser, Christian, 54
Stainton, Henry, 121, 200
"stamp collectors," 97

Standard Natural History, The
    (Cassino), 120
Staten Island, N.Y., 70–1, 107, 109
Staten Island Railroad, 70, 73
Staudinger, Otto, 46–7, 60, 61, 104–5,
    110, 189, 193, 199, 213, 261, 265,
    270, 277
  as dealer, 198–200, 202, 210, 215,
    217, 220, 221, 222, 264
  death of, 264
Staudinger, Paul, 210
Staudinger collection, 200
Steinbach, José, 266
Stevenson, Robert Louis, 214–15
Stevens's Auction Rooms, 196
Stratton-Porter, Gene, 42, 241
Strecker, Anna Kern, 53, 56
Strecker, Eveline, 264
Strecker, Ferdinand, xxiv, 53–4, 56
Strecker, Herman, vii, 2–4, 47, 53, 71,
    82, 86, 96, 98, 115, 117, 122,
    126–7, 133, 152, 168, 181–2,
    189–90, 196, 197, 198, 202, 213,
    234, 242, 248, 252, 257, 260–2,
    267, 282
  acclaim for, 63–4
  accusations of thievery against,
    100–1, 108–10, 127, 262–3
  as agnostic, 52–3, 87
  artistic heritage and skill of, 53–4,
    63–4, 104, 109, 130, 145, 240,
    251, 263
  on beauty, 187–9
  collection of, xvi, 51, 54, 57–64,
    79, 104–6, 217, 223, 226–7, 232,
    264–5
  as commercial collector, 222–5
  covetousness of, 106
  death of, 264, 274
  early years of, 53–6
  as embodiment of democratic
    collecting, 64–5
  exotics of, 216–18, 222–4
  in feud with Grote, 67, 80–1,
    99–106, 108–10, 112, 113, 263,
    274
  literature of, 130–2, 143, 146
  marriages of, 56–7, 59, 264

  and Neumogen, 199, 222, 260–1
  in nomenclature debate, 86–7, 89–90
  passion and zeal of, 58, 61–2, 64–6,
    104, 250–1, 262
  personal misfortunes of, 53, 56–8,
    107–10
  poverty of, 59–60, 105–6, 109, 110,
    193, 199, 264
  Scudder vs., 60, 99–102
  and Staudinger, 60, 61, 200
  as stonecutter, xvi, xxiv, xxv, 53–4,
    103, 104, 109, 222, 263–4
  in troubled friendship with
    Neumoegen, 65–6, 104–6, 110,
    113
  W. H. Edwards vs., 99–103
Strecker, Louisa Roy, 56, 59
Strecker, Paul, 264
Stretch, Richard, 17, 63, 234
suburbanization, 254
sugaring, 66, 79, 96–7
sulphurs, yellows (*Coliadinae*), xxi, 3,
    4, 87, 112, 136, 150, 237, 238,
    275
  see also specific species
swallowtails (*Papilionidae*), 14, 15, 24,
    49, 58, 64, 82, 99, 112, 120, 122,
    158, 163, 167, 172, 199, 207, 217,
    221–3, 233, 237, 272, 277
  in hierarchy dispute, 5, 44, 90–1,
    123, 147, 277
  see also specific species
swarming, 167, 169
sweetheart underwing (*Catocala
    amatrix*), 54
Swinton, A. H., 125
Sydney Museum of Natural History, 29
symbiosis, 176–8, 207
"Systematic Revision of American
    Butterflies" (Scudder), 90

tanning business, 17
Tappan, Arthur, 13
taxidermy, 26, 54, 65, 127, 204, 218,
    235
technology
  advancements in, xxiii, xxvi, 226–54
  adverse impact of, 239–46

in butterfly collecting, 231–9
in color reproduction, 241–2
featured at Chicago Fair, 227–31
in publishing, 235–9
Teinopalpidae, 277
temperature, effect on butterflies of, 25,
    157–60, 169, 252
Tepper, Fred, 62, 70–1, 91, 98, 103,
    254
textbooks, 116, 123
Thatcher, Roland, 118
Thaumantis, 172, 206
Thayer, Abbott, 171–2
Thoas swallowtail (*Papilio thoas*), 262
Thompson, Elizabeth, 93, 140
Thoreau, Henry David, 16, 41, 71, 133
tiger swallowtails, 130, 134, 150, 158,
    179
Townshend, Tyler, 266
*Treatise on Some of the Insects
    Injurious to Vegetation* (Harris),
    41, 143
Trimen, Roland, 170
Tring (museum), 194, 215–16, 236
Troides minos, 83
tropicals, 7, 60, 83, 106, 169–70, 172,
    193
Tutt, James, 85
Twain, Mark, 282
type specimens, 108, 130

Uganda Railway, 205, 265–6, 272,
    273
United States
    changing scientific culture of, 6–7,
        20–1, 36, 67, 85
    economic priorities of, 6, 7, 12, 16,
        21, 73, 112, 125, 182, 189–90,
        230–1, 247–54
    industrialization of, 230–1
    landscape of, xvii, 5, 70–1, 150, 231,
        254
    natural history tradition in, xv–xvi,
        xviii–xix, xxiii
    rise of collecting in, xxiii–xxvi,
        3–49
Urania fulgens, 6–7
Uraniidae, 72

Vanderbilt, Cyrus, 73
Vanessa io, 125
Veblen, Thorsten, 162–3
Venezuela, xxi, 39, 236, 262
viceroys, 173, 174, 179, 248
Villa Sphinx, 199
Voyage of a Naturalist (Darwin), 19
Voyage up the River Amazon (W. H.
    Edwards), 11

Wailly, Alfred, 197–8, 200
Walden Pond, 41
Wallace, Alfred Russel, xxi, 12, 21, 64,
    78, 82, 120, 157, 169, 171, 179,
    188, 189, 193, 196–7, 225, 245,
    252, 282
Walsh, Benjamin, xxiv, 21–2, 25, 32,
    37–9, 43, 76–7, 168
Walsingham, Lord, 125, 215
Walters, William, 122
Wanamaker, John, 76
Wandering Hawk Moth, 72
Ward, Lester, 230
Warren, Edward, 234
Washington, Mount, 40, 44, 48, 78,
    141, 157
Watkins & Doncaster, 196, 215, 216,
    217
Weaver, Willis, 203
Weeks, Archibald, 160, 162
Weidemeyer, John, 13, 50–1, 52
Weismann, August, 16, 156–7, 159,
    177, 245, 252
Werneburg, Adolf, 186
Western Pennsylvania University
    (University of Pittsburgh), 235
western tiger swallowtail (*Papilio
    rutulus*), 4, 28, 139
West Virginia, xxv, 32, 229
    ecological damage to, 244–7
    W. H. Edwards's land in, 11–24, 28,
        107, 140, 169
Westwood, John O., 64
White, Gilbert, 10
White Mountains, N.H., 40, 49, 57,
    118, 119, 141, 147, 167
Whitfield, Annie, 135
Whitfield, William, 135

Whitman, Sarah Wyman, 187, *187*
Whitman, Walt, xviii, 18, 118,
    132–3
Wigglesworth, W., 4
wildlife preserves, 246–7
Wilhelm II, Kaiser, 200, 212, 264
Williams College, 10, 33–4, 40
William Seymour Edwards Oil
    Company, 244
Williston, Samuel, 163–4
Wilson, O. E., 84
wings, 5, 71, 121, 172, 174, 179, 180,
    183, 188
  veins of, 5, 147, 155, 159, 276–7
Winter, William, 258
Wood, Charles, 54
Wood-Mason, James, 170

wood nymphs, 44, 127, 134, 136, 158
Woodworth, Charles William, 147
Worster, Donald, 246
Wortman, J. L., 271
Wright, William Greenwood, 82,
    136–9, 156, 232, 255, 281–3

Xerces blue, 181

Yankee butterfly collectors, 3–49
Yosemite, 31

zebra heliconian, 134
zebra swallowtail, 3, 14, 15, 25, 45,
    156–7, 245
Zeller, Philipp, 17, 100
Zimmerman, Karl, 50